Experimental
Hematology
Today 1979

Experimental Hematology Today 1979

Edited by
Siegmund J. Baum
G. David Ledney

With 123 illustrations

 Springer-Verlag   New York   Heidelberg   Berlin

Siegmund J. Baum
Chairman, Experimental Hematology Department
Armed Forces Radiobiology Research Institute
Defense Nuclear Agency
Bethesda, Maryland 20014

G. David Ledney
Head, Immunology Division
Experimental Hematology Department
Armed Forces Radiobiology Research Institute
Defense Nuclear Agency
Bethesda, Maryland 20014

9 8 7 6 5 4 3 2 1

ISBN-13: 978-1-4612-6181-0   e-ISBN-13: 978-1-4612-6179-7
DOI: 10.1007/978-1-4612-6179-7

Preface

Experimental Hematology Today 1979 again attempts to present significant scientific discoveries of the past year. It contains papers presented at the Seventh Annual Meeting of the International Society for Experimental Hematology held in Chicago, Illinois, 1978. The manuscripts were selected and reviewed by the local scientific committee and the editors of the yearbook. The book is divided into six parts, in an attempt to represent equally the various facets of basic sciences and clinical applications. The first part entitled "Studies of Hematopoietic Stem Cells" is chaired by Dr. T. J. MacVittie, Bethesda, Maryland. It deals with discussions of very recent new findings of stem cell physiology. This is followed by Part II entitled "Myelopoiesis" and headed by Dr. S. J. Baum, Bethesda, Maryland. In this division experimenters relate to physiological properties of granulocyte and macrophages, particularly to the anatomical and functional heterogeneity of these cells. Part III is led by Dr. W. Fried, Chicago, Illinois, and is entitled "Erythropoiesis and Megakaryocytopoiesis." It summarizes the new insights obtained in red cell and platelet production. Part IV is designated as "Transplantation Immunology." It is led by Dr. P. J. Tutschka, Baltimore, Maryland. It deals with the intricate mechanisms of immune reactions involved in bone marrow transplantation. Part V is chaired by Dr. M. M. Bortin, Milwaukee, Wisconsin, and is entitled "Bone Marrow Transplantation." It is comprised of a series of papers dealing with new approaches and evaluations of clinical bone marrow transplantation. Part VI is headed by Dr. D. E. Harrison, Bar Harbor, Maine, and is entitled "Animal Models of Clinical Conditions." It discusses the proper utilization of animal models for the study of specific human hematological diseases. Attempts were made to represent all disciplines involved in the studies of experimental and clinical hematology. Consequently this yearbook again should be of extreme value to experimental and clinical scientists.

ACKNOWLEDGMENT

The International Society for Experimental Hematology wishes to acknowledge the generous assistance provided by the following agencies of the United States government: Defense Nuclear Agency; Armed Forces Radiobiology Research Institute; Department of Energy.

Contents

Contents

Contents

Contributors

John W. Adamson, Divisions of Hematology and Oncology, Department of Medicine, University of Washington, Fred Hutchinson Cancer Research Center, and Veterans Administration Hospital, Seattle, Washington

Aftab Ahmed, Department of Immunology, Naval Medical Research Institute, Bethesda, Maryland

Siegmund J. Baum, Experimental Hematology Department, Armed Forces Radiobiology Research Institute, National Naval Medical Center, Bethesda, Maryland 20014

Jerry Becker, Oncology Research Laboratory, Division of Oncology, Albany Medical College, Albany, New York 12208

B. Belohradsky, Universitätsklinik, 8000 Munich, West Germany

R. Benner, Department of Cell Biology and Genetics, Erasmus University, P.O. Box 1738, Rotterdam, The Netherlands

W. B. Bias, The Johns Hopkins Oncology Center, Baltimore, Maryland

Mortimer M. Bortin, International Bone Marrow Transplant Registry, and The May and Sigmund Winter Research Laboratory, Mount Sinai Medical Center, Milwaukee, Wisconsin 53233

H. G. Braine, The Johns Hopkins Oncology Center, Baltimore, Maryland

N. H. C. Brons, Department of Cell Biology and Genetics, Erasmus University, P.O. Box 1738, Rotterdam, The Netherlands

D. Brookoff, Medical Research Center, Brookhaven National Laboratory, Upton, New York 11973

J. Bryan, Section of Cancer Biology, Division of Radiation Oncology, Mallinckrodt Institute of Radiology, Washington University School of Medicine, St. Louis, Missouri 63110

Hillary Brzyski, Oncology Research Laboratory, Division of Oncology, Albany Medical College, Albany, New York 12208

H. Burlington, Department of Physiology, Mt. Sinai School of Medicine, City University of New York, New York, New York 10029

Melinda Cahan, UCLA School of Medicine, Center for the Health Sciences, Los Angeles, California 90024

Richard A. Cahill, Department of Immunology, Naval Medical Research Institute, Bethesda, Maryland

Martin J. Cline, UCLA School of Medicine, Center for the Health Sciences, Los Angeles, California 90024

Peter F. Coccia, Departments of Pediatrics, Laboratory Medicine and Pathology, and Therapeutic Radiology, University of Minnesota, Minneapolis, Minnesota 55455

E. P. Cronkite, Medical Research Center, Brookhaven National Laboratory, Upton, New York 11973

D. A. Crouse, Department of Anatomy, University of Nebraska Medical Center, Omaha, Nebraska 68105

H. Joachim Deeg, Divisions of Hematology and Oncology, Department of Medicine, University of Washington, Fred Hutchinson Cancer Research Center, and Veterans Administration Hospital, Seattle, Washington

K. A. Dicke, Department of Developmental Therapeutics, University of Texas System Cancer Center, M. D. Anderson Hospital and Tumor Institute, Houston, Texas

G. J. Elfenbein, The Johns Hopkins Oncology Center, Baltimore, Maryland

Ger van den Engh, Radiobiological Institute TNO, 151 Lange Kleiweg, 2280 HV Rijswijk, Netherlands

John L. Fahey, UCLA School of Medicine, Center for the Health Sciences, Los Angeles, California 90024

Stephen Feig, UCLA School of Medicine, Center for the Health Sciences, Los Angeles, California 90024

E. J. Freireich, Department of Developmental Therapeutics, University of Texas System Cancer Center, M.D. Anderson Hospital and Tumor Institute, Houston, Texas

W. Fried, Division of Hematology–Oncology, Michael Reese Medical Center, Chicago, Illinois 60616

T. E. Fritz, Division of Biological and Medical Research Argonne National Laboratory, Argonne, Illinois 60439

Robert P. Gale, UCLA School of Medicine, Center for the Health Sciences, Los Angeles, California 90024

E. O. Gaston, Departments of Medicine and Radiology, The Committee of Immunology, and The Franklin McLean Memorial Research Institute, The University of Chicago, Chicago, Illinois 60637

E. Gerard, Institut de Pathologie Cellulaire, Hôpital de Bicêtre, 94270 Kremlin Bicêtre, France

S. Go, Institute for Cancer Research and Department of Pathology, Medical School, Osaka University, Kita-ku, Osaka, 530 Japan

David W. Golde, UCLA School of Medicine, Center for the Health Sciences, Los Angeles, California 90024

Eugene Goldwasser, Department of Biochemistry, and The Franklin McLean Memorial Research Institute, University of Chicago, Chicago, Illinois 60637

J. Gordon, Department of Surgery, McGill University, Montreal, Quebec, Canada

Theodore C. Graham, Divisions of Hematology and Oncology, Department of Medicine, University of Washington, Fred Hutchinson Cancer Research Center, and Veterans Administration Hospital, Seattle, Washington

Peter R. Graze, Department of Medicine, Division of Hematology-Oncology, UCLA School of Medicine, Center for the Health Sciences, Los Angeles, California 90024

C. W. Gurney, Departments of Medicine and Radiology, The Committee of Immunology, and The Franklin McLean Memorial Research Institute, The University of Chicago, Chicago, Illinois 60637

R. J. Haas, Universitätsklinik, 8000 Munich, West Germany

Anne W. Hamburger, Section of Hematology and Oncology, Department of Internal Medicine, University of Arizona College of Medicine, Tucson, Arizona 85724

W. D. Hankins, Department of Medicine, Veterans Administration Hospital, Nashville, Tennessee 37203

C. M. Harper, Departments of Anatomy and Radiology, University of Nebraska Medical Center, Omaha, Nebraska 68105

Alan W. Harris, The Walter and Eliza Hall Institute of Medical Research, Royal Melbourne Hospital, P.O. Box 3050, Victoria, Australia

D. E. Harrison, The Jackson Laboratory, Bar Harbon, Maine 04609

Susanne Hasthorpe, Radiobiological Institute TNO, Lange Kleiweg 151, Rijswijk, The Netherlands

K. Hatanaka, Institute for Cancer Research and Department of Pathology, Medical School, Osaka University, Kita-ku, Osaka, 530 Japan

Winston Ho, UCLA School of Medicine, Center for the Health Sciences, Los Angeles, California 90024

Grover M. Hutchins, Bone Marrow Transplant Unit, Oncology Center, John Hopkins University School of Medicine, Baltimore, Maryland 21205

N. N. Iscove, Basel Institute of Immunology, 4005 Basel, Switzerland

G. Janka, Universitätsklinik, 8000 Munich, West Germany

R. K. Jordan, Departments of Anatomy and Radiology, University of Nebraska Medical Center, Omaha, Nebraska 68105

H. Kaizer, The Johns Hopkins Oncology Center, Baltimore, Maryland

Shogo Kano, Division of Clinical Immunology, Department of Medicine, Jichi Medical School, Minamikawachi-machi, Tochigi-ken 329-04, Japan

John H. Kersey, Departments of Pediatrics, Laboratory Medicine and Pathology, and Therapeutic Radiology, University of Minnesota, Minneapolis, Minnesota 55455

Mary B. Kim, Section of Hematology and Oncology, Department of Internal Medicine, University of Arizona College of Medicine, Tucson, Arizona 75624

Tae H. Kim, Departments of Pediatrics, Laboratory Medicine and Pathology, and Therapeutic Radiology, University of Minnesota, Minneapolis, Minnesota 55455

Yukihiko Kitamura, Institute for Cancer Research, Medical School, Osaka University, Hakanoshima 4-chome, Kita-ku, Osaka, 530 Japan

H. J. Kolb, Institüt für Hamatologie der GsF, Landwehrstr. 61, 8000 Munich 2, West Germany

T. A. Kost, Department of Medicine, Veterans Administration Hospital and Vanderbilt University School of Medicine, Nashville, Tennessee 37203

M. J. Koury, Department of Medicine, Veterans Administration Hospital and Vanderbilt University School of Medicine, Nashville, Tennessee 37203

S. B. Krantz, Department of Medicine, Veterans Administration Hospital and Vanderbilt University School of Medicine, Nashville, Tennessee 37203

William Krivit, Department of Pediatrics, Laboratory Medicine and Pathology, and Therapeutic Radiology, University of Minnesota, Minneapolis, Minnesota 55455

Kazuo Kubota, Division of Hemopoiesis, Institute of Hematology, Jichi Medical School, Minamika-wachi-machi, Tochigi-ken 329-04, Japan

Seymour H. Levitt, Departments of Pediatrics, Laboratory Medicine and Pathology, and Therapeutic Radiology, University of Minnesota, Minneapolis, Minnesota 55455

Lloyd Lininger, Department of Mathematics, State University of New York, Albany, New York 12222

Eva Lotzová, Department of Developmental Therapeutics, University of Texas System Cancer Center, M.D. Anderson Hospital and Tumor Institute, Houston, Texas

Nancy Lyddane, UCLA School of Medicine, Center for the Health Sciences, Los Angeles, California 90024

J. W. Lyon, Departments of Medicine and Radiology, The Committee of Immunology, and The Franklin McLean Memorial Research Institute, The University of Chicago, Chicago, Illinois 60637

T. J. MacVittie, Experimental Hematology Department, Armed Forces Radiobiology Research Institute, Bethesda, Maryland 20014

B. A. Malcolm, Departments of Medicine and Radiology, The Committee of Immunology, and The Franklin McLean Memorial Research Institute, The University of Chicago, Chicago, Illinois 60637

H. Matsuda, Institute for Cancer Research and Department of Pathology, Medical School, Osaka University, Kita-ku, Osaka, 530 Japan

K. B. McCredie Department of Developmental Therapeutics, University of Texas System Cancer Center, M.D. Anderson Hospital and Tumor Institute, Houston, Texas

W. L. McLellan, Section of Cancer Biology, Division of Radiation Oncology, Mallinckrodt Institute of Radiology, Washington University School of Medicine, St. Louis, Missouri 63110

Mark Miani, Oncology Research Laboratory, Division of Oncology, Albany Medical College, Albany, New York 12208

Yasusada Miura, Division of Hemopoiesis, Institute of Hematology, Jichi Medical School, Minamika-wachi-machi, Tochigi-ken 329-04, Japan

Hideaki Mizoguchi, Division of Hemopoiesis, Institute of Hematology, Jichi Medical School, Minamikawachi-machi, Tochigi-ken 329-04, Japan

L. E. Mobraaten, The Jackson Laboratory, Bar Harbor, Maine 04609

L. Muesse, Department of Developmental Therapeutics, University of Texas System Cancer Center, M.D. Anderson Hospital and Tumor Institute, Houston, Texas

A. Nakeff, Section of Cancer Biology, Division of Radiation Oncology, Mallinckrodt Institute of Radiology, Washington University School of Medicine, St. Louis, Missouri 63110

B. Netzel, Universitatsklinik, 8000 Munich, West Germany

Mark E. Nesbit, Departments of Pediatrics, Laboratory Medicine and Pathology, and Therapeutic Radiology, University of Minnesota, Minneapolis, Minnesota 55455

C. Nissen, Division of Hematology, Department of Internal Medicine, Kantonsspital, 4031 Basel, Switzerland

W. P. Norris, Division of Biological and Medical Research, Argonne National Laboratory, Argonne, Illinois 60439

C. .J. O'Hara, Division of Biological Research, Ontario Cancer Institute, and Division of Hematology, Department of Medicine, Toronto General Hospital, University of Toronto, Toronto, Ontario, Canada

Gerhard Opelz, UCLA School of Medicine, Center for the Health Sciences, Los Angeles, California 90024

D. G. Osmond, Department of Anatomy, McGill University, Montreal, Quebec, Canada

Lester J. Peters, M.D., M.D. Anderson Hospital, Dept. of Radiotherapy, 6723 Bertner, Houston, Texas 77030

G. B. Price, Division of Biological Research, Ontario Cancer Institute, and Division of Hematology, Department of Medicine, Toronto General Hospital, University of Toronto, Toronto, Ontario, Canada

Norma K. C. Ramsay, Departments of Pediatrics, Laboratory Medicine and Pathology, and Therapeutic Radiology, University of Minnesota, Minneapolis, Minnesota 55455

U. Reincke, Medical Research Center, Brookhaven National Laboratory, Upton, New York 11973

I. Rieder, III. Medizinische Klinik Grobhadern der Universität, 8000 Munich, West Germany

Alfred A. Rimm, International Bone Marrow Transplant Registry, and The May and Sigmund Winter Research Laboratory, Mount Sinai Medical Center, 950 North Twelfth Street, Milwaukee, Wisconsin 53233

H. V. Rodt, Institut für Hämatologie der GsF, Landwehrstr. 61, 8000 Munich 2, West Germany

John C. Ruckdeschel, Oncology Research Laboratory, Division of Oncology, Albany Medical College, Albany, New York 12208

Sydney E. Salmon, Section of Hematology and Oncology, Department of Internal Medicine, University of Arizona College of Medicine, Tucson, Arizona 85724

Jean E. Sanders, The Fred Hutchinson Cancer Research Center, Seattle, Washington 98104

George W. Santos, Bone Marrow Transplant Unit, Oncology Center, Johns Hopkins University School of Medicine, Baltimore, Maryland 21205

Gregory Sarna, UCLA School of Medicine, Center for the Health Sciences, Los Angeles, California 90024

T. M. Seed, Division of Biological and Medical Research, Argonne National Laboratory, Argonne, Illinois 60439

Kenneth W. Sell, National Institute of Allergy and Infectious Diseases, National Institute of Health, Bethesda, Maryland

L. L. Sensenbrenner, The Johns Hopkins Oncology Center, Baltimore, Maryland

Saul J. Sharkis, Oncology Center, Johns Hopkins University Medical School, Baltimore, Maryland

J. G. Sharp, Departments of Anatomy and Radiology, University of Nebraska Medical Center, Omaha, Nebraska 68105

M. Shimada, Institute for Cancer Research and Department of Pathology, Medical School, Osaka University, Kita-ku, Osaka, 530 Japan

K. H. Shumak, Division of Biological Research, Ontario Cancer Institute, and Division of Hematology, Department of Medicine, Toronto General Hospital, University of Toronto, Toronto, Ontario, Canada

E. L. Simmons, Departments of Medicine and Radiology, The Committee of Immunology, and The Franklin McLean Memorial Research Institute, The University of Chicago, Chicago, Illinois 60637

Robert Sparkes, UCLA School of Medicine, Center for the Health Sciences, Los Angeles, California 90024

B. Speck, Division of Hematology, Department of Internal Medicine, Kantonsspital, 4031 Basel, Switzerland

Rainer Storb, Divisions of Hematology and Oncology, Department of Medicine, University of Washington, Fred Hutchinson Cancer Research Center and Veterans Administration Hospital, Seattle, Washington 98108

Fumimaro Takaku, First Department of Medicine, Jichi Medical School, Minamikawachi-machi, Tochigi-ken, 329-04, Japan

Mary Territo, UCLA School of Medicine, Center for the Health Sciences, Los Angeles, California 90024

Alan Tesler, UCLA School of Medicine, Center for the Health Sciences, Los Angeles, California 90024

S. Thierfelder, Institut für Hämatologie der GsF, Landwehrstr. 61, 8000 Munich 2, West Germany

D. V. Tolle, Division of Biological and Medical Research, Argonne National Laboratory, Argonne, Illinois 60439

Beverly J. Torok-Storb, Hematology Research Laboratory, Veterans Administration Hospital, Seattle, Washington 98108

Barbara Trask, Radiobiological Institute TNO, 151 Lange Kleiweg, 2280 HV Rijswijk, The Netherlands

Peter J. Tutschka, Bone Marrow Transplant Unit, Oncology Center, Johns Hopkins University School of Medicine, Baltimore, Maryland 21205

F. A. Valeriote, Section of Cancer Biology, Division of Radiation Oncology, Mallinckrodt Institute of Radiology, Washington University School of Medicine, St. Louis, Missouri 63110

Jan Visser, Radiobiological Institute TNO, 151 Lange Kleiweg, 2280 HV Rijswijk, The Netherlands

O. Vos, Department of Cell Biology and Genetics, Erasmus University, P.O. Box 1738, Rotterdam, The Netherlands

Richard I. Walker, Experimental Hematology Department, Armed Forces Radiobiology Research Institute, National Naval Medical Center, Bethesda, Maryland 20014

E. B. Watkins, Departments of Anatomy and Radiology, University of Nebraska Medical Center, Omaha, Nebraska 68105

Paul L. Weiden, Divisions of Hematology and Oncology, Department of Medicine, University of Washington, Fred Hutchinson Cancer Research Center and Veterans Administration Hospital, Seattle, Washington 98108

Wieslaw Wiktor-Jedrzejczak, Department of Nuclear Medicine, Postgraduate Center, Military School of Medicine, 00-909 Warsaw, Poland

Drew Winston, UCLA School of Medicine, Center for the Health Sciences, Los Angeles, California 90024

Cathy M. Wise, Divisions of Hematology and Oncology, Department of Medicine, University of Washington, Fred Hutchinson Cancer Research Center and Veterans Administration Hospital, Seattle, Washington 98108

E. A. J. Wolters, Department of Cell Biology and Genetics, Erasmus University, P.O. Box 1738, Rotterdam, The Netherlands

S. K. Wright, The Johns Hopkins Oncology Center, Baltimore, Maryland

Lowell S. Young, UCLA School of Medicine, Center for the Health Sciences, Los Angeles, California 90024

Gary Van Zant, Department of Biochemistry, and The Franklin McLean Memorial Research Institute, University of Chicago, Chicago, Illinois 60637

Lyda Vellekoop, M.D., M.D. Anderson Hospital, Department of Developmental Therapeutics, 6723 Bertner, Houston, Texas 77030

PART I

Studies of Hemopoietic Stem Cells

T. J. MacVittie

1

Alterations Induced in Murine Hemopoietic Stem Cells Following a Single Injection of *Corynebacterium parvum*

T. J. MacVittie

Corynebacterium parvum is one of several agents shown to have marked anti-tumor activity (10,20). The tumor growth-inhibiting activity has been associated with the ability of *C. parvum* to stimulate the mononuclear phagocyte system. Considerable evidence has implicated the activation of macrophages as nonspecific effector cells in the *C. parvum*-elicited response of the host to the presence of tumor. This implication is supported by the marked stimulation of marrow and spleen-derived granulocyte-macrophage precursors (CFU-c) in normal (2,3,6,7,-24) and tumor-bearing mice (4,23) following injection of such agents as *C. parvum,* BCG, and glucan. Fisher et al. (5) have also shown that the macrophages cloned from the responsive CFU-c in mice bearing tumors are potentially cytotoxic toward tumor cells.

We have recently observed marked changes in proliferation and migration of both the CFU-c and the macrophage colony-forming cell (M-CFC) within the hemopoietic and lymphoid organs following an injection of *C. parvum* (13). Our results revealed the potential of M-CFC mobilization and amplification within the extramedullary organs and tissue spaces as well as the changes induced in the hemopoietic system.

The increased production of CFU-c and M-CFC implied a significant effect, direct or indirect or both, upon the pluripotent stem cell. Although there is a lack of data concerning the response of the hemopoietic stem cell (CFU-s) to such agents as *C. parvum,* BCG, and glucan, two recent reports have dealt with certain aspects of the stem cell response. Maruyama (16) has shown a significant increase in the endogenous CFU-s within 24 hr following *C. parvum* injection. The stimulation extended for at least 7 days and was characterized by a rapid movement of endogenous CFU-s into the S phase. Gordon et al. (9) have shown a stimulatory effect of *C. parvum* and BCG on both bone marrow- and spleen-derived exogenous CFU-s that were being exposed to continuous wholebody irradiation. They indicated that BCG and *C. parvum* can maintain hemopoietic precursor cell levels above the irradiated values for up to 5 weeks. Although these reports indicated sustained stimulation of stem cells, the temporal response of the stem cell population within bone marrow and its mobilization, migration, and proliferation within the spleen are unknown.

This report describes the alterations induced within the exogenous stem cell populations derived from the bone marrow, spleen, and peripheral blood as well as the endogenous stem cell for 25 days following a single injection of *C. parvum*. It is essential to know the status of the hemopoietic stem cell in designing combined modality therapy as well as in understanding the mechanism of action involved in nonspecific stimulation of the mononuclear phagocyte system (25) by immunoadjuvants.

MATERIALS AND METHODS
Cell Suspensions

Femoral bone marrow (BM), spleens (SPL), and peripheral blood leukocytes (PBL) were obtained from 8- to 12-week-old male or female mice of the strain B6D2F1/Cum BR (Cumberland View Farms, Clinton, Tenn.). An appropriate number of BM, SPL, and PBL were suspended in McCoy's 5a medium with 25 mM Hepes buffer and 5% fetal calf serum (McCoy's 5a/5 FCS). Blood was obtained by cardiac puncture of mice under ether anesthesia.

The animals were maintained on a 6:00 A.M. to 6:00 P.M. light–dark cycle. Wayne Lab-Blox and acidified (pH 2.5) water were available *ad libitum*. All mice were acclimated to laboratory conditions for 2 weeks. During this time, the mice were examined and found to be free of lesions of murine pneumonia complex and of oropharyngeal *Pseudomonas* sp.

Stem Cell Assays and Seeding Efficiency

Endogenous colony-forming cells (15) were counted on the spleens of control and experimental mice (8 to 10 per group), 9 days after 550 rad total-body irradiation from the AFRRI ^{60}Co source at a dose rate of 153 rad/min. Exogenous colony-forming cells (CFU-s) were determined by the spleen colony assay of Till and McCulloch (22). The seeding efficiency (f-fraction) of pooled BM and SPL samples from four to six mice was determined as described by Siminovitch et al. (21). The primary irradiated recipients were injected with 2×10^6 nucleated BM cells or 2×10^7 nucleated SPL cells and sacrificed 2 hr later; spleen suspensions were subsequently prepared for injection into secondary irradiated recipients. Secondary recipients (8 to 10) were each injected with approximately one-fifth of a spleen and sacrificed 9 days later.

Proliferative State of Exogenous CFU-s

The proliferative status of CFU-s derived from marrow and spleen were assayed using hydroxyurea (Sigma Chemical Co., St. Louis, Mo.). The hydroxyurea (900 mg/kg) was diluted in saline and injected intravenously 2 hr prior to sacrifice of control and experimental groups. Hydroxyurea, a rapidly metabolized drug, is selectively lethal for cells in DNA synthesis. The decrease in the number of CFU-s per femur or spleen 2 hr after hydroxyurea injection allowed us to estimate the percentage of these cell populations synthesizing DNA at the time of administration.

Corynebacterium parvum

Corynebacterium parvum (Burroughs Wellcome Co., Research Triangle Park, N.C.) strain 6134 was kindly supplied by Dr. Richard Tuttle, at 7 mg (dry weight) per milliliter suspension of washed, formalin-killed organisms in saline solution. The organisms were injected intraperitoneally in a 0.2 ml volume (1.4 mg dry weight) per mouse.

Statistical Analyses

The data presented represent the mean values ($\pm$SE) of at least four replicate experiments. Cell suspensions from four to six mice were pooled for the determination of each datum point within each replicate experiment. The two-tailed Student's t test was used to determine the statistical significance of mean values.

RESULTS
Marrow, Spleen, and Peripheral Blood Cellularity after Injection of Mice with C. parvum

Femoral and splenic cellularity were markedly affected by a single injection of *C. parvum* (Table 1.1). Marrow cellularity decreased to values approximately 50% of control within 48 hr and remained at that level through 14 days and then rose to control levels by 20 days. Splenic cellularity, however, increased significantly ($p < .005$) within 4 days after injection of *C. parvum*, rose to peak cellularity (364% of control) at 17 days, and then began to fall, although it remained at levels higher than control

TABLE 1.1 Alterations in Bone Marrow, Spleen, and Peripheral Blood Cellularity[a] following a Single Injection of 1.4 mg *Corynebacterium parvum*, CN 6134

NO. DAYS CONTROL: C. parvum	BONE MARROW 2.10 ± 0.18	SPLEEN 16.47 ± 1.05	PERIPHERAL BLOOD LEUKOCYTES 0.41 ± 0.09
1	1.46 ± 0.36[b]	19.08 ± 0.7	0.39 ± 0.06
2	0.97 ± 0.20[b]	25.58 ± 3.6	0.61 ± 0.08
3	0.80 ± 0.18[b]	22.10 ± 1.7	0.62 ± 0.17
4	1.17 ± 0.15[b]	27.80 ± 2.7[b]	0.64 ± 0.11
7	1.06 ± 0.25[b]	37.90 ± 1.9[b]	0.69 ± 0.07
10	1.11 ± 0.19[b]	49.13 ± 4.2[b]	0.58 ± 0.12
14	0.80 ± 0.15[b]	68.46 ± 13.1[b]	0.57 ± 0.10
17	1.33 ± 0.40[b]	69.16 ± 15.2[b]	0.57 ± 0.07
20	2.35 ± 0.36	44.90 ± 4.4[b]	0.44 ± 0.11
23	2.24 ± 0.60	45.20 ± 1.3[b]	0.63 ± 0.15
25	1.73 ± 0.31	33.53 ± 2.1[b]	0.49 ± 0.11
28	2.05 ± 0.19	47.20 ± 3.1	0.49 ± 0.06

[a]Mean values ($\pm$ SEM) of four replicate experiments are 10^7, total nucleated cells per organ (bone marrow, 1 femur) or per milliliter of peripheral blood.

[b]Mean values differ significantly from respective control at $p < .001$ for BM and $p < .005$ for SPL.

levels through the 28th day of observation. Only slight changes were observed in peripheral blood leukocytes.

Concentration of Exogenous CFU-s in Bone Marrow, Spleen, and Peripheral Blood Leukocytes and the Response of Endogenous CFU-s following Injection of Mice with *C. parvum*

Following the injection of *C. parvum*, the concentration of exogenous, BM-derived CFU-s increased to $69/10^5$ nucleated cells within 2 days, maintained that concentration for the 3rd day, and then returned to within normal values by the 4th day (Table 1.2). The concentration of spleen-derived exogenous CFU-s increased from a control value of $3/10^5$ nucleated cells to $8.2/10^5$ cells (256% of control) within 24 hr ($p < .01$), rose to a peak concentration of $20.2/10^5$ cells by the 7th day (631% of control), followed by a decrease and then a second increase in concentration at day 21 to 244% of the control values. The concentration of circulating exogenous CFU-s peaked at 3 days after injection of *C. parvum* and then decreased to within normal values for the remainder of the 21-day observation period.

Endogenous CFU-s increased significantly within 24 hr after injection of *C. parvum* to values 537% of control ($p < .001$) (Table 1.2). This level was maintained throughout the 21-day observation period. In addition, the spleen size increased in association with the increased level of endogenous CFU-s (Table 1.2).

Femoral, Spleen, and Peripheral Blood Content of Exogenous CFU-s following an Injection of Mice with *C. parvum*

The relative increase in exogenous CFU-s through the first 3 days following *C. parvum* injection helped maintain the femoral stem cell content at control levels through the 4th day (Fig. 1.1). Thereafter, the CFU-s diminished in absolute numbers to 25% of control by the 12th day. Femoral content of CFU-s returned to normal levels by the 17th day and remained thereafter at control levels. Splenic content of exogenous CFU-s rose steadily over a 7-day period to 1,300% of control, decreased over the 2nd week to a content nearly 400% of control and rose again through the 3rd week to 700% of control at day 25 after *C. parvum* injection. Exogenous CFU-s per milliliter of blood increased to a level 1,200% of control within 3 days, and then decreased sharply to levels approximately 200 to 300% of control by the 7th day and remained at this level throughout the remaining 18 days of observation.

Seeding Efficiency of Exogenous CFU-s Derived from Bone Marrow and Spleen after Injection of Mice with *C. parvum*

The seeding efficiency of marrow- and spleen-derived CFU-s for control mice was 12.9% ± 1.4% and 7.0% ± 1.2%, respectively (Table 1.3). Three days after *C. parvum* injection, the respective seeding efficiencies of marrow and spleen CFU-s decreased to 6.7% and 2.1%. The diminished efficiency of detecting the CFU-s indicated an even greater increase in marrow and splenic content of CFU-s.

Proliferative State of Exogenous CFU-s Derived from Bone Marrow and Spleen

The proliferative capacity of the exogenous CFU-s was measured in both BM and SPL. At both 3 and 7 days following the injection of mice with *C. parvum*, the majority of the femoral and splenic CFU-s had entered the cell cycle. The cytotoxic drug hydroxyurea reduced the content of femoral CFU-s by 45% ± 9% and 58% ± 9% at 3 and 7 days, respectively (Table 1.4). The splenic content of CFU-s was also reduced significantly by values 53% ± 8% and 41% ± 8% at the 3rd and 7th day after injection of mice with *C. parvum*. Control marrow and spleen CFU-s were reduced by 12% and 20%, respectively.

DISCUSSION

A single injection of *C. parvum* into mice produced marked alterations within the hemopoietic stem cell compartments as measured by both exogenous and endogenous techniques. The femoral content of CFU-s remained within normal values over the 1st week following injection of mice with *C. parvum* and then decreased to values 25% of control by the 12th day. Control levels were reached by the 17th day. Circulating CFU-s were increased 12-fold within 3 days, decreased to levels approximately threefold by the 7th day, and then remained at that elevated level through the 25th day after *C. parvum* injection. The splenic CFU-s showed a marked 13-fold increase in absolute values by the 7th day, then decreased to levels fourfold of control at the 14th day, only to be followed by a second peak at the 23rd day to values sevenfold of control. In addition, a significant five- to eightfold increase in the endogenous CFU-s compartment was observed through 21 days following the injection of *C. parvum*.

It is significant that, following injection of mice with *C. parvum*, the femoral content of CFU-s remained within control values over the first 7 days in spite of a 50% drop in cellularity, which was

TABLE 1.2 Concentration of Exogenous Stem Cells[a] in Bone Marrow, Spleen, and Peripheral Blood Leukocytes and the Endogenous Stem Cell[b] Response following Injection of C. parvum (6134)

Stem Cells	Control	DAYS AFTER C. PARVUM								
		1	2	3	4	7	10	14	17	21
Exogenous										
Bone marrow	35.5 ± 2.2	48.9 ± 3.9[b]	69.3 ± 10.3[b]	69.2 ± 5.6[b]	41.7 ± 3.4	36.2 ± 6.8	31.2 ± 5.1	35.8 ± 2.4	44.0 ± 4.6	57.5 ± 0.8[b]
Spleen	3.2 ± 0.7	8.2 ± 1.7[b]	9.7 ± 1.2[b]	15.8 ± 3.7[b]	19.3 ± 3.1[b]	20.2 ± 6.8[b]	9.2 ± 2.5[b]	3.8 ± 0.6	5.2 ± 0.9	7.8 ± 0.5
Peripheral blood leukocyte	0.9 ± 0.5	1.8 ± 0.3	3.9 ± 0.8[b]	4.2 ± 0.6[b]	2.5 ± 0.3	1.7 ± 0.1	1.2 ± 0.4	1.9 ± 0.2	0.8 ± 0.1	1.46 ± 0.1
Endogenous	4.3 ± 0.55	23.1 ± 2.0[b]	37.4 ± 6.2[b]	32.1 ± 4.2[b]	27.3 ± 3.7[b]	34.5 ± 7.1[b]	28.0 ± 5.0[b]	40.1 ± 4.7[b]	25.0 ± 6.0[b]	27.1 ± 3.1[b]
Spleen weight (mg)	26 ± 2.4	44 ± 3.6	106 ± 8.5	60 ± 5.1	62 ± 3.2	126 ± 6.1	127 ± 4.3	116 ± 5.7	—	—

[a]Exogenous values (± SEM) are means per 10^5 nucleated cells of at least four replicate experiments.

[b]Values are significantly different from control BM, $p < .01$ to .0005; SPL, $p < .05$ to .005; PBL, $p < .025$, Endogenous values are means (± SEM) (number of surface colonies per spleen, eight spleens per group) $p < .0005$. Spleen weight equals mean value (± SEM) of endogenous spleens.

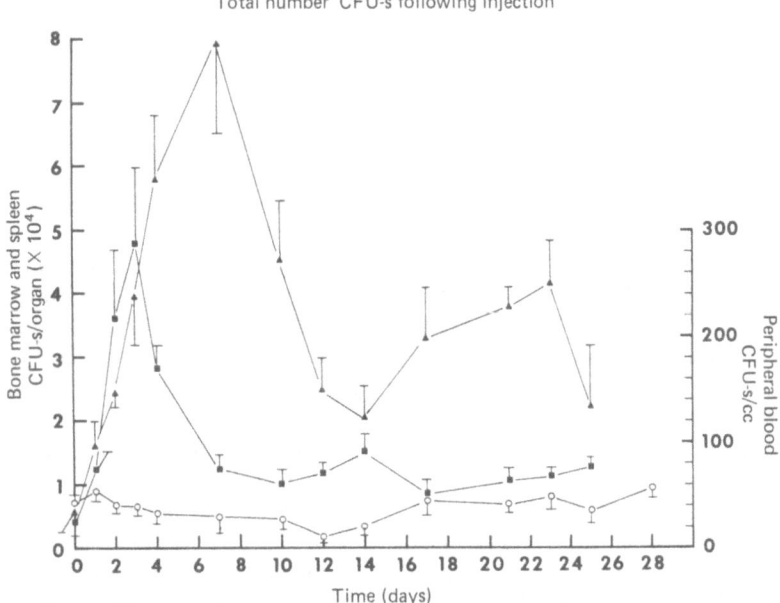

Total number CFU-s following injection

FIGURE 1.1. The number of CFU-s per femur and spleen and per milliliter peripheral blood at various times after the injection of 1.4 mg of *Corynebacterium parvum* (6134) in mice. Values are means (± SEM) of at least four replicate experiments. Bone marrow (o——o), spleen (▲——▲), and peripheral blood (■——■).

maintained through day 14. It is apparent from our data that at least two factors are responsible for the maintenance of CFU-s within control levels. The concentration of CFU-s had increased twofold by 48 to 72 hr, through an apparent differential depletion of non-CFU-s as well as the entrance of a large number of CFU-s into the cell cycle. The hydroxyurea data indicated approximately 45% of the femoral content of CFU-s were in S phase by day 3. This increased proliferation was probably maintained through day 7 when the CFU-s content was again diminished by 51% after exposure to hydroxyurea.

The differential depletion of non-colony-forming cells was also observed within the granulocyte-macrophage colony-forming cells (CFU-c) following the injection of *C. parvum* (8, 13). The concentration of CFU-c was noted to increase at least twofold through the first 2 weeks (13). This effect is somewhat different from that observed following the

injection of endotoxin (11,18,26) and *Bordetella pertussis* (17). The characteristic response to these agents involves a decreased content of marrow CFU-s concomitant with decreased cellularity, mobilization of CFU-s into the circulation, and a variable onset of proliferation ranging from day 1 to day 7. In addition, the decreased seeding efficiency indicated that our observed effect was an underestimate of the actual CFU-s response. A similar drop in seeding efficiency was also observed following treatment with endotoxin by Quesenberry (18) and Fred and Smith (8).

The gradual decrease in femoral CFU-s to 25% of control by day 12 was most likely due to a marked

TABLE 1.3 Seeding Efficiency (%)[a] of Exogenous CFU-s Derived from Bone Marrow and Spleen 3 Days after Injection of *C. parvum*

TISSUE	CONTROL	*C. PARVUM*
Marrow	12.9 ± 1.4	6.7 ± 1.6
Spleen	7.0 ± 1.2	2.1 ± 2.1

[a]Mean percent seeding efficiencies (±SEM) of marrow and spleen-derived CFU-s. Data based on three replicate experiments.

TABLE 1.4 Percent Depression in Exogenous CFU-s per Organ[a] Derived from Bone Marrow and Spleen 3 Days and 7 Days after a Single Injection of *C. parvum*

	% DEPRESSION CFU-S/ORGAN (%)	
TIME	Bone Marrow	Spleen
Control	12 ± 3.1	20 ± 5.6
3 Days	45 ± 8.8	53 ± 6.7
7 Days	58 ± 9.3	41 ± 8.2

[a]Values are means (±SEM) of three replicate experiments. Percent depression in CFU-s per organ are calculated from CFU-s per 10^5 nucleated cells × total nucleated cells per organ = CFU-s per organ. CFU-s/organ (hydroxyurea) CFU-s/organ (saline) × 100 = Percent depression in CFU-s due to hydroxyurea.

differentiative pull from the committed CFU-c (3,4,-7,13) and macrophage colony-forming cell (M-CFC) compartments (13). In another report (13), we emphasized the increased proliferation of femoral and extramedullary CFU-c and M-CFC in association with hyperplasia of the mononuclear phagocyte system in response to injection of mice with *C. parvum.*

The increase in number of circulating CFU-s to 15-fold control within 3 days is most likely the result of the mobilization of femoral CFU-s. This early release of CFU-s from the marrow has also been observed in response to endotoxin (18) and *B. pertussis* (17) as well as to the stress of anemia induced by phenylhydrazine injection (12, 19). In addition to the early mobilization of CFU-s, the peripheral blood level of CFU-s was maintained at approximately threefold of control through the 25th day of observation. This suggested either a continuous release of CFU-s from the marrow, a residual population of migrating CFU-s that are unable to seed within an extramedullary organ, or a continual release of CFU-s from extramedullary organs, such as the spleen.

The CFU-s content within the spleen showed a biphasic response to the injection of *C. parvum:* an early exponential rise in CFU-s over the 1st week, then a decrease followed by a second, smaller rise in CFU-s content over the 3rd week. Both may have been the result of the proliferating resident CFU-s as well as the seeding of circulating CFU-s, with subsequent proliferation. The hydroxyurea technique indicated a large number of CFU-s were proliferating at day 3 and day 7. In addition, determination of the seeding efficiency showed that the splenic CFU-s at day 3 decreased from a control value of 6.7 to 2.1%. As in the case of the femoral CFU-s, this indicated that the observed splenic CFU-s response was an underestimate. The CFU-s proliferation was also associated with a marked increase in spleen size. Foster (7) had previously shown a significant increase in splenic CFU-c content following injection of *C. parvum.* We (13) have recently confirmed such an increase beyond the stem cell in addition to providing evidence for a significant and prolonged rise in the recently detected, M-CFC (14).

Monette et al. (17) have amply demonstrated a similar stem cell response within the spleen of mice injected with *B. pertussis.* Splenic CFU-s increased exponentially and reached levels 12-fold the control value by the fifth day. They also observed an increase in splenic CFU-s entering cell cycle in association with the early rise in absolute number, followed by a resting phase. The response to *C. parvum* was similar in form, although apparently more intense and prolonged in terms of CFU-s prolifera-

tion and amplification within the marrow, spleen, and peripheral blood. Further evidence of a sustained stem cell response was also provided by the data on endogenous CFU-s. We observed a significant five- to eightfold increase in stem cells measured by this technique throughout 21 days following *C. parvum* injection. Maruyama (16) has recently shown that injection of mice with *C. parvum* from 4 hr to 7 days prior to an exposure of 550 rad significantly increased the number of surviving endogenous colony-forming cells. He postulated that the altered radiosensitivity was due to a progression of CFU-s from the resting phase into cell cycle and proliferation. Boggs et al. (1) suggested that the endogenous technique may measure a stem cell population that is predominantly in cell cycle and may therefore be indicative of the cycling portion of exogenous CFU-s. Our data are compatible with this concept.

The data presented in this report illustrated a marked hemopoietic stem cell response following injection of *C. parvum,* a nonspecific stimulant of the mononuclear phagocytic and immune systems. The mechanism(s) involved in this response remains to be elucidated. It appears to be similar in nature to that observed following the use of such agents as endotoxin or *B. pertussis* or the production of severe anemia. It is possible that *C. parvum* exerts a more intense differentiative stress through stimulation of the mononuclear phagocyte system. Another possibility is that *C. parvum* influences the microenvironment within the marrow and spleen and the result is a more prolonged effect on stem cell proliferation, migration, and differentiation. The stem cell is the ultimate source of monocyte and macrophage precursors, the progeny of which serve as anti-tumor effector cells of the mononuclear phagocyte system. Therefore, it is imperative that the integrity of this population remain intact following therapy with cytotoxic agents used in combination with immunoadjuvants. Knowledge of the relative effects on stem cell and progenitor cell populations, as well as their potential for amplification of effector cells, is essential to the timing of combined modality or multi-dose therapy.

SUMMARY

A single injection of *Corynebacterium parvum* into normal mice resulted in marked alterations within the hemopoietic stem cell (CFU-s) populations as measured by both exogenous and endogenous techniques. Exogenous stem cell content was determined for bone marrow, spleen, and peripheral blood. In addition, we measured the fraction of marrow and spleen CFU-s in cell cycle as well as their seeding efficiencies.

Following an injection of mice with *C. parvum*, there was an initial differential depletion of non-CFU-s within the bone marrow. This resulted in a twofold increase in CFU-s concentration and the consequent maintenance of normal CFU-s content during the 1st week. During this period, marrow CFU-s entered cell cycle, their seeding efficiency decreased, and a certain fraction was mobilized into the peripheral circulation. The CFU-s in the circulation peaked at 3 days; then decreased to a value approximately threefold of control by day 7. This elevated level was maintained in the circulation through the next 2 weeks.

The splenic content of CFU-s increased exponentially to peak values 12-fold greater than control by day 7. As with the marrow, a large number of splenic CFU-s had entered the cycle by day 3 and remained in cycle at day 7; they had a significantly decreased seeding efficiency. Stem cell content decreased over the 2nd week and then peaked again over the 3rd week to values sixfold greater than control. The increased stem cell activity was associated with a significant increase in spleen size and a marked five- to eightfold increase in endogenous colony formation from day 1 through day 21. These results indicated an intense and prolonged effect of *C. parvum* on the hemopoietic stem cell populations.

REFERENCES

1. Boggs, S. S., Boggs, D. R., Neil, G. L., and Sartiano, G. Cycling characteristics of endogenous spleen colony-forming cells as measured with cytosine arabinoside and methotrexate. *J. Lab. Clin. Med., 82*:725, 1973.
2. Burgaleta, C., and Golde, D. W. Effect of glucan on granulopoiesis and macrophage genesis in mice. *Cancer Res., 37*:1739, 1977.
3. Dimitrov, N. V., Andre, S., Elioporelos, G., and Halpern, B. Effect of *C. parvum* on bone marrow cell cultures. *Proc. Soc. Exp. Biol. Med., 148*:440, 1975.
4. Fisher, B., Taylor, S., Levine, M., Saffer, E., and Fisher, E. R. Effect of *Mycobacterium bovis* (strain Bacillus Calmette-Guerin) on macrophage production by the bone marrow of tumor-bearing mice. *Cancer Res., 34*:1668, 1974.
5. Fisher, B., Wolmark, N., Coyle, J., and Saffer, E. A. The effect of a growing tumor and its removal on the cytotoxicity of macrophages from cultured bone marrow cells. *Cancer Res., 36*:2302, 1976.
6. Foster, R. S., Jr. Effect of *Corynebacterium parvum* on the proliferative rate of granulocyte-macrophage progenitor cells and the toxicity of chemotherapy. *Cancer Res., 38*:2666, 1978.
7. Foster, R. S., Jr. MacPherson, B. R., and Browdie, D. A. Effect of *Corynebacterium parvum* on colony-stimulating factor and granulocyte-macrophage colony formation. *Cancer Res., 37*:1349, 1977.
8. Fred, S. S., and Smith, W. W. Induced changes in transplantability of hemopoietic colony-forming cells. *Proc. Soc. Exp. Biol. Med., 128*:364, 1968.
9. Gordon, M. Y., Aquado, M., and Blackett, N. M. Effects of BCG and *Corynebacterium parvum* on the haemopoietic precursor cells in continuously irradiated mice: Possible mechanisms of action in immunotherapy. *Europ. J. Cancer, 13*:229, 1977.
10. Halpern, B., Prevot, A. R., Biozzi, G., Stiffel, C., Mouton, D., Morard, J., Bouthillier, Y., and Decreusefound, C. Stimulation de l'activite phagocytaire du reticuloendothelial provoguee par *Corynebacterium parvum. J. Reticuloendothel. Soc., 1*:77, 1964.
11. Hanks, G. S., and Ainsworth, E. J. Endotoxin protection and colony forming units. *Radiat. Res., 32*:367, 1967.
12. Hodgson, G. S. Properties of haemopoietic stem cells in phenyl hydrazine treated mice. *Cell Tissue Kinet., 6*:199, 1973.
13. MacVittie, T. J. Alterations induced in macrophage and granulocyte-macrophage colony-forming cells by a single injection of *Corynebacterium parvum. Cancer Res.* (Submitted)
14. MacVittie, T. J., and Porvaznik, M. Detection of *in vitro* macrophage colony-forming cells (M-CFC) in mouse bone marrow, spleen, and peripheral blood. *J. Cell. Physiol., 97*:305, 1978.
15. Marsh, J. C., Boggs, D. R., Bishop, C. R., Chervenick, P. A., Cartwright, G. E., and Wintrobe, M. M. Factors influencing hematopoietic spleen colony formation in irradiated mice. I. The normal pattern of endogenous colony formation. *J. Exp. Med., 126*:833, 1967.
16. Maruyama, Y., Magura, C., and Feola, J. *Corynebacterium parvum*-induced radiosensitivity and cycling changes of hematopoietic spleen colony-forming units. *J. Nat. Cancer Inst., 59*:173, 1977.
17. Monette, F. C., Morse, B. S., Howard, D., Niskanen, E., and Stohlman, F., Jr. Hematopoietic stem cell proliferation and migration following *Bordetella pertussis* vaccine. *Cell Tissue Kinet., 5*:121, 1972.
18. Quesenberry, P. J., Morley, A., Ryan, M., Howard, D., and Stohlman, F., Jr. The effect of endotoxin on murine stem cells. *J. Cell Physiol., 82*:239, 1973.
19. Rencricca, N. J., Rizzoli, V., Howard, D., Duffy, P., and Stohlman, F., Jr. Stem cell migration and proliferation during severe anemia. *Blood, 36*:764, 1970.
20. Scott, M. T. *Corynebacterium parvum* as an immunotherapeutic anti-cancer agent. *Semin. Oncology, 1*:367, 1974.
21. Siminovitch, L., McCulloch, E. A., and Till, J. E. The distribution of colony-forming cell among spleen colonies. *J. Cell. Comp. Physiol., 62*:327, 1963.
22. Till, J. E., and McCulloch, E. A. A direct measurement of the radiation sensitivity of normal mouse bone marrow cells. *Radiat. Res., 14*:214, 1961.
23. Wolmark, N., and Fisher, B. The effect of a single and repeated administration of *C. parvum* on bone marrow

macrophage colony production in syngenic tumor-bearing mice. *Cancer Res., 34*:2869, 1974.

24. Wolmark, N., Levine, M., and Fisher, B. The effect of a single and repeated administration of *Corynebacterium parvum* on bone macrophage colony production in normal mice. *J. Reticuloendothel. Soc., 16*:252, 1974.

25. vanFurth, R., Cohn, Z. A., Hirsch, J. G., Humphry, J. H., Spector, W. G., and Langevoort, H. L. The mononuclear phagocyte system: A new classification of macrophages, monocytes, and their precursors. *Bull. Wld. Hlth. Org., 46*:85, 1972.

26. Vos, O., Buurman, W. A., and Ploemacker, R. E. Mobilization of haematopoietic stem cells (CFU) into the peripheral blood of the mouse; effects of endotoxin and other compounds. *Cell Tissue Kinet., 5*:467, 1972.

2

Growth Control of Hemopoietic Colony-Forming Cells by Thymidine and Deoxycytidine

Suzanne Hasthorpe
and Alan W. Harris

Thymidine is not an essential nutrient for mammalian cells in culture, although it is rapidly incorporated into DNA when it is added to the culture. This incorporation involves phosphorylation to thymidine triphosphate (dTTP) and then assembly with the other three deoxynucleoside triphosphates into DNA. High levels of dTTP have a negative effect on a number of enzymes instrumental to its own production, but it can also block the action of ribonucleotide reductase on cytidine diphosphate (CDP) (Figure 2.1). Thus, treatment of cells with high concentrations of thymidine inhibits DNA synthesis by inducing a state of deoxycytidine nucleotide starvation (2,18,19). Such inhibition can be specifically reversed by the addition of deoxycytidine, which rapidly restores the intracellular pool of deoxycytidine triphosphate (dCTP) (2,17,23,24).

The concentrations at which thymidine is inhibitory vary somewhat for different cell types, with initial effects commonly occurring between 10 μM and 100 μM and complete inhibition above 100 μM (4). Deoxycytidine reverses the inhibition and allows cell growth to occur in the presence of large amounts of thymidine.

Tritiated thymidine suicide of cells that can form hemopoietic colonies *in vitro* is a method frequently used in experimental hematology to cell cycle status, but information about the sensitivity of such cells to unlabeled thymidine is lacking. In this report, detailed thymidine and deoxycytidine dose-response curves will be presented for erythroid burst-forming units (BFU-e), erythroid colony-forming units (CFU-e) (1,6,13,22), and granulocyte-macrophage colony-forming cells (GM-CFC) (16). In addition, the use of this reversible inhibition system in the study of the kinetics of colony formation in *vitro* will be described.

MATERIALS AND METHODS
Mice
Female BALB/c mice 2 to 3 months of age were used in all experiments.

Culture Technique
Femoral bone marrow cells were cultured in 0.8% methylcellulose containing Dulbecco's modified Eagle's medium (Cat. No. H16, Gibco, Grand Island, N.Y.) supplemented with nonessential amino acids; pyruvate; biotin; Vitamin B_{12}; egg lecithin; Na_2SeO_3 as described by Guilbert and Iscove (8); 0.6 mg/ml human transferrin (Behringwerke); and 100 μM 2-mercaptoethanol. Bovine serum albumin (BSA, Fraction V, Sigma Chemical Co., St. Louis, Mo.) and serum (Flow Laboratories, Irvine, Scotland) were added as follows: for CFU-e, 3% BSA and 5% fetal calf serum (FCS); for BFU-e, 1% BSA

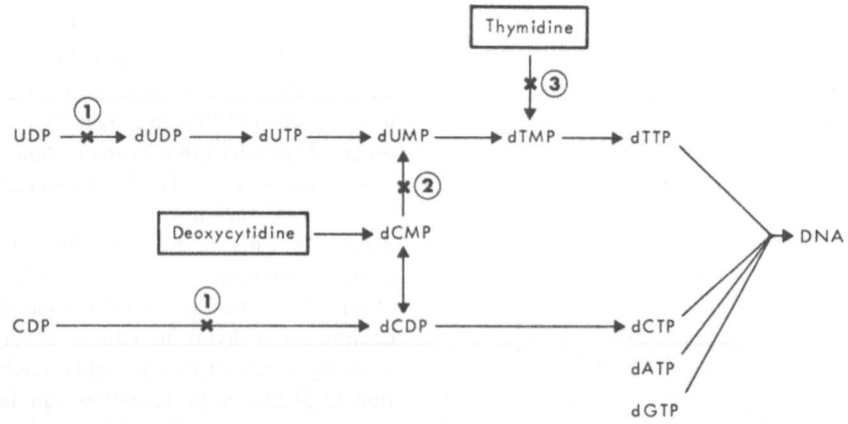

FIGURE 2.1. Pyrimidine deoxynucleotide synthesis pathways leading to DNA replication. Sites of inhibition by dTTP are indicated (**X**) and affect the following enzymes: (1) ribonucleoside diphosphate reductase; (2) deoxycytidylate deaminase; and (3) thymidine kinase. The reversible enzyme reaction is indicated (⟷).

and 10% FCS plus 10% horse serum; for GM-CFC, 1% BSA and 20% FCS. Single batches of FCS (number 2940097) and horse serum (number 402037) were used throughout.

Erythropoietin (step III, Connaught Laboratories, Toronto, Canada) was used at concentrations of 0.25 units/ml of culture for CFU-e and 1 unit/ml for BFU-e. Batch numbers used in these experiments were 3018-10 and 3019-6, which contained 5.9 and 13.9 units of erythropoietin/mg protein, respectively. The GM-CFC were cultured using a preparation of pregnant mouse uterus and embryo extract (3). Cultures of CFU-e were incubated for 2 days, BFU-e were scored at 3 days and 9 days, and GM colonies at 7 days. Incubation was at 37°C in a humidified atmosphere of 10% CO_2 in air.

Deoxynucleosides

Both thymidine (Sigma Chemical Co., St. Louis, Mo.) and 2'-deoxycytidine (Boehringer Mannheim, W. Germany) were dissolved in distilled water at concentrations of 10 mM and sterilized by passage through 0.22 μM Millipore filters. Dilutions were made in Hanks' balanced salt solution, buffered to pH 7.4 with HEPES, and then added to the cultures.

RESULTS

Effects of Thymidine on Colony Formation

The results of a detailed study of colony formation by bone marrow cells in the presence of various concentrations of thymidine are shown in Figure 2.2. A striking sensitivity of CFU-e was apparent, with substantial inhibition occurring even at thymidine concentrations below 10 μM. The BFU-e were less sensitive, with the 3-day BFU-e sensitivity being intermediate between CFU-e and the 9-day BFU-e.

Colony formation by GM-CFC was somewhat less thymidine sensitive.

The CFU-e, in addition to displaying the highest sensitivity, also showed a reduced dose-response rate curve. The colony numbers rose only gradually toward control levels as the thymidine concentration was lowered, indicating a considerable heterogeneity in thymidine responsiveness within the bone marrow CFU-e population. A more extensive dose-response curve in Figure 2.3 demonstrates a significant inhibition of colony development even at submicromolar concentrations of thymidine.

Tests were performed to ensure that the differences in thymidine sensitivity among the various cell types were intrinsic and not due to differences in the serum component of the culture media employed. The various types of progenitor cells under study could not all form colonies efficiently in a single culture medium and had restricted serum requirements; thus, serum intercomparisons were limited. Results obtained by culturing CFU-e in medium containing 5 or 10% FCS are shown in Figure 2.3. The difference in serum concentration had no substantial effect on CFU-e sensitivity to thymidine. Results of thymidine sensitivity assays on GM-CFC grown in medium containing 20% FCS or a mixture of 10% FCS with 10% horse serum also indicated that GM colony numbers were not markedly affected (Table 2.1) by the different sera.

Effect of Deoxycytidine on Colony Growth Inhibition by Thymidine

To determine whether inhibition by thymidine was due to its ability to inhibit deoxycytidine nucleotide synthesis, the effect of deoxycytidine in cultures containing an inhibitory concentration of thymidine (100 μM) was tested. Deoxycytidine restored colony

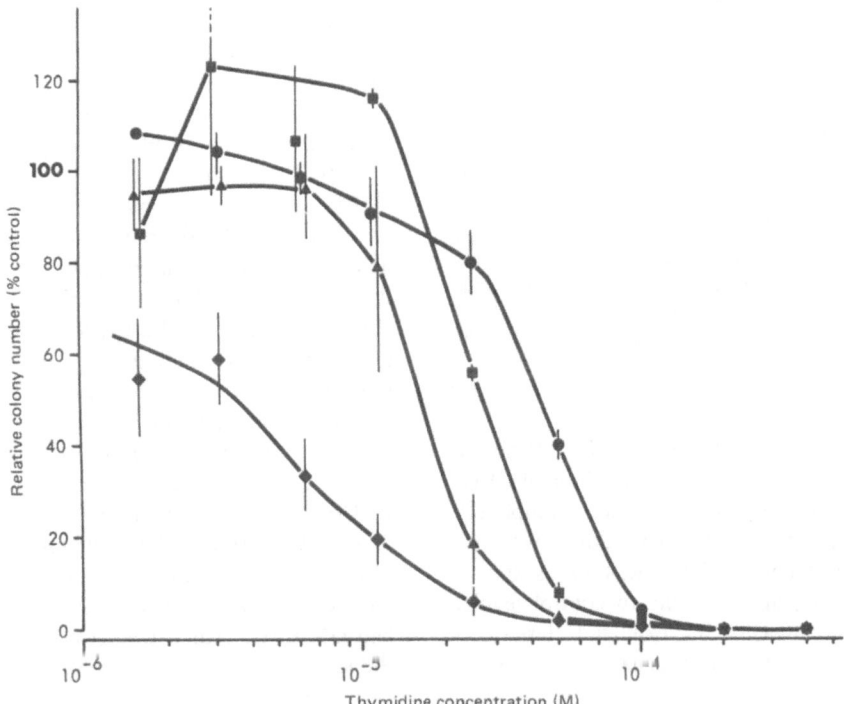

FIGURE 2.2. The effect of thymidine on colony formation. Colony numbers at various concentrations are indicated as follows: CFU-e (♦——♦); 3-day BFU-e (▲——▲); 9-day BFU-e (■——■); GM-CFC (●——●). The data from four experiments were used to construct dose-response curves. Means and standard errors for control numbers were CFU-e, 596 ± 75; 3-day BFU-e, 20.3 ± 2.8; 9-day BFU-e, 15 ± 2.0; GM-CFC, 72 ± 10/10⁵ bone marrow cells.

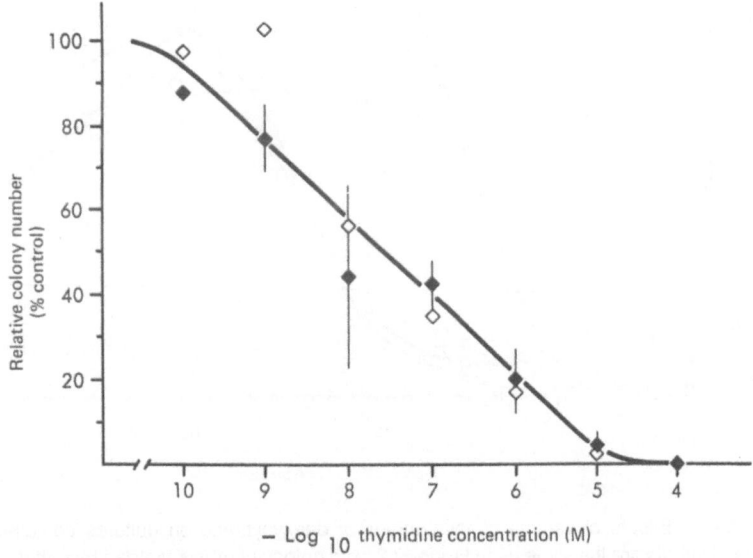

FIGURE 2.3. Titration of thymidine for inhibition of CFU-e in cultures containing 5 (♦——♦) and 10 (◊——◊) % FCS. For cultures containing 5% FCS, mean colony number from three experiments was 370± 124 CFU-e and for two experiments using 10% FCS, 550 ± 82 CFU-e/10⁵ marrow cells.

TABLE 2.1 Effect of Serum Type on Inhibition of Granulocyte-Macrophage Colony Formation by Thymidine[a]

SERUM	THYMIDINE CONCENTRATION (μM)					
	0	100	50	10	5	1
Fetal calf (20%)	79	3	22	80	77	93
Fetal calf (10%) + horse (10%)	81	4	30	71	81	87

[a]Mean colony number (triplicate cultures) is shown for GM-CFC cultures containing different concentrations of thymidine in the presence of fetal calf serum or a mixture of fetal calf and horse serum. Bone marrow cells were plated at a concentration of 10^5/ml culture.

numbers (Figure 2.4), but with differing efficiency for the various types of colonies. Inhibition of CFU-e colony formation by thymidine was completely overcome by about 0.4 μM deoxycytidine, whereas a tenfold higher concentration was necessary to restore erythroid burst numbers to control levels. Colony formation by GM-CFC exhibited a less steep deoxycytidine dose-response curve, with control

levels being attained around 1 μM. Furthermore, CFU-e colony formation was increased above control numbers by deoxycytidine even without added thymidine.

Colony-forming Cell Survival after Treatment with Thymidine

The survival of colony-forming cells in the presence of 100 μM thymidine, and their ability to achieve colony formation upon addition of 10 μM deoxycytidine is shown at various times after the cells were placed in culture (Figure 2.5). Here, CFU-e survival declined only slightly up to 12 hr of thymidine exposure, but between 12 and 24 hr the survival rate fell rapidly; GM-CFC survival was similarly affected in this period. On the other hand, a notable deviation from this pattern was seen with the 9-day BFU-e. Burst formation remained above 50% for at least 24 hr, with complete inhibition requiring 96 hr of thymidine exposure. The rate of decline in 9-day BFU-e numbers was also lower than that of CFU-e and GM-CFC, demonstrating a major difference in the kinetics of loss of colony-forming capacity.

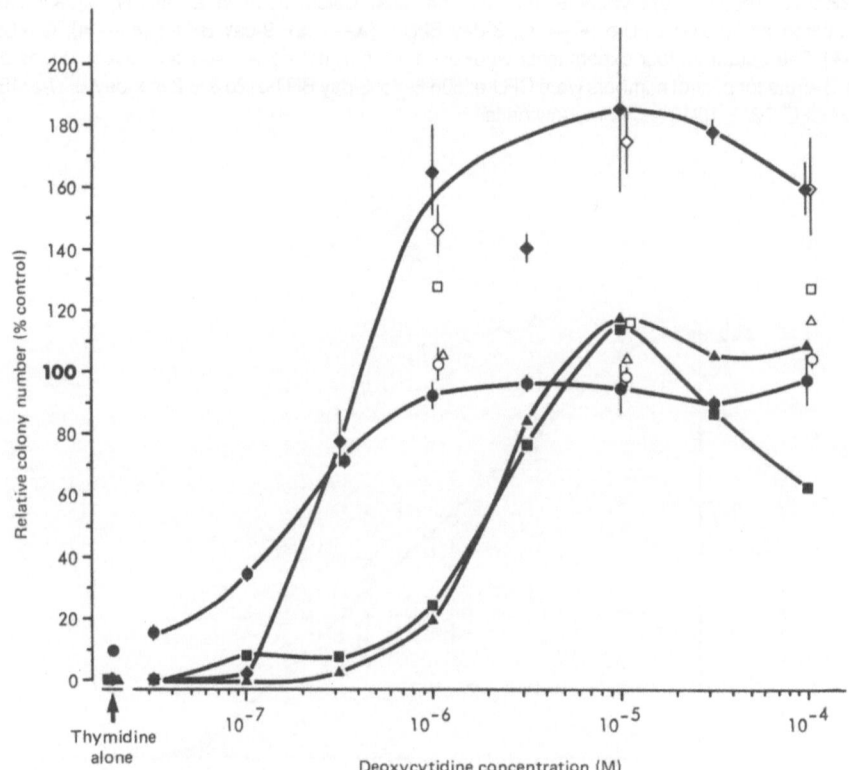

FIGURE 2.4. Effects of various concentrations of deoxycytidine on cultures containing 100 μm thymidine. Symbols are the same as in Figure 2.2, with colony numbers yielded by cultures containing 100, 10, or 1 μM deoxycytidine alone also shown (open symbols). Mean and standard error values are shown for duplicate CFU-e and GM-CFC experiments. Control colony numbers (100% levels) from cultures containing no added thymidine or deoxycytidine were as follows: CFU-e, 446 ± 109; 3-day BFU-e, 29; 9-day BFU-e, 18.5; GM-CFC, 87 ± 3 colonies/10^5 bone marrow cells.

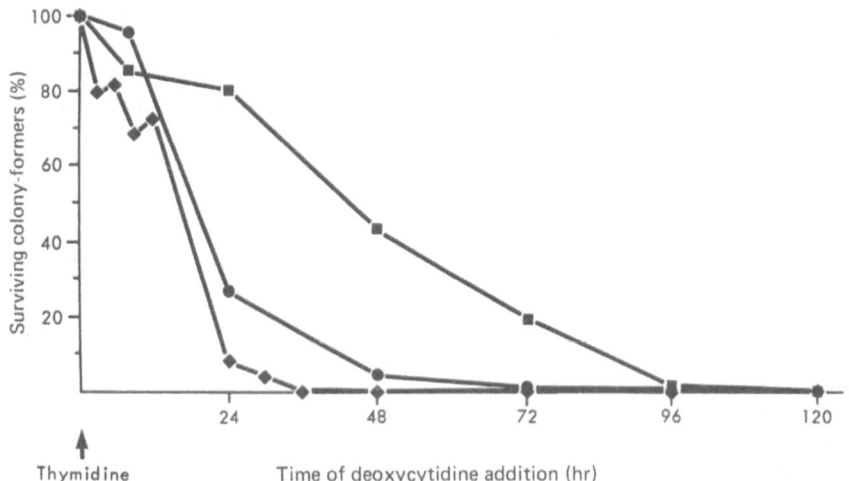

FIGURE 2.5. The effect of delayed deoxycytidine (10μM) in cultures containing 100 μM thymidine. Percentages of surviving colony-forming cells are shown as follows: CFU-e (♦——♦); GM-CFC (●——●); and 9-day BFU-e (■——■). Control cultures (100%) contained both thymidine and deoxycytidine from the beginning of incubation.

Effects of Delayed Addition of Thymidine to Cultures

For thymidine to affect cell survival, a cell must be in the DNA synthesis (S) phase during the time of exposure. The number of viable colonies will therefore depend on whether the colony-forming cell is in the cell cycle at that time. Most CFU-e from normal bone marrow are actively cycling, about 80% being killed in a tritiated thymidine suicide assay, but only 46% of the GM-CFC are killed (unpublished observations) in the same assay and might, therefore, be expected to exhibit a delay in their response to unlabeled thymidine. Results from an experiment in which thymidine was added to GM-CFC cultures either at the time of culturing or following a 24-hr incubation period are shown in Figure 2.6. No difference was found in the rate of decline of the colony-forming cells. This suggests that the GM-CFC enter S phase shortly after being placed in culture.

DISCUSSION

The quantitative studies of sensitivity to growth inhibition by thymidine reported here were motivated partly by some observations made during tritiated thymidine suicide experiments with CFU-e and partly by previous findings that some types of cells of

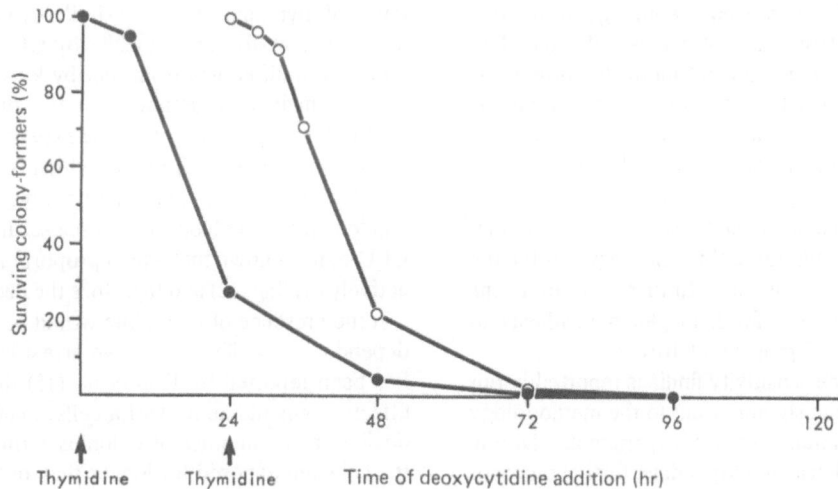

FIGURE 2.6. The effect of thymidine addition at the beginning (●——●) or after 24 hr (o——o) incubation on the numbers of GM colonies formed with subsequent addition of deoxycytidine (10 μM).

the lymphoid cell series are unusually sensitive. Growth and DNA synthesis of thymocytes and T-lymphoma cells are inhibited by concentrations of thymidine as low as 10 to 20 μM (9,11,19,23), whereas several other types of cells are inhibited only by concentrations above 100 μM (2,4,9,11,17,-19). From the present work it appears that the CFU-e is the most sensitive cell type yet found, since concentrations below 1 μM thymidine substantially inhibited colony formation.

The difference in thymidine sensitivity of the colony-forming cells in culture raises an important question about the relationship of the synthetic pathways of thymidine and deoxycytidine nucleotide in these cells. The observation that the three types of erythroid colony-forming cells tend to increase in sensitivity to thymidine with their relative degree of maturity suggests that the pathways may change during erythroid differentiation. These changes may be associated with those reported by others showing comparisons of 9-day BFU-e, 3-day BFU-e, and CFU-e in erythropoietin requirement, cell cycle phase, and physical properties (7,12). The growth of fetal liver CFU-e in the presence of high thymidine concentrations (4×10^{-4}M) (5) and no added deoxycytidine suggests that changes in thymidine sensitivity may also occur during ontogeny.

Differences in sensitivity to growth inhibition by thymidine correlating with the state of developmental maturation have been reported in the lymphoid cell series (9,19). In that series, however, thymidine sensitivity appears to decline with cell maturation.

One aspect of the different sensitivities of CFU-e and BFU-e to thymidine is unclear, however. From the presumed developmental relationship between the cells that form erythroid colonies and the cells that form burst colonies, it would be reasonable to expect that BFU-e progeny would go through a thymidine-sensitive stage of CFU-e–like development during their growth in culture to form burst colonies, but this did not seem to occur. Thymidine concentrations that inhibited CFU-e allowed normal formation of hemoglobin-containing bursts. The cell composition of 9-day BFU-e, at a CFU-e–containing stage, could influence their sensitivity to thymidine, and replating of cultures at this stage may resolve the controversy. Alternatively, further cell divisions may not be required for hemoglobin synthesis to occur in "CFU-e" progeny of BFU-e.

The thymidine sensitivity findings reported in this chapter are obviously important to the methodology of tritiated thymidine suicide experiments. During incubations with tritiated thymidine, further incorporation of labeled nucleoside can be stopped at the required time by the addition of unlabeled thymidine at concentrations of 100 μg/ml (4.1×10^{-4}M) (14).

This concentration is in the range in which DNA synthesis is inhibited, especially in CFU-e cultures. A more recently described method involves washing the cells with thymidine at a concentration of 40 μg/ml and adding all four deoxyribonucleosides (including deoxycytidine) to the final culture medium (12). The present experiments show that deoxycytidine alone is sufficient to overcome any inhibitory effect of the residual thymidine. They suggest, further, that pulses of [^{3}H]thymidine in suicide experiments could be terminated simply by the addition of a large excess of unlabeled thymidine to dilute the label, together with deoxycytidine to counteract the inhibitory effect of the excess thymidine. This may be feasible even when the cells are in a semi-solid culture medium.

The reversal of thymidine inhibition by low concentrations of deoxycytidine indicates that the mechanism of inhibition, i.e., arrest of dCTP synthesis, is the same in all cell types tested, whatever their thymidine sensitivity. The extreme sensitivity of CFU-e and the improvement of CFU-e plating efficiency produced by deoxycytidine even in the absence of added thymidine suggests that the normal rate of dCTP synthesis in these cells is barely sufficient to sustain DNA replication. In the whole animal, CFU-e may draw on extracellular sources of deoxycytidine. Concentrations of deoxycytidine in mammalian serum sufficient to reverse thymidine inhibition have been reported (20,21), although the results presented here suggest that any deoxycytidine present in the serum component of the culture medium could not be utilized by the CFU-e.

The selective inhibition of colony-forming cells, and reversal by deoxycytidine, provides a method to study cell kinetics during colony formation *in vitro*. Addition of deoxycytidine to cultures treated for varying periods of time with completely inhibitory levels of thymidine showed a similarity in the rates of loss of colony-forming capacity by CFU-e and GM-CFC and a difference from that by 9-day BFU-e.

Two main considerations in the time-decay of colony-forming capacity in these experiments are the proportion of cells in S phase, which is immediately affected by thymidine, and the time the cells remain viable while being blocked in S phase. In the case of CFU-e, it is known that a high proportion of cells are actively cycling (12) and therefore the decay of CFU-e in the presence of thymidine would be expected to depend on an ability to survive arrest in S phase. It has been reported by Kim et al. (15) that the toxic effects of thymidine on HeLa cells, manifested by a decline in the number of colonies formed, occur if the cells are exposed for longer than one generation time. On the assumption that this also applies to hemopoietic progenitor cells, the results in Figure 2.5 suggest a generation time of less than 24 hr for

CFU-e. A mean generation time for CFU-e from fetal liver *in vitro* of about 11 hr has been reported previously (5), using time-lapse photography.

Although a smaller proportion of GM-CFC are cycling in normal bone marrow, results in Figures 2.5 and 2.6 show a rapid decline in GM-CFC survival occurring after between an 8 and a 24 hr exposure to thymidine and no change in survival when thymidine addition is delayed by 24 hr. This indicates that the GM-CFC enter the cell cycle very shortly after culture, and the fall in cell survival reflects the length of time they are viable in S phase arrest. From the survival curves, a generation time of between 8 and 12 hr can be predicted for GM-CFC.

The survival curve for 9-day BFU-e in the presence of thymidine is complicated by the small proportion of cells in S phase (20 to 30%) (7,12) in normal bone marrow. But apart from the early 20% drop, a consistent decline in 9-day BFU-e occurs after exposure to thymidine for more than 24 hr, which, in view of the previous discussion, suggests a generation time of over 24 hr. This relatively long generation time would also be consistent with the slow rate of decline in cell survival during exposure to thymidine. An alternative explanation of the findings is that 9-day BFU-e undergo a long lag phase in culture before entering the cell cycle and that the slow decline is due to asynchrony of entry into the cell cycle. Provided the assumptions behind this method can be validated, more detailed experiments of this type should allow precise estimation of the generation times of hemopoietic progenitor cells *in vitro*.

SUMMARY

Addition of thymidine in sufficient concentrations to cultures of mammalian cells inhibits growth by stopping cell division at the S phase of the cell generation cycle. Inhibition of DNA synthesis is due to a restriction of deoxycytidine nucleotide synthesis, and complete reversal of inhibition is achieved by the addition of deoxycytidine.

The assay of thymidine inhibition on colony formation by cells in the murine erythroid and myeloid series revealed a striking sensitivity of erythroid colony-forming units (CFU-e). The 3-day and 9-day erythroid burst-forming units (BFU-e) were sensitive only to progressively higher thymidine concentrations, a sequence that can be correlated with the degree of maturity of the erythroid progenitor cells. The granulocyte-macrophage colony-forming cells (GM-CFC) were the least sensitive of the colony-forming cells tested.

The addition of deoxycytidine to cultures reversed inhibition by thymidine in all cases, although each colony-forming cell exhibited a different responsiveness to different concentrations of deoxycytidine. The CFU-e, in addition to being most sensitive to thymidine, responded to relatively low deoxycytidine levels, which were also found to result in a twofold improvement in plating efficiency, even in the absence of added thymidine. The findings here indicate that consideration of thymidine sensitivity is important in the methodology of tritiated thymidine suicide experiments, especially for CFU-e.

This system of thymidine inhibition and its reversal by deoxycytidine provides a new approach for investigating *in vitro* kinetics of colony-forming cells. A culture generation time for CFU-e of between 12 and 24 hr has been estimated by this method. Similarly, GM-CFC have a predicted generation time of between 8 and 12 hr. But 9-day BFU-e undergo a delayed and more gradual decline in survival when exposed to high thymidine concentrations, which may be explained by a longer generation time (over 24 hr) or an asynchronous entrance into the cell cycle.

ACKNOWLEDGMENTS

The authors thank Dr. D. W. van Bekkum for stimulating discussions and support.

This work was supported by the National Cancer League, Queen Wilhelmina Fund. S. Hasthorpe holds a postdoctoral fellowship from the Netherlands Ministry of Education and Science and is on leave from the Cancer Institute, Melbourne 3000, Australia. A. W. Harris is a Research Fellow of the National Health and Medical Research Council of Australia on sabbatical leave from The Walter and Eliza Hall Institute of Medical Research, Melbourne 3050, Australia.

REFERENCES

1. Axelrad, A. A., McLeod, D. L., Shreeve, M. M., and Heath, D. S. Properties of cells that produce erythrocytic colonies *in vitro*. In Robinson, W. A. ed., *Hemopoiesis in Culture, Second International Workshop,* Washington, D.C.: U.S. Government Printing Office, p. 226, 1974.
2. Bjursell, G. and Reichard, P. Effects of thymidine on deoxyribonucleoside triphosphate pools and deoxyribonucleic acid synthesis in Chinese hamster ovary cells. *J. Biol. Chem., 248*:3904, 1973.
3. Bradley, T. R., Stanley, E. R., and Sumner, M. A. Factors from mouse tissues stimulating colony growth of mouse bone marrow cells *in vitro. Aust. J. Exp. Biol. Med. Sci:, 49*:595, 1971.

4. Cleaver, J. E. *Thymidine Metabolism and Cell Kinetics.* Amsterdam: North-Holland, 1967.

5. Cormack, D. Time-lapse characterization of erythrocytic colony-forming cells in plasma cultures. *Exp. Hematol., 4*:319, 1976.

6. Gregory, C. J. Erythropoietin sensitivity as a differentiation marker in the hemopoietic system: Studies of three erythropoietic colony responses in culture. *J. Cell. Physiol., 89*:289, 1976.

7. Gregory, C. J., and Eaves, A. C. Three stages of erythropoietic progenitor cell differentiation distinguished by a number of physical and biologic properties. *Blood, 51*:527, 1978.

8. Guilbert, L. S., and Iscove, N. N. Partial replacement of serum by selenite, transferrin, albumin and lecithin in haemopoietic cell cultures. *Nature (London), 263*:594, 1976.

9. Harris, A. W. Lymphoid properties expressed by cultured mouse T and B tumour cells: Fc receptor and cellular sensitivity to hydrocortisone and thymidine. In Peeters, H., ed., *Protides of the Biological Fluids, 25th Colloquium.* Oxford: Pergamon Press, p. 601, 1978.

10. Harris, A. W., and Cohn, M. Physiology and genetics of some lymphoid cell functions. In Sterzl, J., and Riha, I., eds., *Developmental Aspects of Antibody Formation and Structure.* Prague: Academia, 1970.

11. Horibata, K., and Harris, A. W. Mouse myelomas and lymphomas in culture. *Exp. Cell Res., 60*:61, 1970.

12. Iscove, N. N. The role of erythropoietin in regulation of population size and cell cycling of early and late erythroid precursors in mouse bone marrow. *Cell Tissue Kinet., 10*:323, 1977.

13. Iscove, N. N., and Sieber, F. Erythroid progenitors in mouse bone marrow detected by macroscopic colony formation in culture. *Exp. Hematol., 3*:32, 1975.

14. Iscove, N. N., Till, J. E., and McCulloch, E. A. The proliferative states of mouse granulopoietic progenitor cells. *Proc. Soc. Exp. Biol. Med., 134*:33, 1970.

15. Kim, J. H., Kim, S. H., and Eidinoff, M. L. Cell viability and nucleic acid metabolism after exposure of HeLa cells to excess thymidine and deoxyadenosine. *Biochem. Pharmacol., 14*:1821, 1965.

16. Metcalf, D. *Hemopoietic Colonies. In Vitro cloning of normal and leukemic cells.* Heidelberg: Springer-Verlag, 1977.

17. Morris, N. R., and Fischer, G. A. Studies concerning the inhibition of cellular reproduction by deoxyribonucleosides. I. Inhibition of the synthesis of deoxycytidine by phosphorylated derivatives of thymidine. *Biochem. Biophys. Acta, 68*:84, 1963.

18. Reichard, P. Control of deoxyribonucleotide synthesis *in vitro* and *in vivo. Adv. Enz. Regulation, 10*:3, 1972.

19. Reynolds, E. C., Harris, A. W., and Finch, L. R. Deoxyribonucleoside triphosphate pools and differential thymidine sensitivities of cultured mouse lymphoma and myeloma cells. *Biochem. Biophys. Acta.* (in press).

20. Schneider, W. C. Deoxyribosides in animal tissues. *J. Biol. Chem., 216*:287, 1955.

21. Sjostrom, D. A., and Forsdyke, D. R. Isotope-dilution analysis of rate-limiting steps and pools affecting the incorporation of thymidine and deoxycytidine into cultured thymus cells. *Biochem. J., 138*:253, 1974.

22. Stephenson, J. R., Axelrad, A. A., McLeod, D. L., and Shreeve, M. M. Induction of colonies of hemoglobin-synthesizing cells by erythropoietin *in vitro. Proc. Nat. Acad. Sci (U.S.), 68*:1542, 1971.

23. Whittle, E. D. Effect of thymidine on deoxyribonucleic acid synthesis and cytidine metabolism in rat-thymus cells. *Biochem. Biophys Acta., 114*:44, 1966.

24. Xeros, N. Deoxyriboside control and synchronization of mitosis. *Nature (London), 194*:682, 1962.

3

Identification of CFU-s by Scatter Measurements on a Light Activated Cell Sorter

Ger van den Engh, Jan Visser, and Barbara Trask

The morphological description of the hemopoietic stem cell [the colony forming unit-spleen (CFU-s) cell] is made more difficult by the low incidence of this cell in hemopoietic tissue. An unequivocal description is only achieved if it is followed by proof that the cell described can regenerate the hemopoietic organs of irradiated mice (CFU-s assay). Therefore, maintenance of cell viability is essential. This precludes the use of the microscope as the sole instrument of investigation.

In previous attempts to describe the stem cell, cell separation techniques to concentrate CFU-s so that the microscope becomes useful (1,10,14) or to derive "morphologic" properties from the study of physical parameters (16) have been used. The first approach assumes a uniform morphology of the CFU-s. If the appearance of the CFU-s changes, for instance, with its cell cycle phase, the search for a unique cell type in stem cell concentrates will not necessarily lead to stem cell identification. The second approach offers the advantage that the cells remain viable. But no physical characteristics, except cell size, are readily translated into a morphologic picture. The analytic application of cell separation has shown that variations in the characteristics of the stem cell do take place (16, 19). To what extent these variations are reflected in cell morphology is uncertain.

This chapter describes an alternative approach to the determination of CFU-s cell morphology. The light-scattering properties of CFU-s were determined in a fluorescence-activated cell sorter (FACS) II. The scatter signals contain information about cell size and internal cell structure (12). By comparing the scatter of CFU-s with that of known cell types, a rough sketch of CFU-s morphology can be derived.

Scatter measurements are also of great help in measuring membrane fluorescence after the staining of cells with fluorescent antibodies. Flow cytometric immunofluorescence measurements are easily disturbed by the background fluorescence of dead cells and small particles. By defining the region of interest by light scatter, immunofluorescence measurements of mouse bone marrow cells can be improved. Such measurements were used to determine the relative H-2 and Thy-1 antigen density on CFU-s.

The Use of FACS II in Light Scatter Measurements

The principle of light-activated cell sorting is shown in Figure 3.1 (5). The cells are centered in a liquid jet that is ejected from a vibrating nozzle. The vibrations cause the buffer stream to break into uniform droplets, which can be charged and subsequently deflected electrostatically. Before the jet breaks into droplets, it passes a laser beam. Light pulses generated by passing cells are registered by

19

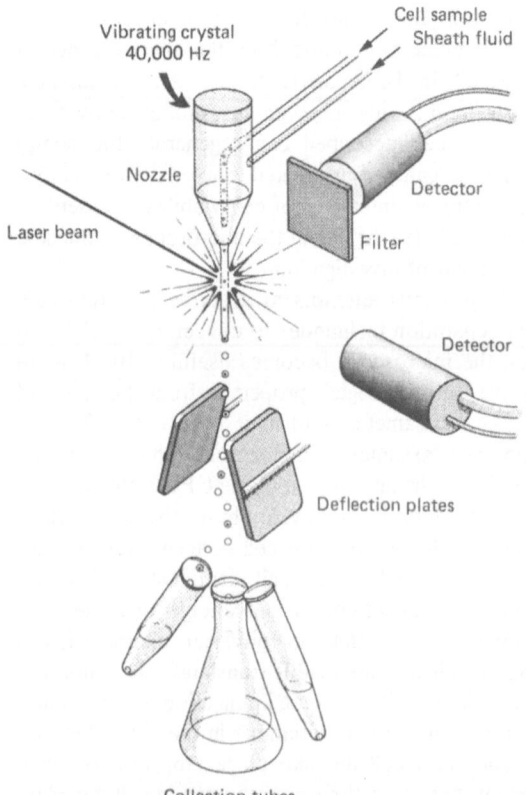

FIGURE 3.1. The principle of a light-activated cell sorter. The cells are suspended in a liquid jet illuminated by a laser beam. Scatter on fluorescence signals are registered by two different detectors. Droplets that contain cells of interest can be changed and deflected into separate tubes.

detectors placed around the point at which the jet intersects the laser beam.

In the sorter used in these studies (FACS II, Becton and Dickinson), two detectors can be used to measure light scatter. One detector is placed along the axis of the incident light. With a fully opened diaphragm, this detector collects light scattered at all angles between 2 and 25 degrees, with respect to the direction of incident light. The signals measured by this detector are indicated as forward light scatter (FLS) intensity. A second detector is placed perpendicularly to the laser beam. In the FACS II, this detector measures fluorescence. If the filter that blocks the laser beam is removed, the detector can be employed to measure light scatter at large angles. These signals are indicated as perpendicular light scatter (PLS) intensity. The PLS detector collects light scattered between 70 to 110 degrees, with respect to the direction of the incident light.

The light scattering properties of small particles are complex and depend upon the size, refractive

index, and internal structure of the particles (3,6,7). The largest scatter signals are registered in the forward direction. These effects are due mainly to diffraction and refraction. So far, no simple theory explains the signals at a discrete angle (6,7,8), but in the design used here, in which the light signals are integrated over all angles between 2 and 25 degrees, a simple relationship may exist. One can assume that the sum of all scatter signals should equal the amount of light intercepted by the particle. The intensity of the scatter signal should be proportional to the cross-sectional area of the particle. Since the FLS detector collects most of the scatter light, this signal can be assumed to be proportional to the square of the particle radius. Here, the loss of light by interference is neglected. Since these effects are most pronounced within the 2 degrees that are shielded on the FLS detector (7), these effects are considered small.

Experimental evidence suggests that the cross-sectional area rule, at least for similarly sized particles with a similar refractive index, is a good first approximation. Fathman et al. (4) reported that the scatter intensity is linearly related to cell volume. These experiments may not have been accurate enough to exclude the possibility that scatter is proportional to the cross-sectional area, however. Mullaney et al. (9) theorized a relationship between scatter and particle size, and Visser (18) showed that the FLS intensity of fluorescent Sephadex spheres (with a refractive index comparable to cells) as measured with the FACS II is proportional to the square of the radius of the spheres. Unpublished data obtained by the author indicates that the cross-sectional area rule holds for several cell types.

Light measured by the PLS detector results from the reflection of light by the particle. Since the FACS II measures light from particles inside a liquid jet, some of the FLS signals will be picked up by the PLS detector as the result of reflections within the liquid. Therefore, for particles that are similarly shaped, a linear relationship between PLS and FLS can be expected. Differences in the extent of the reflective surfaces will cause heterogeneity in the PLS signals. Cells with a relatively large surface area to volume ratio, cells with numerous intracellular particles, or cells with an oddly shaped nucleus, can be expected to give relatively high scatter signals in the perpendicular direction. In this sense, the PLS reflects cell structure.

To illustrate how FLS represents cell size and PLS reflects the complexity of cell structure, Figure 3.2 shows simplified dot diagrams of PLS versus FLS measurements of mouse erythrocytes at various osmolarities. In a high osmotic buffer, the erythrocytes shrank and took on a spiked shape. Such cells are high in PLS and low in FLS. In a low

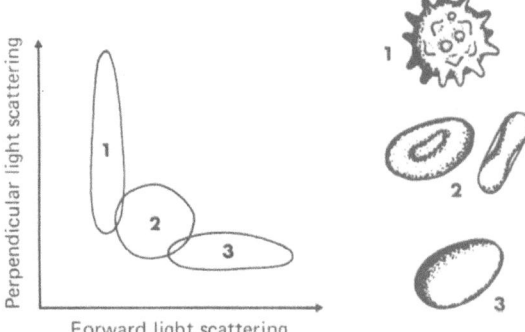

FIGURE 3.2. Perpendicular versus forward light scattering of mouse erythrocytes suspended in buffers of decreasing osmolarity. The spiked shape (1) is high in PLS and low in FLS. As the volume increases and the shape becomes more spherical (2, 3), FLS increases as PLS decreases.

osmotic buffer, the cells swelled and became spherical. Such cells are low in PLS and high in FLS, since the FLS increased with an increase in cell diameter.

MATERIALS AND METHODS

Mice

All the mice used throughout these experiments were (C57BL/Rij x C3H/LAW)F1 and C3H/LAW 8 to 12 weeks of age.

Assay of CFU-s

Hemopoietic stem cells were measured with the CFU-s assay of Till and McCulloch (13).

Antibody Labeling of Cells

Cells were suspended in Hanks/HEPES at a concentration of 10^6 cells/ml and incubated for 30 min with 2,4, dinitro phenol (DNP)-labeled antibody. The cells were washed and then incubated with fluoroscein isothiocyanate (FITC)-labeled rabbit anti-chimpanzee albumin-DNP. After washing, the cells were used for fluorescence measurements. Cells incubated with FITC-rabbit anti-chimpanzee albumin-DNP alone served as a control to establish background fluorescence. The labeling of antisera with DNP and the production of FITC-rabbit anti-chimpanzee albumin-DNP is described elsewhere (18). The papain treatment of antibody-carrying cells was performed as described by Van den Engh and Platenburg (15).

Scatter and Fluorescence Measurements

To eliminate reflections within the liquid jet, the buffer that contained the cells was used as the sheath flow buffer for the scatter measurements. Fluorescence measurements were performed using a Schott SAL 530 interferance filter in combination with a 530-nm glass filter. Filter fluorescence was avoided by placing the glass filter away from the photomulti-

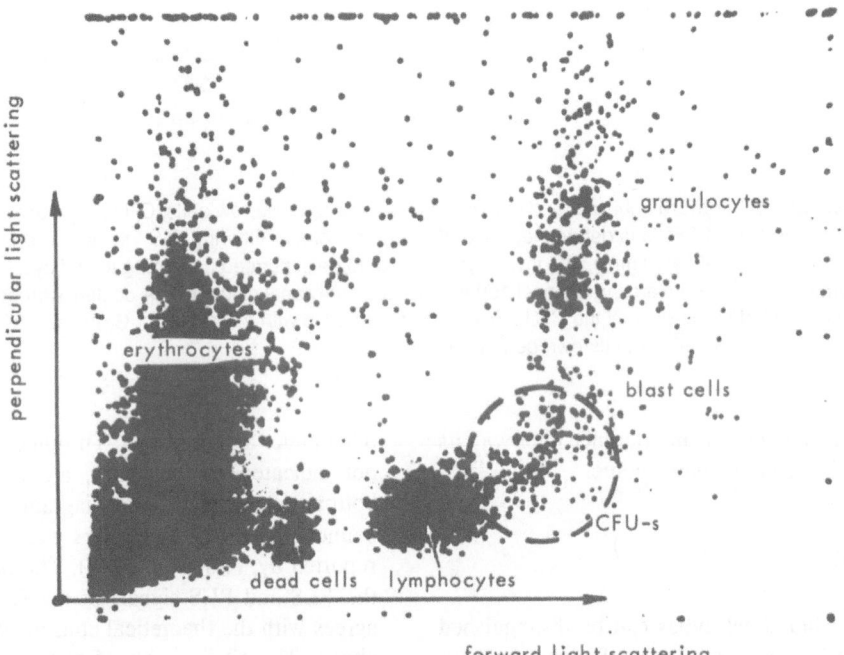

FIGURE 3.3. Perpendicular versus forward light scattering of mouse bone marrow cells. Each point represents the FLS and PLS of one cell. The cells in the clusters are identified by sorting and inspection under the microscope. The encircled area contains approximately 80% of the CFU-s.

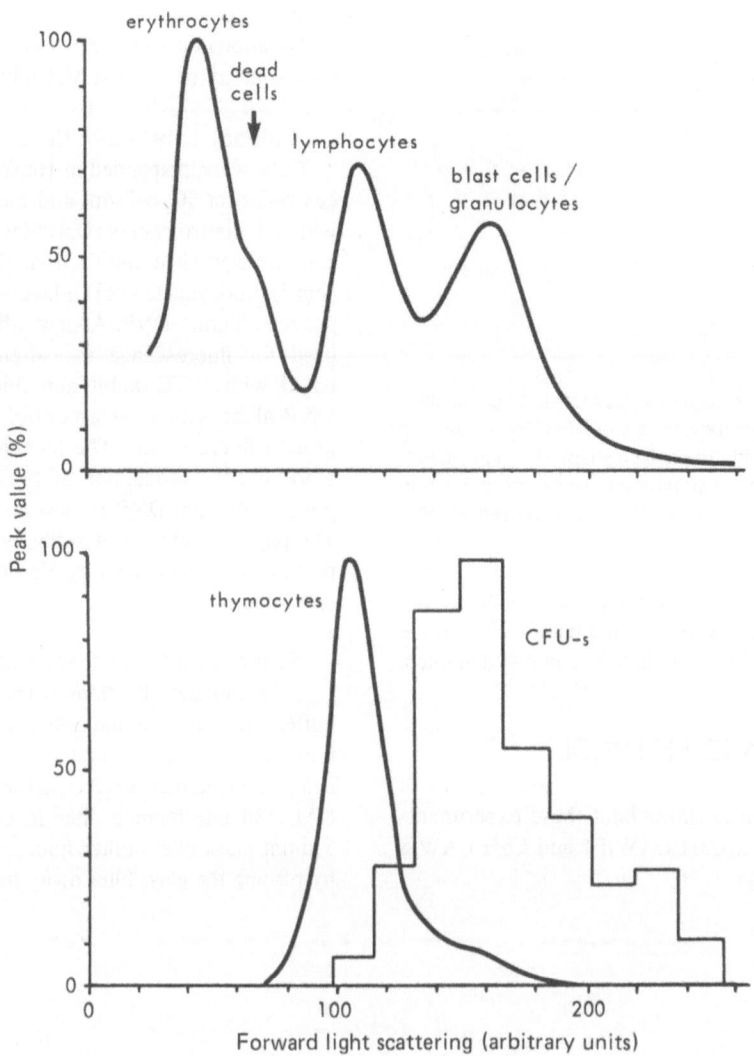

FIGURE 3.4. Forward light scatter distributions of mouse bone marrow cells, CFU-s, and thymocytes. The distribution of CFU-s is determined by deflecting fractions of differing FLS intensity and injecting those cells into irradiated mice. The thymocytes appear to be separated according to cell cycle phase. Thymocytes taken from around channel 150 and subsequently analyzed for DNA content were found to contain 4n DNA. The peak of the CFU-s distribution coincides with the region of G_2 thymocytes. Each distribution is normalized to its own peak value.

plier and by positioning the interference filter with its "mirror side" facing the light source.

RESULTS

The different blood cell types can be distinguished by their light-scattering properties. Figure 3.3 shows a dot diagram of PLS versus FLS signals of mouse bone marrow cells. The cells are grouped in a number of clusters. Cells from each cluster were deflected onto object slides and, after staining, iden-

tified under a microscope. Thrombocytes, which are not indicated in the figure, are found in the left bottom corner of the dot diagram. The differential counts of the various clusters were similar to those reported by Visser et al. (17). The relation between the FLS and PLS signals of the various cell types agrees with the theoretical considerations presented above. The FLS signals of cells increase with cell size. The cells have increasing FLS signals in the sequence platelet, erythrocyte, lymphocyte, granulocyte, and monocyte. The size of the PLS signals depends on cell structure. Spherical cells such as

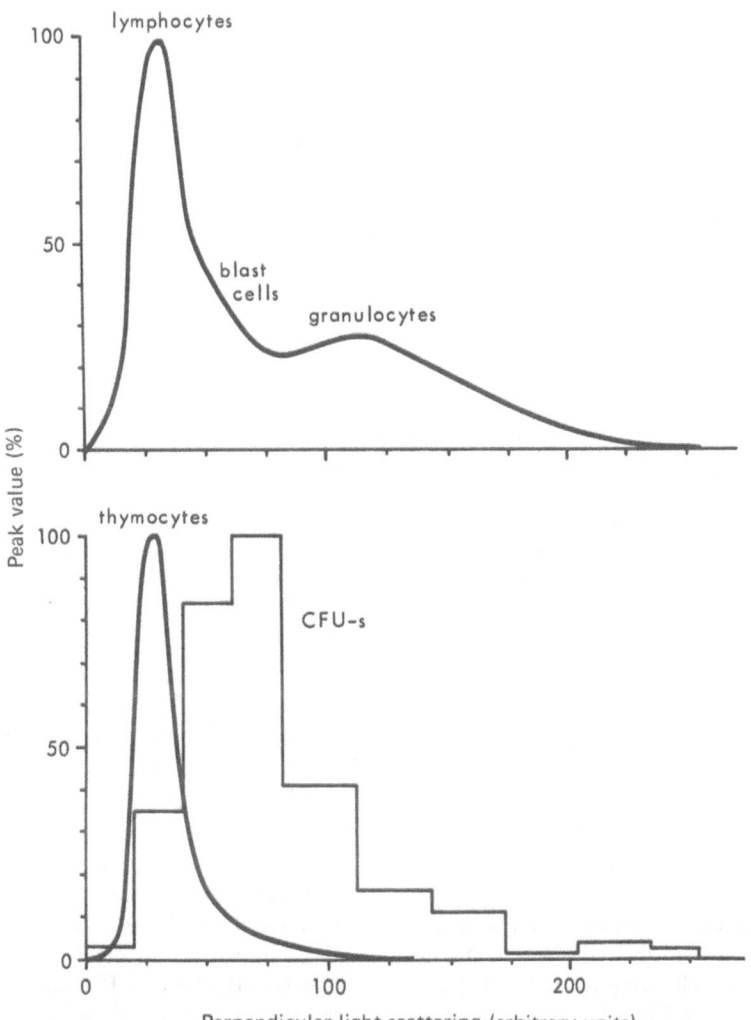

FIGURE 3.5. Perpendicular light-scatter distribution of mouse bone marrow cells, CFU-s, and thymocytes. The distribution of CFU-s is determined by injecting cells of varying scatter intensity into irradiated mice. Thymocytes with a PLS intensity equal to the peak of the CFU-s, when subsequently analyzed for DNA content, were found to contain 4n DNA.

platelets, dead cells, and lymphocytes lie on a straight line through the origin; the slope of the line is determined by the amount of FLS reflected onto the PLS detector. The PLS of cells that deviate from a sphere, have a nonspherical nucleus, and/or have granular inclusions (erythrocytes and granulocytes) is considerably larger than that of lymphocytes and monocytes.

The scattering properties of CFU-s were determined by separating the cells into fractions along the FLS and PLS axes. The fractions were deflected, collected, and injected into irradiated mice for CFU-s determination. The CFU-s distributions that were observed are shown in Figures 3.4 and 3.5. The distributions are plotted with respect to those of

thymocytes and all bone marrow cells. The FLS distribution of thymocytes forms a sharp peak around channel 100. If it is assumed that the FLS signal reflects cross-sectional area, the majority of CFU-s are approximately 1.2 times larger than thymocytes. The size of the thymocytes in the peak of the distribution, as calculated from density and sedimentation rate, is 6 μm (W. Boersma, unpublished results). The diameter of CFU-s can therefore be estimated to be 7.2 μm. The PLS intensity of CFU-s is relatively low. The circled area in Figure 3.2 is the region of highest CFU-s incidence. It contains approximately 80% of the CFU-s. Separation of this region results in a five- to tenfold increase in CFU-s concentration with respect to all nucleated cells.

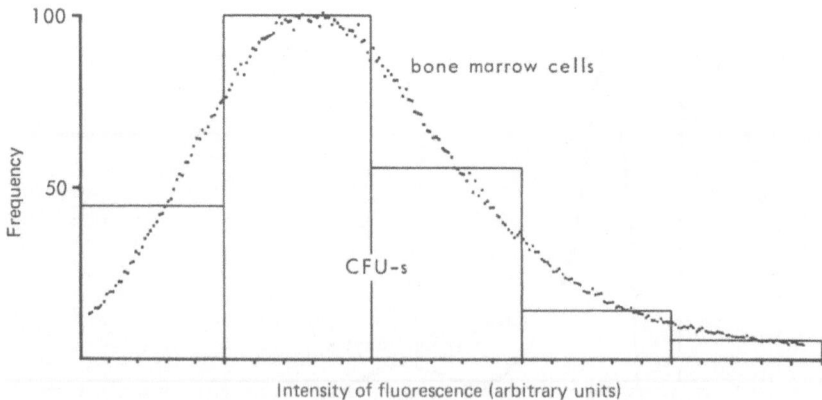

FIGURE 3.6. Fluorescence distribution of CFU-s compared to that of all bone marrow cells of C3H mice after labeling the cells with AKR anti-*C3H* thymocyte serum. Dead cells and erythrocytes are excluded. The anti-Thy-1.2 serum was labeled with approximately five molecules of DNP per Ig molecule. After treatment with anti-Thy-1-DNP, the cells were incubated with a FITC-labeled rabbit anti-chimpanzee albumin-DNP serum. The fluorescence distribution does not appreciably differ from cells treated with fluoresceinated rabbit anti-chimpanzee albumin-DNP alone. Therefore, the fluorescence distribution is largely due to background noise. Thymocytes treated in the same manner show a fluorescence profile that is four to five times higher than control values.

The light scattering properties of bone marrow cells can be used to eliminate dead cells and erythrocytes from the final results. This greatly improves the immunofluorescence measurements on mouse bone marrow cells and allows an analysis of membrane receptors on CFU-s. Figures 3.6 and 3.7 illustrate this. To determine whether Thy-1 and H-2 antigens are present on the CFU-s membrane, mouse bone marrow cells were treated with DNP-labeled anti-H-2 and anti-Thy-1 sera followed by incubation with a FITC-labeled, rabbit anti-chimpanzee albumin-DNP serum (anti-DNP-FITC). After the cells were sorted, they were incubated with papain to eliminate the potentially cytotoxic effect of bound antibody. False positive signals from dead cells were eliminated by gating these cells out on the basis of scatter signals. The filters were carefully chosen such that the fluorescence signals did not interfere with the PLS signals. The fluorescence intensity distribution of bone marrow cells for the two sera are shown in the figures. Binding of antisera to CFU-s was analyzed by fractionating the cells on the basis of FITC fluorescence and determining the CFU-s content of the various fractions. With respect to

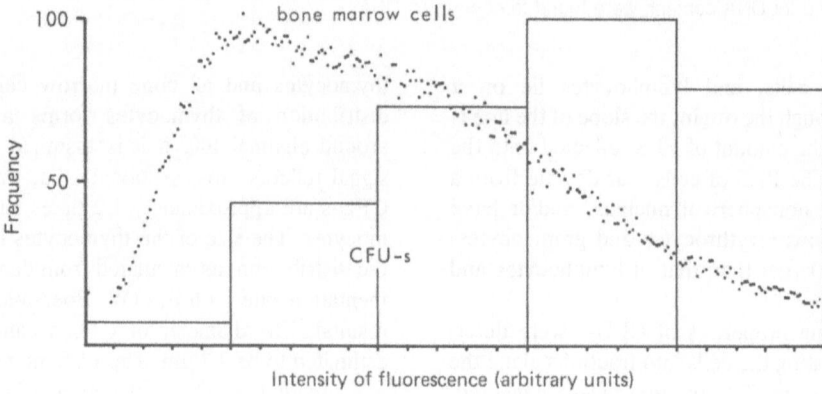

FIGURE 3.7. Fluorescence distribution of CFU-s compared to that of all bone marrow cells of C3H mice after labeling with a anti-H-2^k serum. Dead cells and erythrocytes are excluded. The H-2^k serum was labeled with approximately five molecules of DNP per Ig molecule. After treatment with anti-H-2^k-DNP, the cells were incubated with a FITC-labeled rabbit anti-DNP antiserum. The CFU-s are found among the most intensely labeled bone marrow cells.

Thy-1 labeling, the CFU-s fall in the region of the bulk of bone marrow cells and may be considered to be negative. In a similar experiment with anti-H-2 (Figure 3.7), it was found that FITC-labeled anti-H-2 binds more to CFU-s than the bulk of bone marrow cells. The CFU-s may be considered to have a high H-2 density on the cell surface.

DISCUSSION

The light scattering properties of bone marrow cells provide useful criteria for morphologic identification. Scatter measurements allow the determination of cell size and the analysis of internal structures without the necessity of staining or fixing the cells. When light scattered 2 to 25 degrees is collected (FLS) in the FACS II cell sorter (liquid jet in air), the scatter intensity of small Sephadex spheres is proportional to their cross-sectional area (18). The FLS profiles of thymocytes and bone marrow presented here suggest that the scatter intensities of living cells follow a similar pattern. The scatter signal measured perpendicular to the incident light is due primarily to reflection, which depends on the surface area to the volume ratio of a particle. The study of PLS and FLS signals from living cells allows conclusions to be drawn as to whether the cells are spherical (as are lymphocytes) and whether the cells have an elaborate internal structure.

The overall morphologic characteristics of CFU-s are those of a large lymphocyte. When the FLS and PLS signals of CFU-s are compared to scatter signals of cells of known morphology, the CFU-s appear to be low in PLS. Thus, CFU-s have a spherical shape and few cytoplasmic inclusions. Their FLS signals lie in the region of G_2 thymocytes. Assuming that the cross-sectional area rule holds for cells, the CFU-s measure 7.2 μm in diameter. This size agrees very well with size calculations on the basis of density and sedimentation rate analysis (19).

Light scatter may be used to increase the specificity of immunofluorescence studies. The fluorescence signals from dead cells and erythrocytes greatly distort immunofluorescence distributions. This problem can be eliminated by gating the dead cells and erythrocytes out on the basis of their scatter characteristics. The fluorescence distribution of CFU-s after labeling with anti-H-2 and anti-Thy-1 sera show that CFU-s are among the highest in H-2 density. The CFU-s do not carry detectable amounts of Thy-1. Since it has been demonstrated that CFU-s are negative for Ia (2,11), the H-2 expression must be non-Ia region antigens. These results are in agreement with experiments in which suppression of colony formation by CFU-s is taken as a criterium for the expression of surface antigens (11).

SUMMARY

Light scattered by cells allows information about cell size and cell morphology to be gathered. Experiments are presented in which the intensity of forward light scatter (FLS) angles between 2 to 25 degrees and at perpendicular light scatter (PLS) angles between 70 to 110 degrees of colony forming units-spleen (CFU-s) are measured in a FACS II light-activated cell sorter. The scattering properties are compared to those of known cell types. It is concluded that CFU-s are spherical cells, measuring 7.2 μm in diameter, that do not contain conspicuous cytoplasmic inclusions. Light scatter can be used to eliminate erythrocytes and dead cells from immunofluorescence measurements. Fluorescence measurement of CFU-s in comparison to all bone marrow cells after treatment with anti-H-2 and anti-Thy-1 sera are presented.

ACKNOWLEDGMENTS

This investigation is part of a study on the regulation of hemopoiesis, which is supported by a program grant of The Netherlands Foundation for Medical Research (FUNGO).

B. J. Trask is supported by a NSF training grant.

REFERENCES

1. Barr, R. D., Whang-Peng, J., and Perry, S. Hemopoietic stem cells in human peripheral blood. *Science, 190*:284, 1975.
2. Basch, R. S., Janossy, G., and Greaves, M. F. Murine pluripotential stem cells lack Ia antigen. *Nature, 270*:520, 1977.
3. Brunsting, A., and Mullaney, P. F. Differential light scattering from spherical mammalian cells. *Biophys. J., 14*:439, 1974.
4. Fathman, C. G., Small, M., Herzenberg, L. A., and Weissman, I. L. Thymus cell maturation. II. Differentiation of three "mature" subclasses *in vivo. Cell. Immunol., 15*:109, 1975.
5. Herzenberg, L. A., Sweet, R. G., and Herzenberg, L. A. Fluorescence-activated cell sorting. *Scientific American, 234, 3*:108, 1976.
6. Jovin, T. M., Morris, S. J., Striker, G., Schultens, H. A., Digweed, M., and Arndt-Jovin, D. J. Automatic sizing and separation of particles by ratios of light scattering intensities. *J. Histochem. Cytochem., 24*:269, 1976.
7. Meyer, R. A., and Brunsting, A. Light scattering from nucleated biological cells. *Biophys. J., 15*:191, 1975.
8. Mullaney, P. F., and Dean, P. N. The small angle light scattering of biological cells. Theoretical considerations. *Biophys. J., 10*:764, 1970.
9. Mullaney, P. F., Crowell, J. M., Salzman, G. C.,

Martin, J. C., Hiebert, R. D., and Good, C. A. Pulse-height light scatter distributions using flow-systems instrumentation. *J. Histochem. Cytochem., 24*:298, 1976.

10. Rubinstein, A. S., and Trobaugh, F. E. Ultrastructure of presumptive hematopoietic stem cells. *Blood, 42*:61, 1973.

11. Russell, J. L., and Van den Engh, G. J. The expression of histocompatibility-2 antigens on haemopoietic stem cells. *Tissue Antigens* (in press).

12. Salzman, G. C., Crowell, J. M., Martin, J. C., Trujillo, T. T., Romero, A., Mullaney, P. F., and LaBauve, P. M. Cell classification by laser light scattering: Identification of unstained leukocytes. *Acta Cytologica, 19*:374, 1975.

13. Till, J. E., and McCulloch, E. A. A direct measurement of the radiation sensitivity of normal mouse bone marrow cells. *Radiat. Res., 14*:213, 1961.

14. Van Bekkum, D. W., Van Noord, M. J., Maat, B., and Dicke, K. A. Attempts at identification of hemopoietic stem cell in mouse. *Blood, 38*:547, 1971.

15. Van den Engh, G. J., and Platenburg, M. Suppression of CFU-s activity by rabbit-anti-mouse brain serum can be overcome by treatment with papain. *Exp. Hemat., 6*:627, 1978.

16. Visser, J., Van den Engh, G., Williams, N., and Mulder, D. Physical separation of the cycling and non-cycling compartments of murine hemopoietic stem cells. In Baum, S. J., and Ledney, G. D., eds., *Experimental hematology today,* Springer-Verlag, New York, 1977.

17. Visser, J. W. M., Cram, L. S., Martin, J. C., Salzman, G. C., and Price, B. J. Sorting of murine granulocytic progenitor cells by use of laser light scattering measurements. In A. Desser, et al., eds., *Pulse Cytophotometry*. Ghent: European Press Medikon, 1978.

18. Visser, J., Haaijman, J., and Trask, B. Quantitative immunofluorescence in flow cytometry. In Knapp, et al., eds. *Immunofluorescence and related staining techniques*. Amsterdam: Elsevier/North Holland, 1978.

19. Visser, J., van den Engh, G., Williams, N., and Mulder, D. Buoyant densities and sedimentation rates of the murine haemopoietic pluripotent stem cells. In Bentvelzen, P. et al., eds. *Advances in comparative leukemia research,* 1977. Amsterdan: Elsevier/North Holland, 1978.

4

Are Stem Cells Regulated by Late Erythroid Precursors?

U. Reincke, D. Brookoff,
H. Burlington, E. P. Cronkite,
and E. Gerard

Renewal tissues are characterized by ongoing cell replacement, a function imposed because cells lose their ability to divide while maturing. Differentiation into functional but perishable end cells thus is the hallmark of sustaining stem cell populations. Among renewal tissues, bone marrow is well studied because it is amenable to cell quantifying techniques and its stem cells are assayable by various clone-forming methods (12).

How rapid and efficient is the replacement of maturing bone marrow cells that have been suddenly lost? Experiments relating to this question are usually either performed *in vitro*, and thus devoid of the full complement of regulatory controls, or they employ cytotoxic techniques that injure stem cells along with the differentiated cells. In seeking a means to avoid this nondiscriminatory approach, we developed the method of [^{55}Fe]-erythrocytocide. Intracellular ^{55}Fe causes individual cell death by the emission of cascades of Auger electrons with extremely short path lengths (Table 4.1). The isotope, avidly taken up by erythroid cells, will thus eliminate maturing erythrocyte precursors (19). This produces a discrete and selective effect: The loss of erythroblasts leads to increased erythroid differentiation in a compensatory response that is rapid, specific, sustained, and associated with a continued depletion of pluripotent hemopoietic stem cells.

METHODS

In the present experiment, we enumerated pluripotent stem cells and microscopically classifiable bone marrow cells in ^{55}Fe-injected mice. Cyclotron-produced (22) isotope of high specific activity (> 1 mCi/ μg Fe) was injected intravenously as ferric chloride into female C57B1/6J mice, weighing 18 to 19 g. Thirty mice received 440 μCi each, and 22 mice received an equivalent amount of 0.34 μg cold iron. Bone marrow was obtained after 3, 6, and 18 hr and 1, 2, 3, and 7 days. One hindleg and one foreleg were used to obtain differentials from Giemsa-stained bone marrow smears (1). Cell suspensions were prepared from the opposite leg to determine cellularity and in some cases to estimate the number of pluripotent hemopoietic stem cells (CFU-s). The techniques employed are described in detail elsewhere (21).

RESULTS

The figures show bone marrow differential counts of ^{55}Fe-treated and cold iron-treated mice. The erythropoiesis-enhancing effect of cold iron (13) appears to have caused some increase and oscillation of erythroid cell populations (Figure 4.1). Contrasting,

TABLE 4.1 Distribution and Range of Energy Dissipation of ^{55}Fe (Modified from Ref. 19)

ENERGY DISTRIBUTION ($t_{1/2}$, 2.7 YEARS)				
Radiation	Frequency	Mean Energy	Range in Water (%) 90	50
Auger electrons	95.8% p. dis.	0.889 keV	$\leq 1\mu m$	
K-shell x-rays	4.2% p. dis.	5.961 keV	980 μm;	290 μm

extensive, and sequential changes occurred in the erythroid bone marrow of ^{55}Fe-injected mice. The cytocidal effect was manifested within 18 hr by a significant reduction in polychromatophilic and orthochromatic normoblast numbers (Figure 4.1). Pronormoblasts and basophilic normoblasts were initially unaffected. Between 18 and 24 hr after mice were injected with ^{55}Fe, the pronormoblasts rose sharply from 0.13 ± 0.04 to 0.51 ± 0.19 million (mean and SE) per hindleg. This was followed by a corresponding rise of basophilic normoblasts and then by a normalization of the levels of polychromatophilic and orthochromatic normoblasts. The same sequence occurred in the hindleg and the foreleg, which confirmed that alterations due to ^{55}Fe were similar at different marrow sites.

The fourfold increase of pronormoblasts occurred within 6 hr. A cell cycle time of 3 hr would be required in order to ascribe the increase to mitotic self-replication. This is unlikely, since cell cycle times under 6 hr have not been reported. On the other hand, accumulation of pronormoblasts could result from an inability to complete the maturation process. Since this would cut off production of mature cells, however, it is not compatible with the observed subsequent expansion in all recognizable erythroid compartments. Thus, increased differentiation of unrecognizable precursors into the identifiable pronormoblast pool must account for their increase. Comparable changes were not observed in the granuloid cell series (Figure 4.2), attesting to the specific nature not only of the ^{55}Fe damage but also of the response. This quick and specific course points to a regulatory system in which stem cells are directly responsive to the fate of their differentiated offspring.

Although the number of late normoblasts was normalized after 72 hr, a threefold elevation of pronormoblasts and basophilic normoblasts persisted. Since no corresponding overshoot of late normoblasts occurred, one must conclude that only one-third of the freshly differentiated precursors reached the polychromatic stage. This indicates that, due to reutilization of the radioiron, erythrocytocide continued and was effectively compensated for.

Production of young normoblasts must have orig-

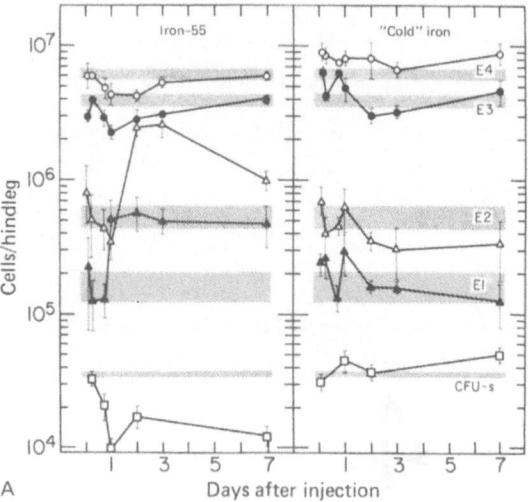
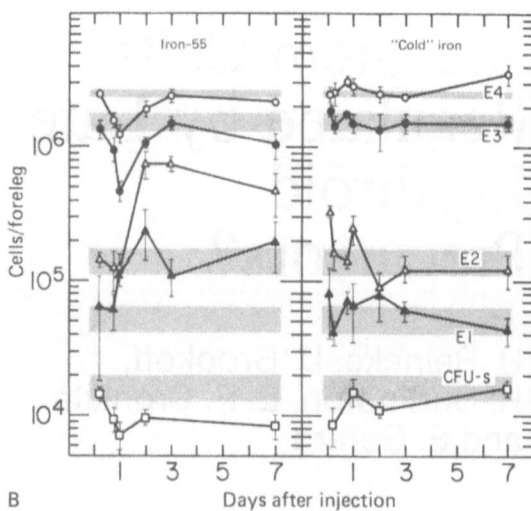

FIGURE 4.1. Erythropoietic precursors and CFU-s in the hindleg (**A**) and foreleg (**B**). Absolute differential counts were obtained by microscopic identification of at least 2,000 cells per bone marrow sample, and multiplication of fractional counts with total nucleated cell count determined in the contralateral leg. Symbols denote mean and standard error for three to six mice: pronormoblasts (E1); basophilic normoblasts (E2); polychromatophilic normoblasts (E3); and orthochromatic normoblasts (E4). Each CFU-s assay was performed with 12 syngeneic recipients and one donor mouse. Each symbol shows the mean and standard error of one assay, corrected for an assumed seeding factor of 0.2. Shaded bands indicate standard error above and below the mean of nine untreated mice for differential counts and five CFU-s assays in untreated mice.

inated from differentiation of progenitors already committed to erythropoiesis, such as BFU-e or CFU-e (12). These were not assayed so that it is not known whether their numbers changed. Pluripotent stem cells, assayed as CFU-s, were significantly

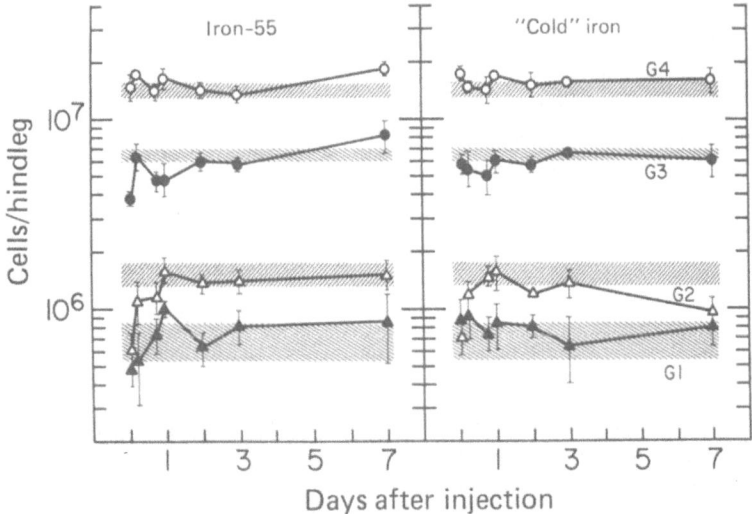

FIGURE 4.2. Granuloid precursors in the hindleg. Symbols are the same as in Figure 4.1; other symbols as follows: myeloblasts (G1); promyelocytes (G2); myelocytes (G3); and metamyelocytes plus banded and segmented granulocytes (G4).

reduced as early as 18 hr after ^{55}Fe injection and remained clearly subnormal throughout the experiment. Their pattern in mice was the reverse of that seen among pronormoblasts. The sudden drop of CFU-s, however, preceded the rapid increase of pronormoblasts by at least 6 hr. Since pluripotent stem cells are not believed to be immediate precursors of pronormoblasts, there is no obvious explanation for the loss of CFU-s as the pronormoblasts emerged. The possibility of ^{55}Fe damage to CFU-s must be considered, but there are good reasons for rejecting this explanation: We measured radiation injury to CFU-s in frozen bone marrow, stored at $-196°C$ in the presence of ^{55}Fe. Survival curves of CFU-s were obtained by frequent assays (20). Significant radiation damage to CFU-s occurred only after weeks of frozen storage. Thus, the precipitous fall of CFU-s as early as 18 hr after ^{55}Fe injection cannot be due to radiation injury. Furthermore, the increase of pronormoblasts in itself shows that at least their immediate precursors could not have been seriously injured. The disappearance of CFU-s must therefore reflect a response to ^{55}Fe erythrocytocide, possibly related to the proliferative stem cell response seen after cytocide with tritiated thymidine (26).

DISCUSSION

This experiment offers evidence for a regulatory mechanism in which hemopoietic stem cells respond to cell loss in the differentiated erythroid cell population. The response is fast, specific, and sustained. It consists of increased production and results in swift compensation of losses of differentiated cells. In the process, pluripotent hemopoietic stem cells are consumed, presumably by rapid differentiation along the erythroid pathway.

Our experimental conditions created a need for differentiation of stem cells. This constitutes a major advantage of the ^{55}Fe method in contrast to procedures that destroy stem cells in order to assess stem cell repopulation kinetics. Direct destruction of stem cells is expected to occur only under extraordinary, usually iatrogenic circumstances such as radiation or drug therapy. Numerous experiments in which stem cell self-repopulation took priority (6, 7) failed therefore to demonstrate swift differentiation. The present experiment clearly displays the regulatory, rather than regenerative, action of stem cells. Possibly this mechanism is not part of daily regulation (2) but is reserved for emergencies in which a sudden interruption of the steady state occurs.

Mediating factors in the differentiation response are undefined. Erythropoietin could be involved. In rodents with arrested erythropoiesis, the hormone generates new pronormoblasts in 12 hr (11, 18) and erythrocytes in 2 to 3 days (24). But loss of pluripotent stem cells is usually not attributed to endogenous erythropoietin action (23), and in the absence of anemia, the erythropoietin-provoking stimulus is unclear.

Are stem cells then responding directly to signals from differentiated daughter populations? A circuit between pluripotent and erythroid cells requires cell communication in the microenvironment (25). Erythrocyte packing density in bone marrow sinuses (5) or the spread of nucleated cells into vacated spaces (9) could play a role as well as a collaboration of stem cells with T-lymphocytes (15,10,17). Nerve fibers (8), drug receptors (4), and chalones (3,14,16) could be involved in transmittal of information. One wonders how cellular characteristics and transmitting factors may compose the microenvironments in which pluripotent stem cells respond to specific requirements in the erythroid, granuloid, or megakaryocytic line.

SUMMARY

Intravenous injection of ^{55}FeCl$_3$ of high specific activity caused cytocide of maturing erythroblasts, which led to increased production of pronormoblasts within 18 to 24 hr. This compensatory differentiation response was preceded by and associated with a decrease of CFU-s. In view of the rapidity and sequence of the hematologic response to ^{55}Fe in the absence of anemia, we suggest that there is an intramedullary, short-range regulatory circuit that monitors the rate of differentiation and draws CFU-s into the erythroid pathway.

ACKNOWLEDGMENTS

We thank Betty Heldman, Mike Makar, and Naomi Pappas for expert technical assistance and Marie Susa for most efficient secretarial services.

Research supported by the U.S. Department of Energy and NIH Grant #HL15685-02. U. R. is a Scholar of the Leukemia Society of America, Inc.

REFERENCES

1. Bessis, M. *Blood Smears Reinterpreted.* New York: Springer Verlag, 1977.
2. Brecher, G. Pluripotent and committed hemopoietic stem cells. A hypothesis. *Nouv. Rev. Franc. d'Hémat., 18*:285, 1977.
3. Bullough, W. S. Mitotic and functional homeostasis: A speculative review. *Cancer Res., 25*:1683, 1965.
4. Byron, J. W. Drug receptors and the haemopoietic stem cell. *Nature New Biology, 241*:152, 1973.
5. Chamberlain, J. K., Weiss, L., and Weed, R. I. Bone marrow sinus cell packing: A determinant of cell release. *Blood, 46*:91, 1975.
6. Chervenick, P. A., and Boggs, D. R. Patterns of proliferation and differentiation of hematopoietic stem cells after compartment depletion. *Blood, 37*:568, 1971.
7. DeGowin, R. L., Hoak, J. C., and Miller, S. H. Erythroblastic differentiation of stem cells in hemopoietic colonies. *Blood, 40*:881, 1972.
8. Fliedner, T. M., Calvo, W., Haas, R., Forteza, J., and Bohne, F. Morphology and cytokinetic aspects of bone marrow stroma. In Stohlman, F., ed., *Hemopoietic Cellular Proliferation.* New York: Grune & Stratton, 1970, p. 67.
9. Folkman, J., and Moscona, A. Role of cell shape in growth control. *Nature, 273*:345, 1978.
10. Goodman, J. W., Basford, N. L., Shinpock, S. G., and Chambers, Z. E. An amplifier cell in hemopoiesis. *Exp. Hemat., 6*:151, 1978.
11. Goldwasser, E. Some molecular aspects of red cell differentiation. In Gordon, A. S., Condorelli, M., and Peschle, C., eds., *Regulation of Erythropoiesis.* Milano: Il Ponte, 1970, p. 227.
12. Gregory, C. J., and Henkelman, R. M. Relationship between early hemopoietic progenitor cells determined by correlation analysis of their numbers in individual spleen colonies. In Baum, S. J., and Ledney, G. D., eds., *Experimental Hematology Today.* New York: Springer-Verlag, 1977, p. 93.
13. Gross, M., and Goldwasser, E. On the mechanism of erythropoietin-induced differentiation. VIII. The effect of iron on stimulated marrow cell functions. *Biochim. Biophys. Acta, 217*:461, 1970.
14. Herman, S. P., Golde, D. W., and Cline, M. J. Neutrophil products that inhibit cell proliferation: Relation to granulocytic "chalone." *Blood, 51*:207, 1978.
15. Jedrzejczak, W. W., Sharkis, S., Ahmed, A., Sell, K. W., and Santos, G. W. Theta-sensitive cell and erythropoiesis: Identification of defect in W/W^v anemic mice. *Science, 169*:313, 1977.
16. Lord, B. I., Shah, G. P., and Lajtha, L. G. The effect of red blood cell extracts on the proliferation of erythrocyte precursor cells, *in vivo. Cell Tissue Kinet., 10*:215, 1977.
17. Nathan, D. G., Chess, L., Hillman, D. G., Clarke, B., Bread, J., Merler, E., and Hausman, D. E. Human erythroid burst forming unit: T-cell requirement for proliferation *in vitro. J. Exp. Med., 147*:324, 1978.
18. Orlic, D., Gordon, A. S., and Rhodin, J. A. G. An ultrastructural study of erythropoietin-induced red cell formation in mouse spleen. *J. Ultrastruct. Res., 13*:516, 1965.
19. Reincke, U., Burlington, H., Cronkite, E. P., and Laissue, J. A. Selective damage to erythroblasts by iron-55. *Blood, 45*:801, 1975.
20. Reincke, U., Brookoff, D., Burlington, H., Cronkite, E. P., Pappas, N., and Zanjani, E. Susceptibility of hematopoietic stem cells (CFU-s) to ^{55}Fe radiation damage. *Radiat. Res., 74*:66, 1978.
21. Reincke, U., Brookoff, D., Burlington, H., and Cronkite, E. P. Forced differentiation of CFU-s by Iron-55 erythrocytocide. *Blood Cells, 5*:1979 (in press).
22. Reincke, U., Brookoff, D., Cronkite, E. P., Hillman, M., Wilcox, D., and Burlington, H. Relevance of specific activity in experimental erythrocytocide by ^{55}Fe. *Experientia* (in press).
23. Schooley, J. C., and Lin, D. H. Y. Hematopoiesis and the colony-forming unit. In Gordon, A. S., Condorelli, M., and Peschle, C., eds., *Regulation of Erythropoiesis.* Milano: Il Ponte, 1970, p. 52.

24. Stohlman, F., Ebbe, S., Morse, B., Howard, D., and Donovan, J. Regulation of erythropoiesis XX. Kinetics of red cell production. *Ann. N.Y. Acad. Sci., 149*:156, 1968.

25. Trentin, J. J. Hemopoietic inductive microenvironments. In Cairnie, A. B., Lala, P. K., and Osmond, D.

G., eds., *Stem Cells of Renewing Cell Populations.* New York: Academic Press, 1976, p. 255.

26. Vassort, F., Winterholer, M., Frindel, E., and Tubiana, M. Kinetic parameters of bone marrow stem cells using *in vivo* suicide by tritiated thymidine or by hydorxyurea. *Blood, 41*:789, 1973.

5

Ly Phenotype and Other T-Cell Antisera Sensitivity of an Anti-Theta Sensitive Cell that Regulates Hematopoiesis

Wieslaw Wiktor-Jedrzejczak,
Aftab Ahmed, Saul J. Sharkis,
Richard A. Cahill,
and Kenneth W. Sell

Hematopoietic and lymphoid cells in adult mice are thought to originate from a common multipotential stem cell, that is, a cell that forms colonies in a conventional CFU-s colony-forming assay (6,13). Our earlier experiments suggested that there are a small number of cells within the marrow, spleen, and thymus that cooperate with hematopoietic stem cells (HSCs) and bear the theta surface antigen (12). These cells, subsequently termed the anti-theta-sensitive regulatory cells (TSRCs), were found (9) to regulate the self-renewal capacity of the HSCs and to direct differentiation along erythropoietic and granulopoietic pathways.

The assay system for the TSRC employs anemic W/W^v mice as marrow recipients. The W/W^v mice possess a genetic HSC deficiency secondary to the mutation in the "w" locus mapped to the XVII linkage group on chromosome 5. Their normal HSC, $+/+$ littermates' bone marrow easily corrects the anemia of the W/W^v mouse with repopulation of hematopoietic tissues with donor cells and, further, forms colonies in anemic mice (7). The treatment of $+/+$ hematopoietic tissue with anti-Thy-1.2 serum and rabbit complement (C') does not destroy the ability of donor HSC to form spleen colonies, but prevents the establishment of normal hematopoiesis in anemic recipients. The addition of normal thymocytes to the anti-Thy-1.2 + C'-treated $+/+$ bone marrow results in a cure of the macrocytic anemia, whereas thymocytes alone cannot form spleen colonies or cure W/W^v recipients (9).

Murine T-cell populations have been characterized by their surface antigens, which are detected by the use of heterologous antithymocyte globulin (ATG) and mouse T-lymphocyte-specific antigen (MTLA), as well as the mouse alloantigen theta (Thy 1). Recent studies for additional cell-surface markers of Thy 1$^+$ cells have revealed that there is considerable heterogeneity among these subpopulations of lymphoid cells (5). Thus, with the use of cytotoxic anti-Ly alloantisera, one can distinguish three major subpopulations of Thy 1$^+$ cells—those that bear the phenotype Ly 1$^+$,2$^+$,3$^+$, those that bear the phenotype Ly 1$^+$,2$^-$,3$^-$, and finally, those that bear the phenotype Ly 1$^-$,2$^+$,3$^+$ (5). Studies in which these cytotoxic alloantisera have been used revealed functional differences between these three lymphoid cell populations. Thus, in general terms, the Ly 1$^+$,2$^+$,3$^+$ cell has been characterized as an amplifier cell; the Ly 1$^+$,2$^-$,3$^-$, the helper T-cell; and the Ly 1$^-$,2$^+$,3$^+$ cell, the cytotoxic and suppressor T-cell. The latter T-cell, the suppressor T-cell, has also been shown to be susceptible to lysis by incubation of cells with alloanti-Ia sera in the presence of C' (2).

Thus, these findings prompted us to examine the surface antigen characteristics of the TSRC for the presence of the various Ly markers and susceptibility to lysis with anti-Ia sera, heterologous ATG, and anti-MTLA. These studies constitute the basis of this report.

MATERIALS AND METHODS
Mice

Six- to 8-week-old male and female WBB6F$_1$ hematologically normal +/+ and anemic W/W^v mice were obtained from the Jackson Laboratory, Bar Harbor, Me. They were derived from C57BL/6-W^v and WB/RE-W/+ parents.

Antisera

Anti-Thy-1.2 serum and rabbit anti-mouse thymocyte globulin were prepared, as previously described (8,12). Anti-Ly sera were made by the method of Shen et al. (10). Anti-Ia^b was selected for the study, as this specificity is carried by the C57Bl/6 parental strain; the Ia specificity for the WB/Re strain, however, remains unknown. The summary of immunizations, absorptions, and titers of the various antisera utilized in this present study is shown in Table 5.1.

In Vitro Antisera Treatment of Cells

Marrow cells were obtained from the femur and tibia of the +/+ mice by flushing the marrow cavity with cold medium. These cells were incubated with the various antisera + C′ in a two-step treatment protocol, as described earlier (12). Dead cells were removed on a Ficoll-Hypaque gradient (sp. gr. 1.094). The cells were then washed twice and injected into groups of five W/W^v recipients. Similarly, aliquots of cells (used as controls) were treated with normal mouse serum (NMS) or normal rabbit serum (NRS) and C′.

The TSRC Assay

For the TSRC assay, 1×10^5 marrow cells treated with the appropriate antiserum + C′ were injected via the tail vein into groups of five $W/^v$ recipients for spleen-colony determinations, as described earlier (7). Simultaneously, 1×10^7 antiserum + C′-treated cells were injected into separate groups of W/W^v recipients because of their capacity to cure the anemia.

Spleen Colony Assay

The number of macroscopic spleen colonies in nonirradiated W/W^v recipients was determined 7 days following transplant of the +/+ marrow cells by the method of Till and McCulloch (11).

Peripheral Blood Examination

Erythrocyte counts and profiles were determined using the Coulter Counter and the Coulter Channelyzer. Because the defective W/W^v mouse produces macrocytes and a normal +/+ mouse produces normocytes, the erythrocyte size profile served as a marker for recognition of either the W/W^v or the +/+ erythrocyte type, respectively. Packed cell volume and the mean corpuscular volume were obtained by routine methods. The W/W^v recipients were considered cured by criteria previously established (9).

RESULTS

The use of cytotoxic antisera for the treatment of bone marrow cells for subsequent evaluation of its ability to cure the anemia of W/W^v mice required an evaluation of its direct cytotoxic effect on spleen colony-forming cells. Experiments were carried out whereby bone marrow from +/+ mice was treated in vitro with optimal concentrations of the various antisera used in these experiments. Aliquots of such antisera + C′-treated bone marrow cells were assayed for their ability to form macroscopic spleen colonies in W/W^v recipient mice. As seen in Table 5.2, aliquots of +/+ bone marrow cells treated with the various antisera + C′ had no effect on the ability of these cells to form macroscopic spleen colonies in W/W^v recipient mice. Controls consisting of no treatment or treatment of bone marrow cells with NMS or NRS + C′ also had no effect on the ability of these bone marrow cells to form macroscopic spleen colonies in irradiated W/W^v recipient mice (17.1 ± 3.1 and 15.9 ± 2.7, respectively), as compared to untreated +/+ bone marrow (18.4 ± 1.6). The data from these groups suggested that the antisera used in these experiments to treat the bone

TABLE 5.1 Summary of Immunizations, Absorptions, and Titers of T-Cell Antisera

SPECIFICITY	DONOR OF THYMOCYTES	RECIPIENT	ABSORPTION	TITER
Anti-Thy 1.2	CBA/J	AKR/J	None	1:3600
Anti-Ly 1.2	CE/J	(C3H)HeJ × DBA/2/F$_1$	DBA/2 thymocytes	1:256
Anti-Ly 2.2	B10.Y/Sn	(C3H)HeJ × BDP/F$_1$	DBA/2 thymocytes	1:64
Anti-Ly 3.2	CE/J	C58	AKR thymocytes	1:128
ATG	BALB/c	Rabbit	Mouse liver powder	1:512
Anti-MTLA	BALB/c	Rabbit	Mouse liver powder, B-cells	1:512

TABLE 5.2 The Effect of Treatment of +/+ Marrow Cells with Various Sera and Rabbit Complement (C') on Their Capacity to Form Spleen Colonies in W/W^v Recipients[a]

TYPE OF SERUM USED TO TREAT +/+ MARROW CELLS IN VITRO	NUMBER OF SPLEEN COLONIES[b] FORMED BY TREATED CELLS (10^5) IN W/W^v RECIPIENTS ($\bar{x} \pm$ SD)
NMS + C'	17.1 ± 3.1
NRS + C'	15.9 ± 2.7
Anti-Thy 1.2 + C'	16.8 ± 2.2
Anti-Ly 1.2 + C'	17.6 ± 4.2
Anti-Ly 2.2 + C'	15.2 ± 3.2
Anti-Ly 3.2 + C'	16.5 ± 1.9
ATG + C'	17.5 ± 2.1
Anti-MTLA + C'	15.9 ± 4.7
Anti-Ia^b + C'	17.2 ± 2.9
No serum	18.4 ± 1.6

[a]Each group consisted of five W/W^v recipient mice.

[b]The number of spleen colonies were determined on day 8 after transfer of cells.

marrow cells *in vitro* had no general cytotoxic effect on the spleen colony-forming precursor cells.

The remainder of the aliquots of cells treated with each antisera + C' was injected into individual groups of W/W^v mice, and the capacity of these cells to cure the W/W^v anemia was assessed starting at 10 days after transplant and every 10 days thereafter until 6 months post-transplant. As seen in Table 5.3, as expected, treatment of the +/+ bone marrow with anti-Thy-1.2 + C' resulted in its inability to cure the anemia of all five W/W injected mice in two separate experiments. Treatment of the +/+ bone marrow cells with anti-Ly-2.2, anti-Ly-3.2, or anti-Ia^b in the presence of C' resulted in the cure of the anemia in 3/3, 4/4, and 4/4 W/W^v recipient mice, respectively, indicating that at least at the dilution used, these antisera have no effect on the spleen-colony forma-

TABLE 5.3 The Effect of Treatment of +/+ Marrow Cells with Various Sera and Rabbit Complement (C') on Their Capacity to Cure W/W^v Anemia

TYPE OF SERUM USED TO TREAT +/+ MARROW CELLS IN VITRO[a]	CURED ANIMALS/ TRANSPLANTED ANIMALS	
	Experiment 1	Experiment 2
NMS + C'	5/5	5/5
NRS + C'	5/5	5/5
Anti-Thy 1.2 + C'	0/5	0/5
Anti-Ly 1.2 + C'	1/5	0/5
Anti-Ly 2.2 + C'	3/3	2/2
Anti-Ly 3.2 + C'	4/4	5/5
ATG + C'	0/4	0/4
Anti-MTLA + C'	0/4	0/3
Anti-Ia^b + C'	4/4	5/5

[a]10^7 antiserum- or normal serum-treated +/+ marrow cells were injected intravenously into nonirradiated W/W^v recipients, and the cure of the anemia was assessed at 2- to 3-week intervals from day 10 to 6 months post-transplant.

tion capacity of the +/+ bone marrow cells. In marked contrast, treatment of the +/+ bone marrow cells with anti-Ly-1.2, ATG, or MTLA in the presence of C' abolished the ability of these cells to cure the anemia in virtually all W/W^v recipients in two separate experiments. Treatment of the +/+ bone marrow cells with either NMS or NRS + C' did not affect its ability to cure the anemia of W/W^v mice.

The data presented in Table 5.3 reflect the cure of the anemia or lack thereof by the criteria for cure of anemic recipients. The results obtained at 10-day intervals were essentially similar up to a period of 6 months, after which the mice were sacrificed.

DISCUSSION

These data extend our previous observations (9,12) on the surface antigen characteristics of the TSRC that are involved in the regulation of hemopoiesis by stem cells in the W/W^v strain of mice. It is shown that these anti-theta-sensitive regulatory cells are sensitive to the cytotoxic action of heterologous ATG, MTLA, and an alloantisera specific for murine helper cells, termed anti-Ly 1 serum, in the presence of C'. Whereas ATG and MTLA are surface markers for all murine T-cells, the Ly 1 marker alone is specific for a subpopulation of murine T-cells constituting less than 1% of the cells in the adult bone marrow. These data suggest that the TSRC is a cell that bears markers recognized by ATG, anti-MTLA, and anti-Ly 1 sera. Further, since the TSRC was refractory to the cytotoxic activity of anti-Ly-2, anti-Ly-3, or anti-Ia^b antisera, these data strongly suggest that the TSRC does not belong or need the cooperation of either amplifier cells (Ly 1$^+$,2$^+$,3$^+$), effector cells (Ly 2$^+$3$^+$), or suppressor cells (Ly 2$^+$,3$^+$,Ia$^+$) (2,3) for its ability to cooperate with stem cells for the cure of the anemia in W/W^v mice. This ability of ATG, anti-MTLA, and anti-Ly 1 antisera to eliminate the TSRC was not secondary to its cytotoxicity for stem cells, as shown by the ability of these antisera-treated cells to form macroscopic spleen colonies equivalent to that obtained with untreated bone marrow cells (Table 5.2).

Since TSRC has been shown to be necessary for HSC renewal (9), one may reason that it has to appear before, or simultaneously with, the appearance of HSC. But the surface phenotype of the TSRC found in the studies reported here appeared to be an Ly 1$^+$,2$^-$,3$^-$ cell. According to Huber et al. (4), the Ly 1$^+$,2$^+$,3$^+$ cell ontogenetically precedes the Ly 1$^+$,2$^-$,3$^-$ cell, the latter cell being the phenotype for the TSRC. Further, we have previously reported that the TSRC is a radioresistant theta-bearing cell. Collectively, these data would suggest that during development either a precursor T-cell, which is Ly

$1^-,2^-,3^-$, is functionally able to perform the role of the TSRC and later acquire the Ly $1^+,2^-,3^-$ surface phenotype and the property of being radioresistant in the adult bone marrow or that there is always a residual small population of Ly $1^+,2^-,3^-$ cells in the bone marrow, or other anatomic sites of the mouse, that functions as the TSRC. Studies on the ability of bone marrow of different age $+/+$ littermate mice before and after treatment with anti-Ly antisera to function as a source of TSRC are currently in progress to ascertain whether a stage in development is achieved whereby TSRC is not sensitive to the cytotoxic action of anti-Ly antisera.

The inability of anti-Ia sera to eliminate the TSRC as the precursor stem cells that form macroscopic spleen colonies is consistent with the recent data of Basch et al. (1) that demonstrates that murine stem cells are Ia$^-$. It also eliminates the possibility of suppressor T-cells being involved in TSRC activity, either directly or indirectly, along with the evidence, as discussed above, that it is not susceptible to the cytotoxic action of anti-Ly 2 serum. Further, the evidence presented here also suggests that the TSRC cannot be an Ly $1^+,2^+,3^+$ suppressor cell, since cells bearing such a phenotype have been reported to act as suppressor cells directly or as amplifier cells to suppressor cells in certain systems.

The studies presented in this communication pose intriguing implications of the cellular requirements for the regulatory mechanisms involved in hemopoiesis and for the self-renewal potential of stem cells. Studies are being undertaken to determine the biologic implications of these findings, using chromosomal markers and other inbred strains of mice as genetic tools.

SUMMARY

Bone marrow cells from hematologically normal $+/+$ mice have the capacity to form spleen colonies and cure the macrocytic anemia of nonirradiated W/W^v mice. In vitro treatment of the bone marrow cells with heterologous antisera directed against thymus-derived T-lymphocytes, alloantisera specific to subpopulations of "T"-lymphoid cells, and Ia antigens results in a preservation of the capacity of these cells to form macroscopic spleen colonies. On the other hand, in vitro treatment of the adult $+/+$ bone marrow cells with heterologous ATG, MTLA, and alloantisera against Ly 1.2 prevents their ability to cure the anemia when they are subsequently injected into W/W^v recipient mice. Similar in vitro treatment of $+/+$ bone marrow cells with anti-Ly 2.2, anti-Ly 3.2, or anti-Ia sera and subsequent injection into W/W^v recipients had no effect; the W/W^v anemia was cured. These results suggest that the TSRC bears the phenotype ATG$^+$, MTLA$^+$, Thy1.2$^+$, Ly $1^+,2^-,3^-$, and Ia$^-$. The implication of these findings is discussed.

ACKNOWLEDGMENTS

We would like to thank Dr. Pawel Kisielow for comments on the manuscript and Mrs. Janie P. Kaczmarowski for editorial assistance.

This work was supported by the Naval Medical Research and Development Command, Research Work Unit No. MR041.02.01.0034. The opinions and assertions contained herein are the private ones of the writers and are not to be construed as official or reflecting the views of the Navy Department or the naval service at large. The experiments reported herein were conducted according to the principles set forth in the Guide for the Care and Use of Laboratory Animals, Institute of Laboratory Resources, National Research Council, Department of Health, Education and Welfare, Pub. No. (NIH) 74-23.

REFERENCES

1. Basch, R. S., Janossy, G., and Greaves, M. F. Murine pluripotential stem cells lack Ia antigens. Nature (London), 270:520, 1977.
2. Beverley, P. C. L., Woody, J., Dunkley, M., Feldmann, M., and McKenzie, I. Separation of suppressor and killer T cells by surface phenotype. Nature (London), 262:495, 1976.
3. Cantor, H., and Boyse, E. A. Functional subclasses of T lymphocytes bearing different Ly antigens. I. The generation of functionally distinct T-cell subclasses in a differentiative process independent of antigen. J. Exp. Med., 141:1376, 1975.
4. Huber, B., Cantor, H., Shen, F-W, and Boyse, E. A. Independent differentiative pathways of Ly 1 and Ly 2,3 subclasses of T cells. J. Exp. Med., 144:1128, 1976.
5. Kisielow, P., Hirst, J. A., Shiku, H., Beverley, P. C. L., Hoffmann, M. K., Boyse, E. A., and Oettgen, H. F. Ly antigens as markers for functionally distinct subpopulations of thymus-derived lymphocytes of the mouse. Nature (London), 253:219, 1975.
6. McCulloch, E. A., Mak, T. W., Price, G. B., and Till, J. E. Organization and communication in populations of normal and leukemic hemopoietic cells. Biochim. Biophys. Acta, 355:260, 1974.
7. McCulloch, E. A., Siminovitch, L., and Till, J. E. Spleen colony formation in anemic mice of genotype W/W^v. Science, 144:844, 1964.
8. Ochiai, T., Ahmed, A., Strong, D. M., Scher, I., and Sell, K. W. Specificity and immunosuppressive

potency of a rabbit antimouse T cell-specific antiserum. *Transplantation, 20*:198, 1975.

9. Sharkis, S. J., Wiktor-Jedrzejczak, W., Ahmed, A., Santos, G. W., McKee, A., and Sell, K. W. Theta-sensitive regulatory cell (TSRC) and hematopoiesis: Regulation of the differentiation of transplanted stem cells in *W/W^v* anemic and normal mice. *Blood, 52*:802, 1978.

10. Shen, F-W, Boyse, E. A., and Cantor, H. Preparation and use of Ly antisera. *Immunogenetics, 2*:591, 1975.

11. Till, J. E., and McCulloch, E. A. A direct measurement of the radiation sensitivity of normal bone marrow cells. *Radiat. Res., 14*:213, 1961.

12. Wiktor-Jedrzejczak, W., Sharkis, S. J., Ahmed, A., Sell, K. W., and Santos, G. W. Theta-sensitive cell and erythropoiesis: Identification of a defect in *W/W^v* anemic mice. *Science, 196*:313, 1977.

13. Wu, A. M., Till, J. E., Siminovitch, L., and McCulloch, E. A. Cytological evidence for a relationship between normal hematopoietic colony forming cells and cells of the lymphoid system. *J. Exp. Med., 127*:455, 1968.

PART II
Myelopoiesis

S. J. Baum

6

The Heterogeneity of Macrophage-Granulocyte Precursor Cell Populations

S. J. Baum

The complex process of *in vivo* regulation of granulopoiesis still remains an enigma (24). Numerous studies utilizing the method of cloning of progenitor cells in semi-solid *in vitro* cultures, however, have revealed important developmental steps that permit some elucidation of the feedback mechanism(s) controlling granulopoiesis (5,14,18). Essentially, the committed granulocytic precursor cell is differentiated from the multipotential stem cells by intrinsic as well as extrinsic mechanisms (3). The early granulocyte precursor cell, known as the colony-forming unit culture (CFU-c) or granulocyte-macrophage colony-forming cell (GM-CFC), is also assumed to be a precursor for monocytes and macrophages (M-CFC) (14). The extreme heterogeneity in size and shape of the agar colonies suggests the presence of several subpopulations. Attempts have been made to describe some of them (11,12,14,15). It has not yet been determined whether all these different granulocyte-macrophage (GM) precursor cells are stimulated by the same colony-stimulating activity (CSA). But the existence of four molecular species of CSA suggests that different subpopulations may be stimulated by different molecular species or possibly by different combinations of molecular species (20). Eventually, the CFU-c reach a point of development amenable to differentiation into recognizable early granulocytes (i.e., myeloblasts), most likely via an intrinsic cellular modification as stimulated by humoral agents. This review is limited to an assessment of the cause and possible meaning of the observed heterogeneity in precursor cell populations and their stimulatory activities. The reader is referred to recent articles for a more comprehensive review of CFU-c production and regulation (3,14,24,28).

The Release of Colony-Forming Unit Culture From the Multipotential Stem Cell Compartment

It is now assumed that multipotential stem cells (CFU-s) can either renew themselves by producing two daughter cells or can differentiate to produce two cells of the committed precursor cell compartment (10). Here, it is assumed that the CFU-s compartment is normally mostly quiescent and maintained at a controlled size by as yet unknown specific regulators and a defined internal environment. Recently, evidence was presented that molecules produced by adult leukocytes in inflammatory exudates may increase CFU-s activity and, possibly, may be involved in the release of cells from the most primitive compartment (4). But since these agents also stimulate an increase in CFU-c activity and the release of these cells into the maturation compartment, it is possible that the reduced CFU-c compartment stimulates differentiation of CFU-s into CFU-

c. This could be accomplished by a direct, cell-to-cell stimulatory interaction or by less inhibitory action upon CFU-s by the reduced CFU-c compartment. It was also suggested that cell membranes may exert regulatory influences (20) either by components separated from cell surfaces or by configurational changes that would alter the characteristics of physical contact between adjacent cells.

Regulation of the Committed Stem Cell Compartment Colony-Forming Unit Culture

It has been postulated that the CFU-c represent an important buffer population between the CFU-s and the functional granulocytes or macrophages in the three-tiered structure of hemopoietic cell populations (20). This may possibly protect the relatively small CFU-s population against depletion through excessive demand for mature cells. Furthermore, since it appears that the primitive precursors represent a heterogeneity of cells with multiple sites for stimulation, they provide the potential for a highly sensitive and precise cellular delivery system. Another possible system to prevent excessive cellular expansion is inherent in the process of differentiation that appears to limit proliferation. It has been indicated that abnormal proliferation usually originates in the CFU-s prior to its loss of self-renewal and the onset of the differentiation processes (1,8).

Of the many interesting findings concerning GM progenitor cells reported by Metcalf's laboratory, the observation on the heterogeneity of the size and shape of GM colonies generated by those cells is possibly of greatest significance (14,15). This may indicate that the ancestral CFU-s form of these cells may also have been a heterogeneous group of cells. These cells apparently remain in distinct precursor compartments and may be selectively responsive to the different combinations of the basic regulatory factors also termed colony-stimulating activities (CSA). Recently, it has been reported, at least for the macrophages, that these cells could be divided into subpopulations with distinct functional properties (29). At present, it is not clear whether the observed heterogeneity is related to differences in degree of maturation or differentiation, or possibly, whether it is indicative of multiple origins (i.e., CFU-s). Five subpopulations of neutrophil and or macrophage progenitor cells have been reported for the mouse (11,12,14,15,27). These populations form pure neutrophil colonies, mixed neutrophil/macrophage colonies, or pure macrophage colonies. Extensive work on the cellular populations that form only macrophages was completed in our laboratory by MacVittie (12,13). He detected *in vitro* macrophage-forming cells (M-CFC) in bone marrow, spleen, peripheral blood leukocytes, thymus, and lymph nodes. Similar cells were also detected by others in the peritoneal cavity (11,27). The bone marrow- and spleen-derived M-CFC required pregnant mouse uterine extract (PMUE) for colony formation, whereas those derived from peripheral blood leukocytes could be stimulated by either PMUE or L-cell conditioned medium. All the colonies formed contained only macrophages, they required a lag period of 13 to 18 days prior to initiation of colony formation, and therefter they proliferated at markedly slow rates. All the M-CFC colonies showed a marked ability to survive in culture in the absence of PMUE. In contrast, the CFU-c perish rapidly under similar culture conditions. Heterogeneity in the cell populations, however, is again demonstrated by the observation that M-CFC derived from bone marrow or spleen survived significantly longer in the absence of PMUE than did cellular colonies obtained from peripheral blood leukocytes or thymus (13).

Control of Colony-Forming Unit Culture Homeostasis

Current available information appears to indicate that differentiation of multipotential stem cells into CFU-c is under the control of the microenvironment, although humoral influences have not been completely ruled out (4). Differentiation and proliferation of the committed precursor compartment, however, must be mainly under the control of specific humoral factors (16). In culture, the CFU-c are only produced in the presence of certain molecular species, collectively termed CSA or feeder cells, which produce these molecules. In the mouse, feeder cells are obtainable from a variety of tissues (2,17,26) for the production of CSA. Human urine can initiate culture growth in the mouse (28). But all mouse-derived CSA proved to be incapable of stimulating human cells. Human cells can only be activated by human CSA. Human CSA has been obtained from peripheral blood, bone marrow, embryo kidney cells, placenta, and leukocytes (6,9,20). Three molecular species of CSA having molecular weights of 93,000, 36,000, and 14,700 Daltons have been isolated from human leukocytes. These are nondialyzable and are usually referred to as high molecular weight (HMW) CSA. A fourth species is dialyzable, has a molecular weight of only approximately 1,300 Daltons and is consequently called low molecular weight (LMW) CSA. It has been determined that the HMW and LMW CSA differ not only in structure but also in biologic activity. The three species of HMW CSA were found primarily in fractions containing cell membranes (22). It was proposed that membrane-associated molecules are in equilibrium with free surface molecules to create an environment favorable to the proliferation of granulocyte progenitors (20). The main support for this comes from the observation that human granulocyte

CSA can be obtained directly from solubilized cell membranes (22). But CFU-c appear to lack, on their cellular membranes, intrinsic immunoglobulins, sheep erythrocyte receptors, and receptors for erythrocyte antibodies. They appear to be essentially null cells (23). It was proposed that they require the presence of peripheral blood mononuclear cells for maximum proliferation. The LMW CSA may be obtained from filtrates or dialysates of leukocyte-conditioned media (LCM) (21) or by incubating 5×10^6 leukocytes in serum-free medium without agar. Enzymatic digestion with trypsin and chymotrypsin indicated that LMW CSA was a peptidyl compound. Biologic heterogeneity was observed for LMW CSA. Materials obtained from leukemic patients differed in stimulatory capability from that of normal individuals, depending on whether cells from normal or leukemic persons were cultured. Generally, LMW CSA does not have the stimulatory capability that HMW CSA has. But when added to CFU-c in the presence of HMW species, LMW CSA appears to increase colony-forming capacity. It was suggested that LMW CSA could be a candidate for a role in the triggering mechanism for proliferation (20).

Feedback Control of Granulocyte-Macrophage Production

Normal granulocyte-macrophage production may be a stochastic event, particularly in the release of CFU-s into the CFU-c compartment. A great deal of the control may be exercised locally in the microenvironment, although humoral stimulators could be involved (14). Conceivably, normal cellular production could be simply controlled by negative feedback (3). But once inflammatory loci are established, humoral signals, presumably from functional leukocytes, may regulate cellular production (4,25). It has also been suggested that increased leukocyte production may possibly be at the expense of other cell lines (7,19). The heterogeneity of precursor cell populations in CSA raises a number of interesting questions that should be of immediate interest to researchers in the field. Research data are needed to determine whether this heterogeneity is randomly produced or whether it is a possible response to variable stimulators released by areas of inflammation depending on the inflammatory condition and the corresponding cellular participation. Is this heterogeneity induced at the CFU-s level or in the committed precursor department? Obviously, a variable response, tailored to specific emergency conditions, could better and perhaps more rapidly normalize a homeostasis disturbed by invading pathogens. It would result in a heterogeneity of functional cell production, each one capable of remedying specific aspects of inflammation and infection.

SUMMARY

At present, most of our knowledge of differentiation and proliferation of the CFU-c has come from *in vitro* observations. Although these studies contributed greatly toward our understanding of granulocyte-macrophage production, we cannot assume that they were conducted under the precise physiological conditions normally present in the whole animal. Possibly, the most important recent finding was the heterogeneity of precursor cell populations and colony-stimulating activities (CSA). There are some suggestions that this heterogeneity originates at the multipotential cell level or at least within the earliest committed precursor cells. The possibility then exists that the differentiation of these cells depends on the character of the stimulatory activities received, presumably, from infectious areas. It is intriguing to speculate that the CSA produced by leukocytes in inflammatory exudates is related to the extent and condition of infection. If this possibility is proven true, it would represent an extremely efficient physiologic feedback system.

REFERENCES

1. Adamson, J. W., Fialkow, P. J., Murphy, S., Prchal, J. F., and Steinmann, L. Polycythemia vera: Stem cell and probable clonal origin of the disease. *N. Eng. J. Med., 295*:913, 1976.

2. Austin, P. E., McCulloch, E. A., and Till, J. E. Characterization of the factor in L-cell conditioned medium capable of stimulating colony formation by mouse marrow in culture. *J. Cell. Physiol., 77*:121, 1971.

3. Baum, S. J. Negative and positive feedback control of the committed granulocytic stem cell compartment. In Baum, S. J., and Ledney, G. D., eds., *Experimental Hematology Today*. New York: Springer-Verlag, 1977, p. 127.

4. Baum, S. J., MacVittie, T. J., Brandenburg, R. T., and Levin, S. G. Stimulatory effects of products from inflammatory exudates upon stem cell production. *Exp. Hemat., 6*:405, 1978.

5. Bradley, T. R., and Metcalf, D. The growth of mouse bone marrow cells *in vitro*. *Aust. J. Exp. Biol. Med. Sci., 44*:287, 1966.

6. Burgess, A. W., Wilson, E. M., and Metcalf, D. Stimulation by human placental conditioned medium of hemopoietic colony formation by human marrow cells. *Blood, 49*:573, 1977.

7. Chikkappa, G., Chauaua, A. D., Chaudra, P., and Cronkite, E. P. Kinetics and regulation of granulocyte precursors during a granulopoietic stress. *Blood, 50*:1099, 1977.

8. Fialkow, P. J., Gantler, S. M., and Yoshida, A. Clonal origin of chronic myelogenous leukemia in man. *Proc. Nat. Acad. Sci. U.S.A., 58*:1468, 1967.

9. Iscove N. N., Senn, J. S., Till, J. E., and McCulloch, E. A. Colony formation by normal and leukemic human marrow cells in culture: Effect of conditioned medium from human leukocytes. *Blood, 37*:1, 1971.

10. Lajtha, L. G. Bone marrow stem cell kinetics. *Seminars Haemat 4*, "Dynamics of hematopoiesis" 293, 1967.

11. Lin, H. S., and Stewart, C. C. Peritoneal exudate cells. I. growth requirement of cells capable of forming colonies in soft agar. *J. Cell. Physiol., 83*:369, 1974.

12. MacVittie, T. J., and McCarthy, K. F. The detection of *in vitro* monocyte-macrophage colony-forming cells in mouse thymus and lymph nodes. *J. Cell. Physiol., 92*:203, 1977.

13. MacVittie, T. J., and Porvaznik, M. Detection of *in vitro* macrophage colony-forming cells (M-CFC) in mouse bone marrow, spleen and peripheral blood. *J. Cellular Physiol., 97*:305, 1978.

14. Metcalf, D. *In vitro* cloning of normal and leukemic cells. In *Hemopoietic Colonies*. Berlin: Springer-Verlag, 1977.

15. Metcalf, D., and MacDonald J. R. Heterogeneity of *in vitro* colony- and cluster-forming cells in the mouse marrow. Segregation by velocity sedimentation. *J. Cell. Physiol., 85*:643, 1975.

16. Metcalf, D., and Moore, M. A. S. Regulation of growth and differentiation in hemopoietic colonies growing in agar. *Ciba Found. Symp., 13*:157, 1973.

17. Pike, B. L., and Robinson, W. A. Human bone marrow colony growth in agar-gel. *J. Cell. Physiol., 76*:77, 1970.

18. Pluznik D. H., and Sachs. L. The cloning of normal "mast" cells in tissue culture. *J. Cell. Comp. Physiol., 66*:19, 1966.

19. Pratt A. G., Emerson, R. J., Levin, S., and Baum, S. J. Granulocytic recovery in the polycythemic dog treated with endotoxin postirradiation. *Rad. Res., 56*:162, 1973.

20. Price, G. B., and McCulloch, E. A. Cell surfaces and the regulation of hemopoiesis. *Sem. Hematol., 15*:283, 1978.

21. Price, G. B., McCulloch, E. A., and Till, J. E. A new human low molecular weight granulocyte colony stimulating activity. *Blood, 42*:341, 1973.

22. Price, G. B., McCulloch, E. A., and Till, J. E. Cell membranes as sources of granulocyte colony stimulating activities. *Exp. Hematol., 3*:227, 1975.

23. Richman, C. M., Chess, L., and Yankee, R. A. Purification and characterization of granulocytic progenitor cells (CFU-c) from human peripheral blood using immunologic surface markers. *Blood, 51*:1, 1978.

24. Robinson, W. A., and Mamgalik, A. The kinetics and regulation of granulopoiesis. *Semin. Hematol., 12*:7, 1975.

25. Scott, R. B., Eanes, R. Z., Cooper, L. W., Higgins, L. L., and Eastman, C. A. Characterization of marrow granulocyte development: Changes in response to inflammatory reactions. *Brit. J. Haematol., 37*:503, 1977.

26. Sheridan, J. W., and Stanley E. R. Tissue sources of bone marrow colony stimulating factor. *J. Cell. Physiol., 78*:451, 1971.

27. Stewart, C. L., Lin, H. S., and Adles, C. Proliferation and colony-forming ability of peritoneal exudate cells in liquid culture. *J. Exp. Med., 141*:1114, 1975.

28. Stohlman, F. Jr., Guesenberry, P. J., Niscanen, E., Morley, A., Tyler, W., Rickard, K., Symann, M., Monette, F., and Howard, D. Control of Granulopoiesis. *CIBA Found. Symp., 13*:206, 1973.

29. Walker, W. S. Functional Heterogencity of macrophages in the induction and expression of acquired immunity. *J. Reticuloendothelial. Soc., 20*:57, 1976.

7

Local Control of Mast Cell Differentiation in Mice

Y. Kitamura, K. Hatanaka,
M. Shimada, S. Go,
and H. Matsuda

Mast cells are detectable in connective tissues of almost every part of the body and are thought to be closely related to fibroblasts and macrophages (17). Fibroblasts originate *in situ* from undifferentiated mesenchymal cells (3.5) whereas macrophages are derived from the bone marrow (3,4,15). Recently, we have shown that tissue mast cells also originate from the bone marrow in irradiated mice (11). A probable route of precursor cells from the bone marrow to the connective tissues of each organ may be the bloodstream (9). But our previous postirradiation bone marrow transplantation studies demonstrated that the appearance rate of donor-type mast cells differed for each organ (11). Thus, mast cells of the skin were of host type 25 weeks after irradiation and bone marrow cell transplantation, whereas most of the mast cells in the intestinal tract were of the donor type 12 weeks after transplantation (11). Thus, there is the possibility that the differentiation of bone marrow-derived precursors of mast cells is under local tissue control. We designed our experiments to examine this possibility using three mutant strains of mice. Mutant beige mice bearing mast cells with giant granules were used as a marker of donor-type cells (2,11). Mutant anemic W/W^v and Sl/Sl^d mice were used as host animals because we have recently found that such genotypes are nearly devoid of tissue mast cells in all parts of the body (7,8). Our experimental results suggest that the differentiation of mast cells may be locally controlled both by humoral regulators and cell-to-cell interactions.

MATERIALS AND METHODS

Mice

Beige (C57BL/6-bg^J/bg^J) mice and their normal litter mates (C57BL/6bg^J/+ and −+/+) were raised in our laboratory using parental stocks obtained from the Jackson Laboratory (Bar Harbor, Me.). The giant granules of mast cells of beige mice are shown in Figure 7.1. The WBB6F$_1$ (WB/Re-W/+ × C57BL/6-W^v/+)−(W/W^d, +/+) and WCB6F$_1$ (WC/Re-Sl/+ × C57BL/6Sl^d/+)−(Sl/Sl^d, +/+) mice were purchased from the Jackson Laboratory or raised in our laboratory. Animals between 2 and 4 months of age were used.

Bone Marrow Transplantation

The method of cell preparation (10) and X-irradiation (12) have been described. Normal C57BL/6 mice (+/+ and bg^J/+) were injected intravenously with the bone marrow cells of beige C57BL/6 mice (bg^J/bg^J) within 3 hr after 800 rad wholebody irradiation. The W/W^v mice were injected with the bone marrow cells of the congeneic +/+ mice without prior irradiation.

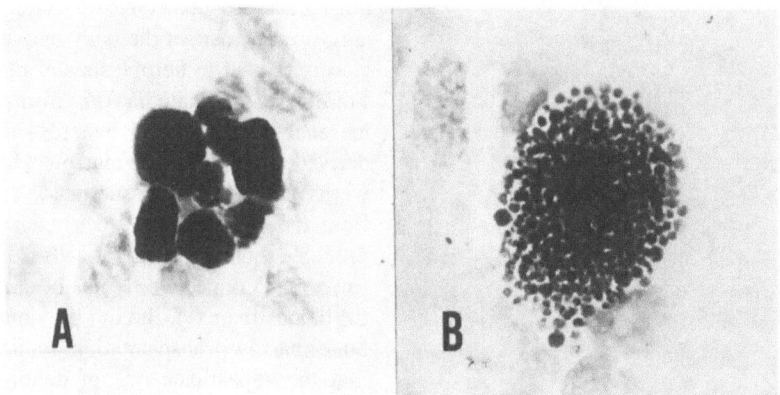

FIGURE 7.1. Mast cells in the stretch preparation of the mesentery stained with toluidine blue. (**A**) A mast cell of the beige C57BL/6 mouse. The size of the metachromatic granules is greatly enlarged and their number decreased. (**B**) A mast cell of the normal C57BL/6 mouse.

Methylcholanthrene Painting

Mast cell production was stimulated by painting 20-methylcholanthrene (0.5% solution in benzene) on the dorsal skin of normal C57BL/6 mice that had been irradiated and reconstituted with the bone marrow cells of beige mice. Starting on the 90th day after bone marrow cell transplantation, the painting was done every 3rd day for 27 days (a total of 10 times) according to the method described by Takeoka et al. (18). The mice were killed 3 days after the last painting (i.e., 120 days after bone marrow cell transplantation). A piece of the painted dorsal skin was removed and compared with a sample of unpainted neutral skin.

Skin Transplantation

A full thickness 25 × 20 mm piece of skin, including the paniculus carnosus, was grafted on the back of a recipient mouse anesthetized with Nembutal. The graft was kept in place by wound clips. Each engrafted mouse was caged separately. To determine the variation of mast cell numbers in each skin, a 25 × 3 mm piece was cut from the skin before grafting. Additionally, a 15 × 3 mm biopsy of the graft was obtained from mice anesthetized with ether. A piece of the grafted skin was also removed at the time of autopsy.

Histology

Tissues (skin, stomach, and cecum) were gently smoothed onto a piece of thick filter paper to keep them flat, fixed in 10% buffered formalin (pH 7.2), and embedded in paraffin. Sections (5 μm thick) were stained with acidified toluidine blue.

Counting of Mast Cells

When normal C57BL/6 mice were transplanted with the bone marrow cells of beige C57BL/6 mice, counts of more than 1,000 mast cells in the skin and more than 100 mast cells in the stomach and cecum of each mouse were obtained to demonstrate the presence or the absence of giant metachromatic granules (magnification, × 1000) (Figure 7.2).

To determine the number of mast cells in a unit length of the skin, the mast cells between the epithelium and the paniculus carnosus were counted at low magnification (× 100). The number of mast cells was divided by the length of each section and expressed as the number of mast cells per centimeter.

RESULTS

Effect of Local Stimulation on Development of Mast Cells

Normal C57BL/6 mice were irradiated and injected with 5 × 10⁶ beige C57BL/6 bone marrow cells. These mice were killed on various days after cell injection, and the skin, stomach, and cecum were examined histologically for the presence of mast cells. Although donor-type mast cells with giant granules appeared in the cecum and the stomach 9 weeks after transplantation, the mast cells of the skin continued to be of the host type up to 28 weeks after transplantation. (Figure 7.3). When methylcholanthrene was painted on the dorsal skin, mast cells of the donor beige-mouse type with giant granules developed at the painted site. In contrast, mast cells in the ventral skin of the same mice continued to be of the host mouse type, with numerous small granules (Figure 7.2).

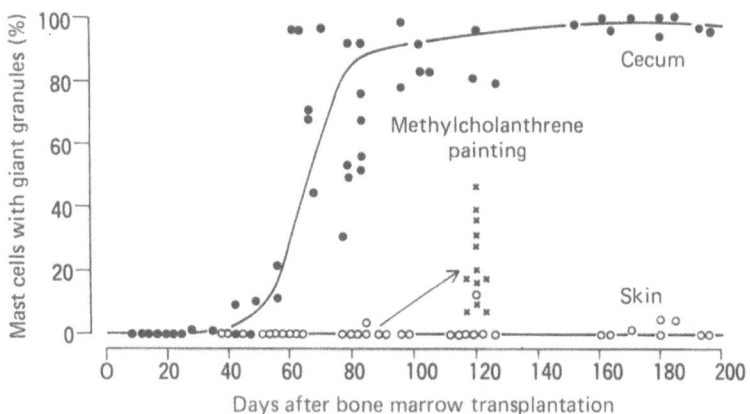

FIGURE 7.2. Appearance of mast cells with giant granules in the normal C57BL/6 mice irradiated and injected with the bone marrow cells of the beige C57BL/6 mice. In the skin of unstimulated mice, mast cells of the beige-mouse origin developed in the cecum (●——●) but not in the skin (○——○). But after methylcholanthrene painting (90 to 117 days), beige-type mast cells also appeared in the skin (x——x).

Inhibitory Effect of Mature Mast Cells on Development of New Mast Cells

As shown in Figure 7.4, the number of mast cells in the skin of W/W^v anemic mice was less than 1% of the number observed in the congeneic +/+ mice. No mast cells were detectable in other tissues of the W/W^v mice. When 2×10^7 bone marrow cells of the +/+ mice were injected into W/W^v mice, the number of mast cells increased not only in the gastrointestinal tract, but also in the skin (8). Fifteen weeks after bone marrow transplantation, the number of mast cells in the skin of W/W^v mice was about 200 times greater than the number of mast cells detected in the skin of control W/W^v mice that had not received the bone marrow cells from +/+ mice (8). The value representing the production of mast cells in the skin

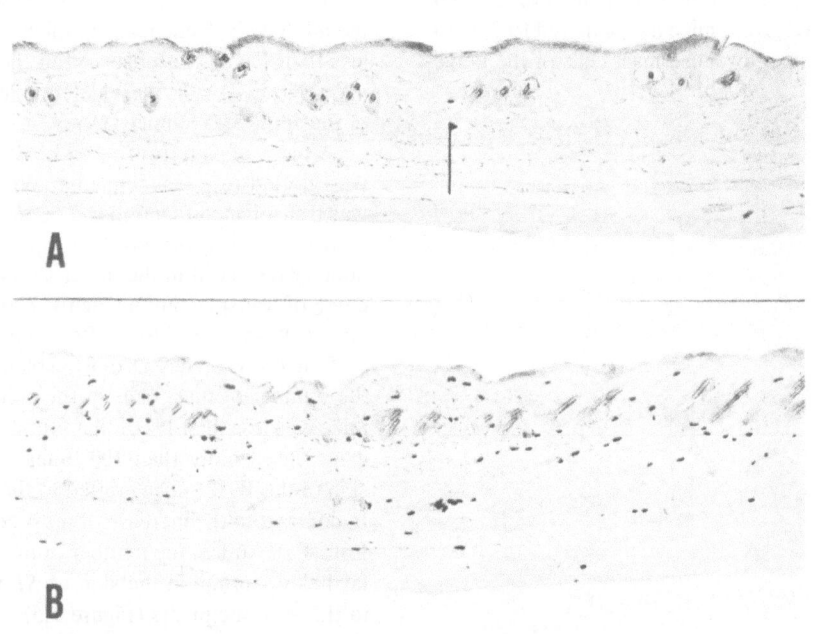

FIGURE 7.3. Decrease in mast cell number in the skin of WBB6F$_1$-W/W^v mice. (**A**) The skin of the W/W^v anemic mouse. Only one mast cell is detectable in this field (arrow). (**B**) The skin of the congeneic +/+ normal mouse.

47

of W/W^v mice by transplanted bone marrow cells was calculated as follows:

Production of Mast Cells
= [Mast Cells in W/W^v
Mice Injected with $+/+$ Cells]
− [No. of Mast Cells in Control W/W^v]

(Number of mast cells were expressed as mast cells per centimeter).

Although irradiation was not necessary for the engraftment of bone marrow cells from $+/+$ mice to W/W^v mice, normal C57BL/6 mice had to be irradiated before injection of the bone marrow cells from the beige C57BL/6 mice to effect successful engraftment. Since the irradiation itself did not reduce the number of host-type mast cells in the skin of normal C57BL/6 mice, the value representing the production of mast cells by the injected bone marrow cells from beige C57BL/6 mice was calculated as follows:

Production of Mast Cells
= [Total Number of Mast Cells
Beige-type mast Cells]
× (Beige-Type Mast Cells
+ Normal-Type Mast Cells)

As shown in Figure 7.5, the production of mast cells by transplanted bone marrow cells from mutant beige was much less in the skin of normal C57BL/6 mice than in the skin of the W/W^v mice. This result could not be attributed to the difference between the number of injected bone marrow cells (i.e., 5×10^6 in the normal C57BL/6 mice as compared to 2×10^7 in the W/W^v mice), because mast cells of the beige-

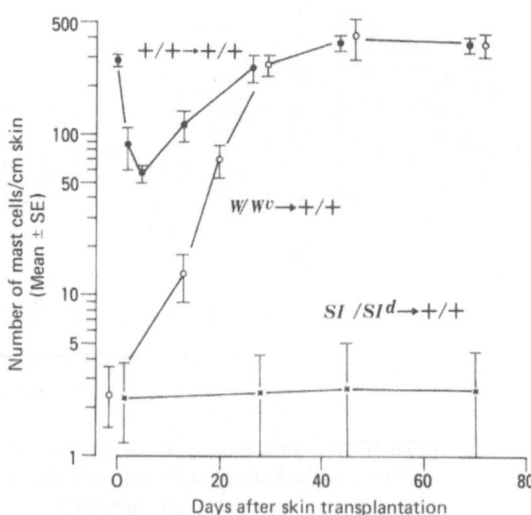

FIGURE 7.5. Number of mast cells in the skin of the $+/+$ or W/W^v or $S1/S1^d$ mice grafted to the congeneic $+/+$ recipients.

mouse type were rarely found in the skin of normal C57BL/6 mice after irradiation and injection with 5×10^7 beige bone marrow cells.

Necessity of the Microenvironment for Mast Cell Development

As compared with the number of mast cells in congeneic $+/+$ mice, the number of these cells in $S1/S1^d$ and W/W^v mice was reduced. In the skin of the adult $S1/S1^d$ mice, the number of mast cells was less than 1% of that observed in the congeneic $+/+$ mice. No mast cells were detectable in other tissues of the adult $S1/S1^d$ mice (7).

A piece of skin from $+/+$, W/W^v, or $S1/S1^d$ mice was grafted to $+/+$ recipients. When the $+/+$ skin was transplanted to another $+/+$ mouse, the number of mast cells in the graft dropped to one-fifth the number observed in the intact skin of $+/+$ mice, on day 5 of transplantation (Figure 7.5). The number of mast cells increased thereafter, and approached pregrafting levels on day 28 of transplantation. Although the number of mast cells in the skin of adult W/W^v mice was less than 1% of that found in the $+/+$ mice, it increased more than 100 times in 28 days in the skin of the W/W^v mice grafted to the $+/+$ recipients. In contrast to the increase of mast cells in skin grafts from W/W^v mice, the number of mast cells remained far below normal in the skin of $S1/S1^d$ mice grafted to the $+/+$ recipients (Figure 7.5).

DISCUSSION

Although the donor-type mast cells appeared in the gastrointestinal tract of normal C57BL/6 mice 9

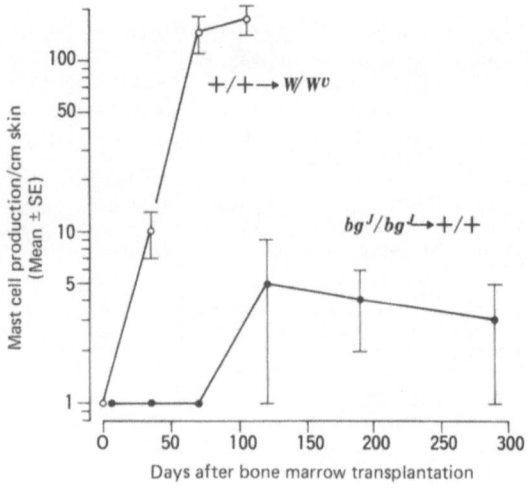

FIGURE 7.4. Production of mast cells in the skin of mice by transplanted bone marrow cells. The W/W^v mice received bone marrow cells from the congeneic $+/+$ mice, and the normal C57BL/6 mice from the beige C57BL/6 mice.

weeks after irradiation and injection of bone marrow cells from the beige C57BL/6 mice, mast cells of the skin continued to be of the host type up to week 28 of transplantation. But when methylcholanthrene was painted on the dorsal skin of the radiation chimeras, mast cells of the donor type appeared at the painted site. Since the donor-type mast cells did not appear in the ventral skin of the same chimeras, development of mast cells seems to be controlled locally. Since we have recently demonstrated by radioautography that division of bone marrow-derived precursors occurred at the site of methylcholanthrene painting (6), we think that some type of stimulator(s) might be produced by connective tissue cells other than mast cells to induce division and maturation of precursor cells after local irritation by methylcholanthrene (Figure 7.6).

The production of mast cells by transplanted bone marrow cells was much smaller in the skin of normal recipients than in the skin of the W/W^v recipients. This difference may be attributed to the genetically determined lack of mast cell production in W/W^v mice. Since inhibitors (chalones) produced by mature granulocytes and erythrocytes were reported to suppress division of granuloid and erythroid precursors, respectively (16), we think that mature mast cells might secrete a "chalone"-like substance that would inhibit the development of mast cells from their bone marrow-derived precursors (Figure 7.6).

Although the number of mast cells is markedly low in both W/W^v and Sl/Sl^d mice (7,8), the possible mechanisms of mast cell production differ in these two mutant strains. Since the number of mast cells in the W/W^v mice can be increased by bone marrow transplantation from the $+/+$ mice, the defect of

mast cell production in W/W^v mice appears to be due to a shortage of bone marrow-derived precursor cells (8). Since the bone marrow cell transplantation from the Sl/Sl^d mice also increased the number of mast cells in W/W^v mice (7), however, the Sl/Sl^d mice seem to have precursors of mast cells in spite of the fact that they have so few tissue adult mast cells. The decrease of mast cells in the Sl/Sl^d mice may be due to a local defect in the tissues in which mast cells differentiate from their precursors, because the number of mast cells in the skin grafted from Sl/Sl^d to $+/+$ mice were under 1% of the number observed in the skin grafted from W/W^v to $+/+$ mice. Since anemia (1,14) and lack of melanocytes (13) in the Sl/Sl^d mice were considered to be due to defects in the cellular microenvironment, the decrease of mast cells in the Sl/Sl^d mice may also be attributable to such defects (Figure 7.6).

SUMMARY

The control mechanisms of mast cell differentiation are unknown. Therefore, beige (bg^J/bg^J C57BL/6 mice with mast cells containing giant granules were used to identify donor-type cells in chimeric W/W^v and Sl/Sl^d mice. These genotypic strains of mice were used as hosts because their tissues are nearly devoid of mast cells. It was observed that mast cells of the beige C57BL/6 mice donor type were barely detectable in the skin of normal irradiated C57BL/6 mice. But when methylcholanthrene was painted on the dorsal skin of these chimeric mice, mast cells of the beige-mouse type appeared in the painted site, although no mast cells of the donor type were detectable in the ventral skin of the same chimeric mice. When W/W^v mice were used as recipients of beige C57BL/6 bone marrow cells, a considerable number of mast cells appeared in the skin without additional treatment. In contrast with this remarkable increase of mast cells in the skin of W/W^v mice grafted into the congeneic $+/+$ mice, no mast cells developed in the skin of Sl/Sl^d mice grafted onto $+/+$ mice. These results suggest that the differentiation of mast cells is controlled locally both by humoral regulators and cell-to-cell interactions.

ACKNOWLEDGMENTS

The authors thank Drs. K. J. Mori, M. Seki, and K. Matsumoto for valuable discussions.

This investigation was supported in part by Grants-in-Aid from the Japanese Ministry of Education, Science and Culture.

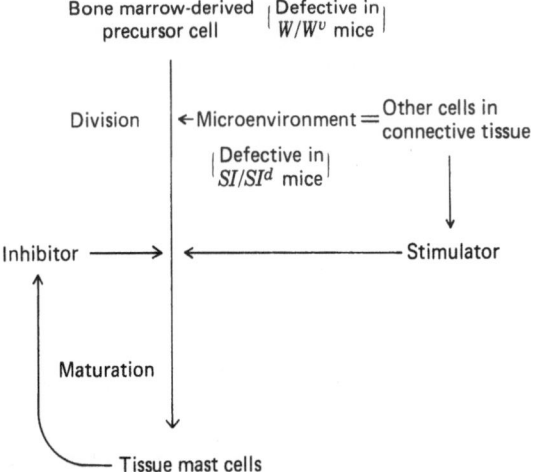

FIGURE 7.6. A scheme showing the probable mechanisms that control the differentiation of bone marrow-derived precursors into tissue mast cells.

REFERENCES

1. Bernstein, S. E. Tissue transplantation as an analytical and therapeutic tool in hereditary anemias. *Am. J. Surg., 119*:448, 1970.

2. Chi, E., and Lagunoff, D. Abnormal mast cell granules in beige (Chediak-Higashi syndrome) mouse. *J. Histochem. Cytochem., 23*:117, 1975.

3. Fialkow, P. J., Jacobson, R. J., and Papayannopoulou, T. Chronic myelocytic leukemia: Clonal origin in a stem cell common to the granulocyte, erythrocyte, platelet and monocyte-macrophage. *Am. J. Med., 63*:125, 1977.

4. Golde, D. W., Burgaleta, C., Sparkes, R. S., and Cline, M. J. Philadelphia chromsome in human macrophages. *Blood, 49*:367, 1977.

5. Greenberg, B. R., Wilson, F. D., Woo, L., and Jenks, H. M. Cytogenetics of fibroblastic colonies in Ph¹-positive chronic myelogenous leukemia. *Blood, 51*:1039, 1978.

6. Hatanaka, K., Kitamura, Y., and Nishimune, Y. Local development of mast cells from bone marrow-derived precursors in the skin of mice. *Blood 53*:142, 1979.

7. Kitamura, Y., and Go, S. Decreased production of mast cells in Sl/Sl^d mice (In press).

8. Kitamura, Y., Go, S., and Hatanaka, K. Decrease of mast cells in W/W^v mice and their increase by bone marrow transplantation. *Blood, 52*:447, 1978.

9. Kitamura, Y., Hatanaka, K., Murakami, M., and Shibata, H. Mast-cell precursors in peripheral blood of mice: Demonstration by parabiosis *(paper submitted)*.

10. Kitamura Y., Kawata, T., Suda, O., and Ezumi, K. Changed differentiation pattern of colony-forming cells in F_1 hybrid mice suffering from graft-versus-host disease. *Transplantation, 10*:455, 1970.

11. Kitamura, Y. Shimada, M., Hatanaka, K., and Miyano, Y. Development of mast cells from grafted beone marrow cells in irradiated mice. *Nature, 268*:442, 1977.

12. Kitamura, Y., Tamai, M., Miyano, Y., and Shimada, M. Development of hematopoietic spleen colonies in nonirradiated mice. *Blood, 50*:1121, 1977.

13. Mayer, T. C., and Green, M. C. An experimental analysis of pigment defect caused by W and Sl loci in mice. *Develop. Biol., 18*:62, 1968.

14. McCulloch, E. A., Siminovitch, L., Till, J. E., Russel, E. S., and Bernstein, S. E. The cellular basis of the genetically determined hematopoietic defect in anemic mice of genotype Sl/Sl^d. *Blood, 26*:399, 1965.

15. Metcalf, D., and Moore, M. A. S. In *Haematopoietic Cells.* Amsterdam: North-Holland, 1971, p. 123.

16. Metcalf, D., and Moore, M. A. S. In *Haemotopoietic Cells.* Amsterdam: North-Holland, 1971, p. 419. Frontiers of Biology ed A. Neuberger & E. L. Tatum

17. Selye, H. *The Mast Cells.* Washington D.C., Butterworths, 1965.

18. Takeoka, O., Ashihara, T., and Tada, N. ³H-Thymidine labelled mast cells in mice treated with 20-methylcholanthrene: Proliferation of precursor cells, their transformation into mast cells and migration of the latter. *Acta Path. Jap., 26*:693, 1976.

8

The Effect of Anti-i on Early Myeloid Progenitor Cells in Human Bone Marrow

C. J. O'Hara, K. H. Shumak, and
G. B. Price

Bone marrow cells incubated for several days in liquid suspension culture with an appropriate stimulator frequently produce more CFU-c than marrow cells assayed without prior incubation. The increase in assayable CFU-c produced by the suspension culture is known as ΔC (2, 11). In the differentiation stage, the cells responsible for ΔC probably fall between CFU-c and the human CFU-s equivalent (2, 8). In support of this hypothesis, it has been shown first that, unlike CFU-c, the ΔC progenitor cells can rarely be detected in the S-phase of the cell cycle (8). Second, two types of stimulatory molecules have been detected in leukocyte-conditioned media. The larger molecules (15,000, 35,000, and 90,000 Daltons) are strong stimulators of human CFU-c but have little or no ΔC effect (6, 9). On the other hand, a class of small molecules (1,200 Daltons) often have a large ΔC effect, but as a class, it is a relatively poor stimulator of CFU-c (6, 12). Third, 4-methyl-histamine, which has been shown by Byron (1) to cause murine CFU-s to start DNA synthesis, has been shown to have a similar effect on the human ΔC progenitors, but to have no detectable effect on the cell cycle phase of human CFU-c (8). This evidence suggests that there are significant similarities between the differentiation stage of the human ΔC progenitors and murine CFU-s.

The i antigen was originally detected on erythrocytes, but it has since been detected on all the types of end cells found in the hemopoietic system (10). We have shown that this antigen is also present on the committed progenitor cells, CFU-e and CFU-c (7). Unlike the end cells of the hemopoietic system, these progenitor cells express detectable amounts of i antigen only during the S phase of the cell cycle.

This chapter describes experiments done to determine whether the i antigen is on the human bone marrow cells that produce the ΔC effect. In particular, we wondered whether the reactions of these cells to anti-i would help to distinguish them from CFU-c.

MATERIALS AND METHODS

Bone Marrow Cells

The bone marrow cells used in this study were obtained from marrow aspirations done for diagnostic purposes in patients with anemia or for staging in patients with solid tumors.

Treatment of Bone Marrow Cells with Anti-i

A detailed description of this procedure has been published (2). Briefly, washed bone marrow buffy coat cells were mixed with anti-i (Den) serum and normal rabbit serum, as a source of complement. In control mixtures, normal human serum was substituted for anti-i or the normal rabbit serum was heat-inactivated (56°C for 30 min). The mixtures were incubated for 30 min at 4°C, to allow the cold agglutinin anti-i to bind, and then for 30 min at 23°C, to permit complement fixation. The cells were then washed and dispensed into culture medium for CFU-c or ΔC assay.

Cell Culture

CFU-c Samples of human bone marrow containing approximately 2×10^5 cells were assayed for CFU-c using the procedure of Iscove et al. (3). The growth medium consisted of α-medium (K.C. Biologicals Inc., Lenexa, Kansas), 40% (v/v) methylcellulose (Dow Chemical Co., Midland, Mich.) in α-medium [0.8 g% (w/v) concentration in final culture medium], 20% (v/v) fetal calf serum (FCS) (Flow Laboratories, Rockland, Md.), and 20% leukocyte-conditioned medium (LCM). The cultures were incubated at 37°C in a water-saturated atmosphere of 10% CO_2 in air. After 14 days, the cultures were examined and colonies having more than 20 cells were counted as CFU-c.

ΔC The procedure described by Wiseman et al. (13) was used to study the ΔC effect. Human bone marrow cells were suspended at about 1.0×10^6 nucleated cells per milliliter in growth medium consisting of α-medium supplemented with 20% FCS and 20% LCM. The cultures were incubated at 37°C in a water-saturated atmosphere of 10% CO_2 in air for 7 days. The cell suspensions were then assayed for CFU-c. In some experiments, the period of liquid suspension culture before the CFU-c assay was varied over a range of 1 to 14 days.

RESULTS

The results of a typical experiment in which the effect of anti-i (Den) on CFU-c and on ΔC are compared are shown in Figure 8.1. The lower two curves in this figure show the number of CFU-c detected in bone marrow assayed without prior liquid suspension culture. The marrow treated with anti-i and complement contained 21 to 37% fewer CFU-c than did the control marrow in which heat-inactivated serum was used instead of active complement. The upper pair of curves, labeled ΔC, show the number of CFU-c found in aliquots of the same bone marrow sample assayed after 7 days in liquid suspension culture. The number of CFU-c in control ΔC cultures was approximately twice the number detected in the original bone marrow aliquot. Treatment of the marrow with anti-i (1:2500) and complement further increased the number of CFU-c ($p < .005$), but this effect of anti-i was not observed in all experiments.

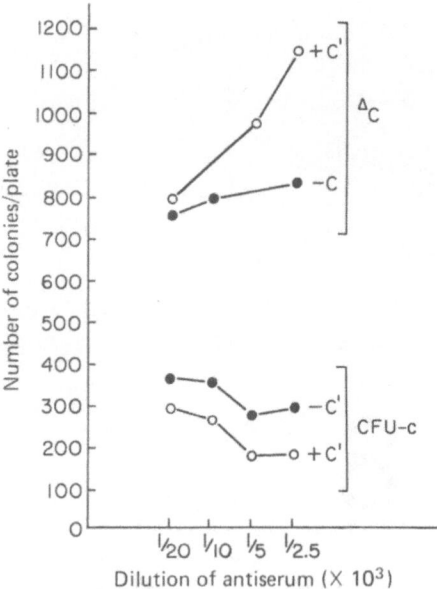

FIGURE 8.1. The effect of anti-i on the ΔC and CFU-c assays. Plates contained approximately 2×10^5 nucleated bone marrow cells. Colony number after treatment with anti-i in the presence of complement (o——o); colony number after treatment with anti-i in the absence of complement (•——•). The lower two curves show the number of CFU-c in bone marrow assayed for CFU-c without prior liquid suspension culture. The upper two curves show the number of CFU-c in bone marrow after 7 days in liquid suspension culture. Points shown are the mean of four replicate plates.

In three of eight marrows in which a significant ΔC was found, treatment with anti-i did not produce an additional increase in the number of CFU-c. In no experiment, however, was the number of CFU-c decreased by treatment with anti-i.

Assays of the effect of anti-i on the cells responsible for ΔC were done on unseparated bone marrow buffy coat cells and on those cells that did not adhere to glass. The effect of anti-i on the size of the ΔC was not affected by the presence or absence of glass-adherent cells (data not shown).

To study further the effect of anti-i on ΔC, liquid suspension cultures of anti-i-treated and control bone marrow were assayed for CFU-c at times ranging from 1 to 14 days. Figure 8.2 shows the results of one such experiment. In both the control and the anti-i-treated bone marrow during the initial few days in culture, the number of CFU-c decreased. Subsequently, the number of CFU-c began to increase and peaked on about day 8 or 9. After this peak, there was a rapid decline in the number of CFU-c in the culture. Bone marrow treated with anti-i (1:100) and complement produced a higher peak in CFU-c than did control marrow, but the time

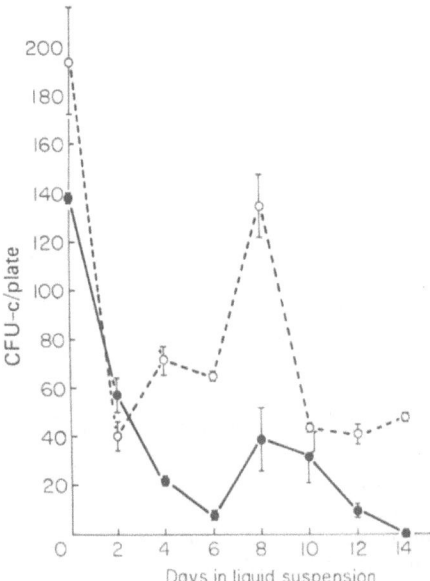

FIGURE 8.2. Kinetics of modification of ΔC culture growth by anti-i treatment. Plates contained approximately 2 × 10⁵ nucleated bone marrow cells. Colony number in marrow after treatment with anti-i (1:100) in the presence of complement (o——o); colony number in marrow after treatment with normal human serum (1:100) in the presence of complement (●——●). Points are the means of two replicate plates: the error bars represent one standard deviation.

at which this peak was reached was similar in treated and control cultures.

DISCUSSION

In previous work (7), we have shown that the progenitor cells CFU-e and CFU-c bear i antigen. The antigen seems to be detectable, however, only during the S phase of the cell cycle.

In the present study, treatment of bone marrow with anti-i and complement enhanced the number of CFU-c produced during ΔC culture in five of eight cases. Figure 8.1 shows that the magnitude of this enhancement was related to the dilution of anti-i serum with which the marrow was treated. Although this enhancement was seen in only five of eight bone marrow specimens in which a significant ΔC was observed, it was not related to the presence or absence of the glass-adherent bone marrow cells removed by the procedure of Messner et al. (5). The kinetic experiments shown in Figure 8.2 show that when anti-i does increase ΔC, this effect is due to an actual increase in the number of CFU-c produced by the cultures, with little effect on the time at which the peak CFU-c numbers are reached in the culture.

It seems unlikely that enhancement of ΔC is due to an interaction between the ΔC-producing cells and anti-i. The expected result of such an interaction would generally be the inactivation of the progenitors. Krogsrud et al. (4) have shown that the ΔC progenitors are no less sensitive to the action of complement than are CFU-e and CFU-c, both of which can be inactivated by anti-i (7). Thus, it appears that the cells responsible for ΔC do not bear i antigen. Since only those CFU-e and CFU-c in the S phase of the cell cycle have been found to be sensitive to anti-i, one possible explanation for the failure of anti-i to react with the ΔC progenitors may be that these cells are seldom out of the G_0 to G_1 phase of the cell cycle. To test this hypothesis, experiments are presently being done using bone marrow specimens from neutropenic patients. Wiseman et al. (13) have suggested that a significantly greater fraction of the ΔC-producing cells from these patients' bone marrows are in the S phase of the cell cycle.

If the ΔC progenitors do lack i antigen, there are at least two other possible explanations for the effect of anti-i on these cells. An i antigen-bearing cell that suppresses ΔC production may be killed by anti i treatment, and thus removal of this cell would allow a larger ΔC. Alternatively, the material released from bone marrow cells killed by anti-i might provide additional nutrients or "stimulators" required for the optimal proliferation of the cells responsible for ΔC. Regardless of the mechanism of action of anti-i on the ΔC progenitors, it is apparent that this action differs from that of anti-i on CFU-c progenitors, providing additional evidence that the ΔC and CFU-c progenitors are not the same cells.

SUMMARY

The number of CFU-c detected in human bone marrow samples decreases if the marrow is treated with anti-i and complement before plating. The number of CFU-c often increases, however, if before plating, the marrow cells are incubated in liquid suspension culture for several days in the presence of an appropriate stimulator. The effect on this increase (ΔC), of treatment of marrow with anti-i before liquid suspension culture, was studied in eight marrow samples. In no case did anti-i decrease the size of the ΔC and in five of the eight cases, the ΔC was greater in marrow that had been treated with anti-i than in untreated marrow. Kinetic experiments showed that, in these cases, the larger ΔC in anti-i-treated marrow was due to an actual increase in the number of CFU-c produced rather than to a change in the time at which peak CFU-c production was reached. The difference in the effect of anti-i on ΔC and CFU-c progenitor cells provides evidence that these cells are not the same.

ACKNOWLEDGMENTS

We wish to thank Ms. K. Benzing and Ms. P. Thompson for excellent technical assistance, Dr. J. S. Senn of the Sunnybrook Hospital, Toronto, for providing samples of bone marrow as well as Dr. J. E. Till for contributing to discussions during the course of these investigations.

This research was supported by the Medical Research Council of Canada, the National Cancer Institute of Canada and the Ontario Cancer Treatment and Research Foundation, and by an Ontario Graduate Scholarship held by C.O.

REFERENCES

1. Byron, J. W. Bone marrow toxicity of metiamide. *Lancet, 2*:1350, 1976.
2. Iscove, N. N., Messner, H., Till, J. E., and McCulloch, E. A. Human marrow cells forming colonies in culture: Analysis by velocity sedimentation and suspension culture. *Ser. Hematol., 5*:37, 1972.
3. Iscove, N. N., Senn, J. S., Till, J. E., and McCulloch, E. A. Colony formation by normal and leukemic human marrow cells in culture: Effect of conditioned medium from human leukocytes. *Blood, 37*:1, 1971.
4. Krogsrud, R. L., Bain, J., and Price, G. B. Serological identification of hemopoietic progenitor cell antigens common to mouse and man. *J. Immunol., 119*:1486, 1977.
5. Messner, H. A., Till, J. E., and McCulloch, E. A. Interacting cell populations affecting granulopoietic colony formation by normal and leukemic human marrow cells. *Blood, 42*:701, 1973.
6. Niho, Y., Till, J. E., and McCulloch, E. A. Granulopoietic progenitors in suspension culture: A comparison of stimulating cells and conditioned media. *Blood, 45*:811,1975.
7. O'Hara, C. J., Shumak, K. H., and Price, G. B. The i antigen on human myeloid progenitors. *Clin. Immunol. Immunopathol., 10*:420, 1978.
8. Price, G. B., and Krogsrud, R. L. Use of molecular probes for detection of human hemopoietic progenitors. In *CSH Symposium on the Differentiation of Normal and Neoplastic Hemopoietic Cells.* Cold Spring Harbor Laboratory: Cold Spring Harbor, New York (in press).
9. Price, G. B., Senn, J. S., McCulloch, E. A., and Till, J. E. The isolation and properties of granulocyte colony stimulating activities. *Biochem. J., 148*:209, 1975.
10. Pruzanski, W., and Shumak, K. H. Biologic activity of cold-reacting autoantibodies. *N. Eng. J. Med., 297*:538, 583, 1977.
11. Sutherland, D. J. A., Till, J. E., and McCulloch, E. A. Short-term cultures of mouse marrow cells separated by velocity sedimentation. *Cell Tissue Kinet., 4*:479, 1971.
12. Till, J. E., Price, G. B., Senn, J. S., and McCulloch, E. A. Cell interactions in the control of hemopoiesis. In Drewinko, B., and Humphrey, R. M., eds., *Growth Kinetics and Biochemical Regulation of Normal and Malignant Cells.* Baltimore: Williams & Wilkins, 1976, p. 223.
13. Wiseman, L. L., Senn, J. S., Miller, R. G., and Price, G. B. Stem cell characterization of neutropenia: Velocity sedimentation and mass culture analysis. *Br. J. Cancer, 34*:46, 1976.

9

Hematologic Contributions to Increases in Resistance or Sensitivity to Endotoxin

Richard I. Walker

Infectious complications are often associated with hematologic disorders. Frequently, it is the endogenous (i.e., intestinal) gram-negative flora of the compromised host that invade normally sterile tissues. Control of such infections requires an adequate inflammatory response by various components of the hematologic system. This host response to microorganisms is very similar to that induced by endotoxin. Endotoxin is a characteristic lipopolysaccharide component of the cell walls of gram-negative bacteria and is of major biologic and medical significance for three reasons:

One, endotoxin is an important factor in many common bacterial infections that occur when hematologic defenses are compromised. The inflammatory response initiated by endotoxin is suited to the trapping and destruction of microorganisms associated with it. But if this response is too prolonged or if an essential component is missing (i.e., neutropenia), the defensive response may become detrimental to the host (71).

Two, endotoxin is a regulator of host defenses. Endogenous (intestinal) endotoxin can enhance granulopoiesis (40,41,76) and nongranulocyte-mediated antimicrobial activity (72). Furthermore, the toxin enhances tissue regeneration (37,54) and the antitumor (9), antiviral (5,35), and antiparasitic (64) activities of macrophages. Endotoxin can also be used to alter donor cell populations with consequent reduction in graft versus host disease (GvHD) (63,65).

Three, subclinical endotoxemias may occur in normal humans (19). The long-term consequences of these events on metabolic and degenerative diseases have not been adequately evaluated, but when the membrane (17) and nuclear (81) affinities of the toxin, plus its *in vitro* capability to alter metabolism of cells (4), is considered it is obvious that this area should be studied further.

Successful clinical treatment of gram-negative sepsis as well as possible therapeutic uses of endotoxin necessitate a knowledge of toxin action in the host so that the inflammatory response can be controlled. It has been hypothesized that the interaction of endotoxin with the hematologic system is important in the pathogenesis of the toxin. This concept is consistent with previous reports on the sequence of events in the elimination of microorganisms and endotoxin:

1. Platelets interact directly with bacteria (12) or purified endotoxin (18,62). Aggregates resulting from these interactions may signal the host of the presence of microorganisms.
2. Leukocytes [particularly polymorphonuclear cells (PMN)] can respond to this signal and participate in a cooperative antibacterial action with

the platelets. Platelets promote phagocytosis and destruction of bacteria by phagocytes following platelet-bacterial aggregation (12,58). Unlike macrophages, PMN have minimal degradatory effects on ingested endotoxin (22).

3. Macrophages, and particularly those in the liver, eventually ingest most injected endotoxin (7,32). Therefore endotoxin must be transported to these cells for further processing. Processing of antigenic materials by macrophages has been reviewed recently (3,15).

A critical event in endotoxemia may be the *persistence* of the toxin and its effects (i.e., inflammation) in the bloodstream rather than delivery to the reticuloendothelial system. Recently, sustained phagocytic activity by marginating leukocytes was associated with the development of endotoxin shock in primates (2). Several earlier studies are consistent with the idea that there are detrimental consequences if endotoxin or endotoxin-induced inflammatory foci persist in the microcirculation:

1. Endotoxin given intramuscularly persists in low titer in the bloodstream over longer periods of time than when it is administered by other routes (48). When endotoxin-injected animals die, their livers contain endotoxin in amounts associated with sublethal challenges.

2. Endotoxin inoculated directly into the hepatic circulation of an animal does not induce mortality nor does it cause the hepatic damage seen when the same dose is injected via the systemic circulation (46,55).

3. Toxic doses of endotoxin are cleared from the blood more slowly than are smaller doses (7). Furthermore, animals made tolerant to the lethal effects of endotoxin remove large doses of endotoxin from the blood faster than normal animals (7,32).

4. Reticuloendothelial blockade impairs particle clearance and increases the sensitivity of animals to injected endotoxin (30).

Our studies of hematologic responses to endotoxin to be described below can be grouped under two general approaches: One, determination of causes for increased sensitivity to endotoxin seen in irradiated animals, and two determination of causes for increased resistance to endotoxin seen in tolerant animals.

Sensitivity of Irradiated Mice to Endotoxin

An early (days 2 to 5) and a late (day 7 onward) increase in sensitivity to injected endotoxin [0.3 mg of *Salmonella typhosa* endotoxin (Difco) was injected intraperitoneally into mice] can be seen when B6CBF1 mice are exposed to radiation in the hemopoietic death range (Figure 9.1). This biphasic

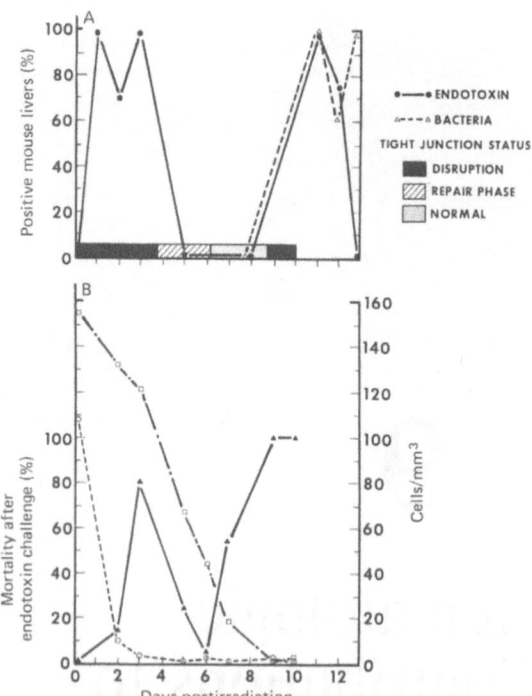

FIGURE 9.1. (**A**) Correlation of per cent of mouse livers positive for endotoxin or bacteria with the status of tight junctional permeability barriers in the intestine after lethal radiation (1,000 rad, ^{60}Co). These data can also be correlated with decreases in platelets (□——□) or granulocytes (o——o) and the biphasic increases in per cent mortality (▲——▲) following challenge with 0.3 mg of *Salmonella typhosa* endotoxin (**B**).

pattern correlates well with the presence of microbial agents in host tissues (71,73,74):

Early Phase Endotoxin, but not bacteria, was detected in livers of B6CBF1 mice as early as 24 hr after irradiation and remained detectable through day 3 (74). During this period, the permeability barriers (tight junctions or zonula occludens) between the intestinal epithelial cells (73) were disrupted, and we feel that it is likely that endotoxin can leave the gut by this route (69,73; M. Porvaznik, personal communication). During this period of aseptic endotoxemia, granulocyte numbers reach their nadir (day 3), an event that has been associated with decreased resistance to endotoxin (1,70,73,78). Disruption of tight junctions seems to become a life-threatening event in combination with granulocytopenia. Loss of granulocytes may contribute to the progressively larger amounts of endotoxin that have been found in the blood of mice between days 1 and 3 after irradiation (43). When relatively small amounts of the toxin are injected, the combined endotoxemia could be lethal.

Late Phase Although tight junctions are intact at 7 days after irradiation (78,79) and no endotoxin is detectable in mouse livers at 7 days (74), sensitivity to injected endotoxin began to increase at this time, reaching 100% mortality at days 9 to 10 (73). Increased sensitivity to endotoxin seen at this time may be due in part to two factors not present during the early phase: thrombocytopenia and bacterial invasion of host tissue.

Thrombocytopenia can reduce the efficiency of endotoxin clearance from the bloodstream (18), as well as permit formation of petechial hemorrhages in the small bowel (10). This latter event, coupled with the breakdown of intermicrobial antagonisms in the intestine (66,68), could lead to the invasion of normally sterile host tissue by enteric microorganisms (Figure 9.2). Challenge with a normally sublethal dose of endotoxin at this time, in addition to disrupting tight junctions (79), can reduce the effectiveness of the remaining antibacterial defenses (23,34) and thereby contribute to mortality. This concept is consistent with our previous observation that subclinical infections induced by hyperbaric stress are associated with a reduced LD_{50} for mice injected with endotoxin (28).

The importance of intestinal microorganisms to endotoxin sensitivity during this phase of radiation injury is indicated by studies using mice treated orally with poorly absorbed antibiotics. These animals were much more resistant than conventional mice 7 days after irradiation to challenges with normal endotoxins (70,71) or attenuated endotoxins (27, 79). Furthermore, we recently observed that endotoxin attenuated with chromium chloride (60) initiates disruption of tight junctional complexes (79), thereby providing a route by which non-attenuated endotoxin could enter host tissue.

There is no evidence to suggest a macrophage contribution to increased sensitivity to endotoxin after irradiation. Macrophages are known to be radioresistant (10), and they apparently retain *in vivo* antibacterial activity after radiation (75). Hepatic uptake of ^{51}Cr-labeled endotoxin is slightly reduced by day 12 after irradiation. But this does not seem to

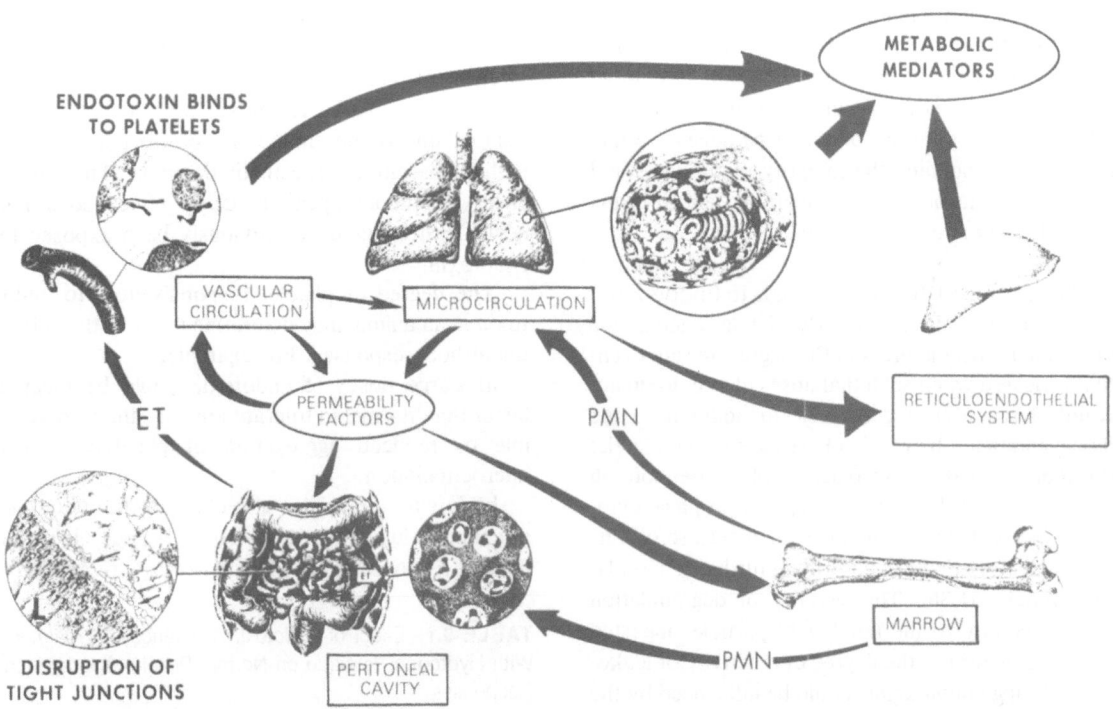

FIGURE 9.2. Graphic representation of the major components of host response to endotoxin. Endotoxin (flat ribbons attached to platelets in enlargement) can cause inflammatory congestion within the microcirculation. Here, mast cells degranulate, cells escape through walls of capillaries, and various inflammatory mediators are generated by leukocytes phagocytizing aggregates. Some of these mediators disrupt intercellular barriers in the intestine (arrows), thereby permitting more endotoxin (ET) from the lumen (L) to enter the circulation. Polymorphonuclear leukocytes (PMN) from the marrow may contribute to microcirculatory and intraperitoneal clearance. If a sublethal amount of endotoxin is administered, the microcirculation will eventually be cleared as the toxic materials enter the reticuloendothelial system.

correlate with the increased sensitivity to endotoxin (75) and possibly is a consequence of thrombocytopenia (18,77). It is possible that the cytotoxic effects of endotoxin on macrophages may be enhanced after irradiation due either to generation of mediators at other sites (61) or to direct presentation of the toxin to the macrophage. This latter alternative is based on a recent report that macrophage cytotoxicity is influenced by the biochemical form of the endotoxin (29).

Irradiation injury to hematologic defenses is important clinically, due to its use in conjunction with bone marrow transplantation. When B6CBF1 mice were grafted with 5×10^6 CBA spleen cells after 850 rad x-radiation, microorganisms were detectable in the liver in less than 24 hr, compared to 8 or 9 days in mice who were only irradiated (68,73). Furthermore, when challenged 3 days after irradiation and grafted with ordinarily sublethal amounts of endotoxin, all animals undergoing GvHD died (70). The rapid bacterial invasion in these animals is probably due to the damage of intestinal epithelium by the grafted cells.

The importance of control of intestinal microorganisms in irradiated hosts is indicated by the studies described above. Additional work is necessary, however, to determine the most suitable means to effect this control. Oral antibiotic therapy and protective environments may be useful, but are complicated and may render the subject even more susceptible to colonization by potential pathogens. Some means of maintaining the integrity of the intestinal barrier and the normal intermicrobial antagonisms within the gut should be studied further.

Resistance of Tolerant Animals to Endotoxin

Endotoxin induces an early platelet aggregation in the microvasculature, but the aggregates are eventually cleared when sublethal doses of endotoxin are administered (67) (Figure 9.2). In addition to the damage initiated by mediators released from platelet thrombi (33), leukocytes reaching platelet-endotoxin aggregates in the microcirculation begin phagocytosis of the particles, with subsequent release of various mediators that can contribute further to circulatory shock (51,36). The severity of degranulation relates directly to the number of particles ingested (53), and therefore, the degree of reactivity of leukocytes during endotoxemia could be influenced by the amount of platelet damage that has occurred.

When animals are made resistant to endotoxin by repeated challenges with sublethal amounts of the substance, certain adaptations occur that can reduce the persistence of platelet aggregates in the microcirculation.

Platelet Resistance to Endotoxin Damage Platelets taken from rabbits made tolerant to the lethal effects

of endotoxin aggregate faster than normal in the presence of endotoxin, but they can disaggregate, in contrast to platelets from normal animals whose platelets clump more slowly but irreversibly *in vitro* (77). Reversible platelet aggregation in tolerant animals, which could facilitate passage through the microcirculation to the reticuloendothelial system, has also been observed *in vivo* (B. Urbaschek, personal communication). These characteristics of platelets from tolerant animals are due to a humoral factor in the plasma (77) and seem to be specific for endotoxin-induced aggregation, leaving unaffected that aggregation initiated by such other factors as ADP or collagen (Table 9.1).

A pivotal role for platelets in the inflammatory response to endotoxin is indicated by our observations that leukopenia is a dose-independent phenomenon (similar degrees of leukopenia were obtained in dogs given doses of endotoxin ranging from 0.4 μg/kg to over 2 mg/kg), whereas thrombocytopenia was seen only in animals given high-dose challenges with endotoxin. Neutropenia is a complement-mediated response (42,49,50), but thrombocytopenia involves direct platelet-endotoxin interactions that increase aggregation with increasing dose (J. R. Fletcher, personal communication).

Induction of tolerance can alter the dose responsiveness of platelets. Platelets in normal dogs challenged with 4 μg/kg endotoxin do not decrease in number unless the animals have also been treated with the barbiturate anesthetic thiamylal. This synergistic thrombocytopenia-inducing effect does not occur if the dogs have previously been exposed to endotoxin.

The degree of platelet responsiveness to endotoxin in an animal may determine many of the subsequent host responses. For example,

a) Large doses of endotoxin could be cleared faster than normal in tolerant animals due to reversible or reduced aggregation of platelets in the microcirculation.

b) *In vitro* responses of leukocytes to endotoxin are much slower than *in vivo* responses (14), but platelet responses to endotoxin *in vitro* are rapid. *In*

TABLE 9.1 Effect of Endotoxin Tolerance or Treatment With Hydrogen Peroxide on Normal Platelet Aggregation Responses

INDUCER OF AGGREGATION	ENDOTOXIN TOLERANCE	HYDROGEN[a] PEROXIDE
Endotoxin	Rapid and Reversible	None
ADP	None	Reduced
Collagen	None	Reduced

[a]Three microliters of 3% hydrogen peroxide were added to the platelet suspension approximately 100 sec before the inducer of aggregation.

vivo leukocytes may be responding to platelet aggregates.

c) Dogs receiving large doses of endotoxin do not exhibit pyrogenic tolerance as they do when smaller challenges are given (20). This may be explained in part by a resistance of platelets to small doses of endotoxin such as we observed in dogs treated with thiamylal before sublethal challenges. In tolerant animals, larger doses of endotoxin induce formation of aggregates that can be phagocytized with a consequent release of endogenous pyrogen.

d) Recently, we found a more rapid return of platelet numbers toward normal following an initial thrombocytopenia in dogs challenged a second time with a large dose (2 mg/kg) of endotoxin, and this return could contribute to the general appearance of well-being in these animals.

e) Inhibition of leukocyte migration seen in animals (13,25,26) challenged with increasing amounts of endotoxin may be mediated in part by platelet-endotoxin interactions. This possibility is based on our observation (56) that, at certain higher concentrations of endotoxin, human platelets significantly reduce leukocyte chemotactic responses *in vitro*.

f) Cortisone prevents the irreversible aggregation of dog platelets induced by endotoxin (47) and in this way may help obviate lethal effects of the toxin. The protective role of cortisone has also been associated with inhibition of complement-activated generation of neutropenia (49), but our observation that sublethal amounts of endotoxin cause neutropenia (margination) comparable to that induced by toxic doses suggests that the neutropenia per se is not the major factor in determining lethality.

Leukocyte Adaptations in Resistant Animals There is no evidence that leukocytes, as well as platelets, can become resistant to endotoxin in tolerant animals. Instead, marrow in tolerant animals becomes hyperplastic after treatment with endotoxin (11,52,-72) (Figure 9.2). This adaptation permits mobilization of leukocytes that could contribute to resistance to the biologic effects of endotoxin in two ways.

a) Inflammatory mediators such as can be released in the microcirculation of endotoxin-challenged animals increase the permability of the intestine to endotoxin produced by enteric microorganisms (8,16), apparently through disruption of some tight junctional barriers between epithelial cells (69). If this occurs unchecked, a low titer of endotoxin will persist in the bloodstream (8), with additional hematologic injury to the host. Both tolerant (26) and genetically resistant (44) animals can mobilize neutrophils to the peritoneal cavity more rapidly than can normal animals. This could be a defense against endotoxin from the intestine, since neutrophil accumulation in the peritoneal cavity has been shown to

protect mice against the lethal consequences of intraperitoneal challenges with endotoxin (44,78).

It is not known whether platelets are also involved in leukocyte-endotoxin interactions in the peritoneal cavity as well as in the microcirculation. Platelets are motile (39), however, and could conceivably enter the peritoneum where they have been shown to be capable of protecting rats against intraperitoneal challenges with endotoxin (18).

b) In addition to protection against intestinal endotoxin, rapid mobilization of leukocytes following a challenge with endotoxin could enhance the clearance of the endotoxin-platelet aggregates in the microcirculation and further reduce aggregation caused by host factors (i.e., collagen, ADP). Granulocytes are known to phagocytize platelet aggregates efficiently (57). By using oxygen consumption measurements (an index of phagocytosis), we have detected vigorous responses from rabbit granulocytes incubated with endotoxin-platelet aggregates. During phagocytosis, granulocytes release hydrogen peroxide (6,38), which can limit further thrombosis after endotoxin is trapped (Table 9.1).

The importance of an early leukocytosis in protecting dogs challenged with endotoxin has recently been indicated (80). These studies used multiple injections of small amounts of endotoxin to achieve leukocytosis, so the possibility remains that other components of the host defense system may also have adapted to the toxin. But if granulocytes do have a role in controlling endotoxemia, this property may be useful in testing the *in vivo* effectiveness of leukocyte preparations purified or preserved by various methods. Instead of the multiple transfusions usually required to treat experimental infections, it may be sufficient to use a single transfusion against endotoxemia.

Although additional factors may also be involved, our recent studies with the genetically resistant C3H/HeJ mice support the contention that cellular adaptations, such as those just described for platelets and leukocytes, contribute to survival. We found that the resistant mouse strain quickly clears cellular foci of inflammation from the pulmonary microcirculation, whereas the sensitive C3HeB/FeJ strain mice were unable to do so. Such persisting inflammation is consistent with our observation that tight junction disruption was present in mice dying from endotoxin (75).

Macrophage Adaptations in Endotoxin-Resistant Animals Macrophages, particularly in the liver, accumulate and degrade endotoxin (7,22). Therefore, if accelerated particle clearance from the microcirculation is to benefit the host, macrophage adaptations to handle the increased phagocytic load would be expected. In tolerant mice, these cells phagocytize particles more rapidly than normal, but they release less of the mediators that can contribute to the inflammatory process (61).

One aspect of macrophage resistance that should be studied concerns the possibility that the microvascular response may yield mediators toxic to the macrophage (61). A result of tolerance may be the diminished release of cytotoxins in the microvasculature, thereby preserving the functional status of macrophages. It is also possible that macrophages in tolerant animals are more resistant to the cytotoxins to which they may be exposed during endotoxemia.

Resistance seen in tolerant animals seems to depend on adaptations that limit the release of mediators from hematologic cells rather than refractoriness to the mediators themselves. Thus, tolerant or genetically resistant animals respond normally to injections of the mediators themselves (24,31,45). Furthermore, non-hematologic-mediated reactions, such as adrenocortical stimulation, are not altered in tolerant animals (20,21).

In short, the object of hematologic adaptations to endotoxin is to prevent persistence of the toxin of inflammatory foci induced by it within the circulation. This limits the generation of potentially harmful levels of mediators by inflammatory cells. Persistence can be prevented by increasing granulocytic reserves and enhancing resistance of platelets and macrophages to the biologic effects of endotoxin.

SUMMARY

Humoral and cellular elements in the blood interact with endotoxin and each other to initiate an inflammatory response such as would be directed against gram-negative bacteria (Figure 9.2). These agents, and particularly platelets, can bind to the endotoxin and undergo reactions to facilitate delivery of particles to the reticuloendothelial system, where they are subsequently degraded.

A vulnerable stage of this process is the failure of the circulating components of the host defense system to remove all toxin from the circulation. Endotoxin that persists in the vasculature (probably in platelet aggregates in the microcirculation) will allow the inflammatory response to continue, which involves release of mediators and ultimately depletes or damages cellular and metabolic reserves and thereby contributes to death. Furthermore, some of these mediators may disrupt permeability barriers in the intestine and thereby permit additional toxin to enter the circulation.

Irradiation compromises resistance to endotoxin in a biphasic manner. In the early phase, sensitivity can be associated with endotoxin escape from the intestine through disruptions in tight junctional com-

plexes and depletion of leukocyte reserves. In the late phase, thrombocytopenia and invasion of host tissues by opportunistic pathogens can contribute to increased sensitivity to endotoxin.

Resistance to endotoxin involves adaptations to facilitate resolution of cellular aggregates in the microcirculation and to make macrophages of the reticuloendothelial system refractory to large loads of toxic materials. One specific adaptation to eliminate persistence of inflammatory foci is increased granulocytic reserves in the marrow, so that the leukocytes can be mobilized rapidly to the microcirculation and the peritoneal cavity to perform clearance functions. Another adaptation is the elaboration of a humoral substance that increases platelet resistance to the aggregating effects of endotoxin.

REFERENCES

1. Altura, B. M., and Hershey, S. G. Endotoxemia in the rat: Role of blood coagulation and leukocytes in the progression of the syndrome. *Circulat. Shock, 1*:113, 1974.
2. Balis, J. U., Rappaport, E. S., Gerber, L., Fareed, J., Buddingh, F., and Messmore, H. L. A primate model for prolonged endotoxin shock. Blood-vascular reactions and effects of glucocorticoid treatment. *Lab. Invest., 38*:511, 1978.
3. Bona, C. A. Fate of endotoxin in macrophages: biological and ultrastructural aspects. *J. Infect. Dis., 128*:S74, 1973.
4. Bradley, S. G., and Howe, D. B. Perturbation by bacterial lipopolysaccharide of the metabolic processes of human cells in continuous culture. *J. Reticuloendoth. Soc., 20*:35, 1976.
5. Campbell, J. B., and White, S. L. A comparison of the prophylactic and therapeutic effects of poly I:C and endotoxin in mice infected with Mengo virus. *Canad. J. Micro., 22*:1595, 1976.
6. Canoso, R. T., Rodvien, R., Scoon, K., and Levine, P. H. Hydrogen peroxide and platelet function. *Blood, 43*:645, 1974.
7. Carey, F. J., Braude, A. I., and Zelesky, M. studies with radioactive endotoxin. III. The effect of tolerance on the distribution of radioactivity after intravenous injection of *Escherichia coli* endotoxin labeled with Cr^{51}. *J. Clin. Invest., 37*:441, 1958.
8. Cardis, D. T., Ishiyama, M., Woodruff, P. W., and Fine, J. Role of the intestinal flora in clearance and detoxification of circulating endotoxin. *J. Reticuloendoth. Soc., 14*:513, 1973.
9. Carswell, E. A., Old, L. J., Kassel, R. L., Green, S., Fiore, N., and Williamson, B. An endotoxin-induced serum factor that causes necrosis of tumors. *Proc. Nat. Acad. Sci. (U.S.A.), 72*:3666, 1975.
10. Casarett, A. P. *Radiation Biology.* Englewood Cliffs: Prentice-Hall, 1968.
11. Chervenick, P. A., and Boggs, D. R. Granulocytic hyperplasia and induction of tolerance in response to

chronic endotoxin administration. *J. Reticuloehdoth. Soc.*, 1:288, 1971.

12. Clawson, C. C. Modification of neutrophil function by platelets. In Baldini, M. G., and Ebbe, S., eds., *Platelets: Production, Function, Transfusion and Storage.* New York: Grune & Stratton, 1978, p. 287.

13. Cluff, L. E. Studies of the effect of bacterial endotoxins on bacterial leukocytes. II. Development of acquired resistance. *J. Exp. Med., 98*:349, 1953.

14. Collins, R. D., and Wood, W. B., Jr. Studies on the pathogensis of fever. VI. The interaction of leukocytes and endotoxin *in vitro. J. Exp. Med., 110*:1005, 1959.

15. Cruchaud, A. The role of mononuclear phagocytes in the defense against infections and malignancies. In Baum, S., and Ledney, G. D., eds., *Experimental Hematology Today.* New York: Springer-Verlag, 1978, p. 119.

16. Cuevas, P., and Fine, J. Production of fatal endotoxic shock by vasoactive substances. *Gastroenterology, 64*:285, 1973.

17. Donlon, M. A., and Walker, R. I. Adenyl cyclase activity of mouse liver membranes after incubation with endotoxin and epinephrine. *Experientia, 32*:179, 1976.

18. Das, J., Schwartz, A. A., and Folkman, J. Clearance of endotoxin by platelets: Role in increasing the accuracy of the limulus gelatin test and in combatting experimental endotoxemia. *Surgery, 74*:235, 1974.

19. DuBose, D., Lemaire, M., Brown, J., Wolfe, D., and Hamlet, M. Survey for positive *Limulus* amoebocyte lysate test in plasma from humans and common research animals. *J. Clin. Microbiol., 7*:139, 1978.

20. Egdahl, R. H., and Melby, J. C. Induced tolerance of dogs to endotoxin: Correlation of febrile response and adrenal function. *J. Clin. Invest., 37*:891, 1958.

21. Egdahl, R. H., Melby, J. C., and Spink, W. W. Adrenal cortical and body temperature response to repeated endotoxin administration. *Proc. Soc. Exp. Biol. Med., 101*:369, 1959.

22. Filkins, J. P. Blood endotoxin inactivation after trauma and endotoxicosis. *Proc. Soc. Exp. Biol. Med., 151*:89, 1976.

23. Flynn, J., and McEntegart, M. G. Gonococcal enhancement of staphylococcal virulence for the mouse. *J. Med. Microbiol., 6*:371, 1973.

24. Freedman, H. H. Further studies on passive transfer of tolerance to pyrogenicity of bacterial endotoxin. The febrile and leucopenic responses. *J. Exp. Med., 112*:619, 1960.

25. Fruhman, G. J. Factors influencing neutrophil mobilization. In Gordon, A. S., ed., *Regulation of Hematopoiesis*, Vol 2. New York: Appleton-Century-Crofts, 1970, p. 873.

26. Fruhman, G. J. Endotoxins and leukocyte mobilization. *J. Reticuloendoth. Soc., 12*:62, 1972.

27. Galley, C. B., Walker, R. I., Ledney, G. D., and Gambrill, M. Evaluation of biological activity of attenuated endotoxin in mice. *Exp. Hemat., 3*:197, 1975.

28. Gillmore, J. D., and Walker, R. I. Evidence of bacteremia and endotoxemia in mice undergoing hyperbaric stress. *Undersea Biomed. Res., 4*:67, 1977.

29. Glode, L. M., Jacques, A., Mergenhagen, S. E., and

Rosenstreich, D. L. Resistance of macrophages from C3H/HeJ mice to the *in vitro* cytotoxic effects of endotoxin. *J. Nat. Cancer Inst., 119*:162, 1977.

30. Greisman, S. E., Carozza, F. A., Jr., and Hills, J. D. Mechanisms of endotoxin tolerance. I. Relationship between tolerance and reticuloendothelial system phagocytic activity in the rabbit. *J. Exp. Med., 117*:663, 1963.

31. Herion, J. C., Walker, R. I., and Palmer, J. G. Relation of leukocyte and fever responses to bacterial endotoxin. *Am. J. Physiol., 199*:809, 1960.

32. Herring, W. B., Herion, J. C., Walker, R. I., and Palmer, J. G. Distribution and clearance of circulating endotoxin. *J. Clin. Invest., 42*:79, 1963.

33. Jorgensen, L., Hovig, T., Rosusell, C., and Mustard, J. F. Adenosine diphosphate-induced platelet aggregation and vascular injury in swine and rabbits. *Am. J. Pathol., 61*:161, 1970.

34. Kaplan, J. E., and Saba, T. M. Low-grade intravascular coagulation and reticuloendothelial function. *Am. J. Physiol., 234*:H323, 1978.

35. Kojima, Y. Sites of interferon production in rabbits induced by bacterial endotoxin. *Kitasato Arch. Exp. Med., 43*:35, 1970.

36. Lefer, A. N., and Glenn, T. M. Role of the pancreas in the pathogenesis of circulatory shock. *Adv. Exp. Med. Biol., 239*:311, 1972.

37. Leibovich, S. J., and Roxx, R. The role of the macrophage in wound repair. *Am. J. Pathol., 78*:71, 1975.

38. Levine, P. H., Weinger, R. S., Simon, J., Scoon, K. L., and Krinsky, N. I. Leukocyte-platelet interaction. Release of hydrogen peroxide by granulocytes as a modulator of platelet reactions. *J. Clin. Invest., 57*:955, 1976.

39. Lowenhaupt, R. W., Glueck, H. I., Miller, M. A., and Kline, D. L. Factors which influence blood platelet migration. *J. Lab. Clin. Med., 90*:37, 1977.

40. MacVittie, T. J., and Walker, R. I. Canine granulopoiesis: Alterations induced by artificial suppression of gram-negative bacterial flora. *Exp. Hemat.* (in press).

41. MacVittie, T. J., and Walker, R. I. Endotoxin-induced alterations in canine granulopoiesis: Colony-stimulating activity, colony-forming cells in culture, and growth of cells in diffusion chambers. *Exp. Hemat. 6*:613, 1978.

42. McCall, C. E., DeChatelet, L. R., Brown, D., and Lachmann, P. New biological activity following intravascular activation of the complement cascade. *Nature, 249*:841, 1974.

43. Meter, I. D. Endotoxemia of lethally irradiated animals. *Radiobiologica, 14*:98, 1974.

44. Moeller, G. R., Terry, L., and Snyderman, R. The inflammatory response and resistance to endotoxin in mice. *J. Immunol., 120*:116, 1978.

45. Moore, R. N., Goodrun, K. J., Couch, R. E., Jr., and Berry, L. J. Elicitation of endotoxemic effects in C3H/HeJ mice with glucocorticoid antagonizing factor and partial characterization of the factor. *Infect. Immun., 19*:79, 1978.

46. Mori, K., Matsumoto, K., and Gans, H. On the *in vivo* clearance and detoxification of endotoxin by lung and liver. *Ann. Surg., 177*:159, 1973.

47. Nelson, Wm. R. *In vitro* inhibition of endotoxin induced platelet aggregation with hydrocortisone sodium succinate (Solu-Cortef). *Scand. J. Haematol., 15*:35, 1975.

48. Noyes, H. E., McInturf, C. R., and Blahuta, G. J. Studies on distribution of *Escherichia coli* endotoxin in mice. *Proc. Soc. Exp. Biol. Med., 100*:65, 1959.

49. O'Flaherty, J. T., Craddock, P. R., and Jacob, H. S. Mechanism of anti-complementary activity of corticosteroids *in vivo:* Possible relevance to endotoxin shock. *Proc. Soc. Exp. Biol. Med., 154*:206, 1977.

50. O'Flaherty, J. T., Showell, H. J., and Ward, P. A. Neutropenia induced by systemic infusion of chemotactic factors. *J. Immunol., 118*:1586, 1977.

51. Pennington, D. G., Hyman, A. L., and Jaques, W. E. Pulmonary vascular response to endotoxin in intact dogs. *Surgery, 73*:246, 1973.

52. Quesenberry, P., Devon, J., Ryan, M., and Stohlman, F., Jr. Endotoxin tolerance and granulopoiesis. *Am. Soc. of Hemat. (Dallas)*, p. 56, 1975.

53. Rogers, D. E. Host mechanisms which act to remove bacteria from the bloodstream. *Bact. Rev., 24*:50, 1960.

54. Rossolini, A., Cellesi, C., and Barberi, A. Effects of *E. coli* O 127 endotoxin on regenerating rat liver. *Bull. Istituto Sieroterapico Milanese, 54*:487, 1976.

55. Rutenburg, S., Skarnes, R., Palmerio, C., and Fine, J. Detoxification of endotoxin by perfusion of liver and spleen. *Proc. Soc. Exp. Biol. Med., 125*:455, 1967.

56. Sheil, J. M., and Walker, R. I. Evidence for inhibition of the leukocyte chemotactic response to endotoxin by human platelets. *Toxicon* (in press).

57. Shirasawa, K., and Chandler A. B. Phagocytosis of platelets by leukocytes in artificial thrombi and in platelet aggregates induced by andenosine diphosphate. *Am. J. Phathol., 63*:215, 1971.

58. Smith, S. B. Platelets in host resistance: *In vitro* interactin of platelets, bacteria and polymorphonuclear leukocytes. *Blut, 25*:104, 1972.

59. Snyder, S. L. and Walker, R. I. Inhibition of lethality in endotoxin challenged mice treated with zinc chloride. *Infect. Immunol., 13*:990, 1976.

60. Snyder, S. L., Walker, R. I., MacVittie, T. J., and Sheil, J. M. Biologic properties of bacterial lipopolysaccharides treated with chromium chloride. *Canad. J. Micro., 24*:495, 1978.

61. Snyder, S. L., Eklund, S. K., and Walker, R. I. Release of B-glucoronidase from peritoneal macrophages of normal and endotoxin-tolerant mice. *Infect. Immun.* In press.

62. Springer, G. F., and Adye, J. C. Endotoxin-binding substances from human leukocytes and platelets. *Infect. Immun., 12*:978, 1975.

63. Thomson, P. D., Rampy, P. A., and Jutila, J. W. A mechanism for the suppression of graft-vs-host disease with endotoxin. *J. Immun., 120*:1340, 1978.

64. Tizard, I. R., and Ringleberg, C. P. The effect of bacterial adjuvants on *Trypanosoma lewisi* infections in rats. *Folia Parasitologica* (Praha), *22*:323, 1975.

65. Truitt, R. L., Rose, W. C., Rimm, A. A., and Bortin, M. M. Graft versus leukemia. VIII. Selective reduction in antihost reactivity without loss of antileukemic reac-

tivity by treatment of donor mice with lipopolysaccharide. *Exp. Hemat., 6*:488, 1978.

66. Vincent, J. G., Veomett, R. C., and Riley, R. F. Relation of the indigenous flora of the small intestine of the rat to postirradiation bacteremia. *J. Bact., 69*:38, 1955.

67. Urbaschek, B. Addendum—the effects of endotoxins in the microcirculation. In Kadis, S., Weinbaum, G., and Ajl, S. J., eds., *Microbial Toxins. V. Bacterial Endotoxins.* New York: Academic Press, 1971, p. 261.

68. Walker, R. I. The contribution of intestinal endotoxin to mortality in hosts with compromised resistance. *Exp. Hemat., 6*:172, 1978.

69. Walker, R. I., and Porvaznik, M. J. Disruption of the permeability barrier (zonula occludens) between intestinal epithelial cells by lethal doses of endotoxin. *Infect. Immun., 21*:655, 1978.

70. Walker, R. I., and Sheil, J. M. Contributions of granulocytopenia to endotoxin sensitivity of mice irradiated or undergoing graft-versus-host reaction. *Exp. Hemat., 4*:329, 1976.

71. Walker, R. I., and Snyder, S. L. Endotoxemia in irradiated mice. In Rosenberg, P., ed., *Toxins: Animal, Plant and Microbial.* New York: Pergammon Press, 1978, p. 933.

72. Walker, R. I., and MacVittie, T. J. Hematologic adaptations of dogs to endotoxin. *Exp. Hemat.* (under review).

73. Walker, R. I., and Porvaznik, M. J. Assoiation of leukopenia and intestinal permeability with radiation-induced sensitivity to endotoxin. *Life Sciences* (in press).

74. Walker, R. I., Ledney, G. D., and Galley, C. G. Aseptic endotoxemia in radiation injury and graft-vs-host disease. *Rad. Res., 62*:242, 1975.

75. Walker, R. I., Moon, R. J., Alm, P. F., and Ledney, G. D. Bactericidal activity in conventional or decontaminated mice undergoing GVHD or radiation-induced injury. *Experientia, 15*:1527, 1976.

76. Walker, R. I., MacVittie, T. J., Sinha, B. L., Ewald, P. E., Egan, J. E., and McClung, G. L. Antibiotic decontamination of the dog and its consequences. *Lab. Animal Science, 28*:55, 1978.

77. Walker, R. I., Shields, L. J., and Fletcher, J. R. Platelet aggregation in rabbits made tolerant to endotoxin. *Infect. Immun., 19*:919, 1978.

78. Walker, R. I., Snyder, S. L., Sobocinski, P. Z., McCarthy, K. F., and Egan, J. E. Possible association of granulocyte mobilization to the peritoneal cavity with $ZnCl_2$-induced protection against endotoxin. *Canad. J. Microbiol., 24*:834, 1978.

79. Walker, R. I., Porvaznik, M. J., Egan, J. E., and Miller, A. M. Hageman factor activation in mice challenged with attenuated endotoxin. *Experientia* (under review).

80. White, G. L., Archer, L. T., Beller, B. K., Holmes, D. D., and Hinshaw, L. B. Leukocyte response and hypoglycemia in superlethal endotoxic shock. *Circ. Shock, 4*:231, 1977.

81. Zlydaszyk, J. C., and Moon, R. J. Fate of ^{51}Cr-labeled lipopolysaccharide in tissue culture cells and livers of normal mice. *Infect. Immun., 14*:100, 1976.

10

Suppression of Erythroid Differentiation by Colony-Stimulating Factor

Gary Van Zant and
Eugene Goldwasser

Under normal conditions, erythropoiesis and granulopoiesis occur simultaneously in the bone marrow, and moderate stimulation of one line of differentiation does not affect the other line. A large perturbation of the erythropoietic steady state, however, has been shown to affect the rate of granulopoiesis. A greatly increased demand for cells, for example, was found to increase erythropoiesis and concomitantly to decrease granulopoiesis, suggesting competition for a common stem cell (6).

The primary inducer of erythropoiesis, erythropoietin (epo), has been purified, and its action *in vivo* and *in vitro* has been studied extensively (4,12). An analogous inducer of granulopoiesis has not been unequivocally identified. Colony-stimulating factor (CSF), which induces the growth of granulocyte and macrophage (G-M) colonies *in vitro,* however, is a candidate granulopoietin (15). According to a widely accepted model of hemopoietic differentiation, these inducers act on unipotent stem cells specific for each line of differentiation, but do not have an effect on the pluripotent stem cell, which, according to this scheme, bears a parent-progeny relationship to the unipotent cells (10). In an alternative model, inducers act on the pluripotent stem cell itself and thereby regulate hemopoietic differentiation(4).

Using marrow cells *in vitro,* we have put to experimental test these two hypotheses concerning the cellular locus of action of epo and CSF. It is implicit in the unipotent stem-cell hypothesis that the presence of the inducer of one differentiation pathway should not have a short-term effect on the other pathway or on the pluripotent stem cell compartment.

Our results do not support this latter hypothesis, but do suggest instead that epo and CSF share a common target cell, probably the pluripotent stem cell. The addition of epo to primary cultures of mouse marrow caused an increase in the number of colony forming unit-spleen (CFU-s) by the 2nd day of incubation; the increment was maximal at day 7 (17). Moreover, the increase was due solely to an increase in cells forming erythroid spleen colonies. At the same time, epo caused a decrease in non-erythroid spleen colonies, demonstrating not only that it had a positive effect on erythroid function, but also that it caused the number of cells available for non-erythroid differentiation to decrease. In agreement with these findings, we recently showed that epo suppressed G-M colony formation induced by CSF and, conversely, that CSF suppressed epo-stimulated hemoglobin synthesis by rat marrow cells in culture (18). In the work reported here, we studied the simultaneous effects of epo and CSF on marrow cells under conditions that permit erythroid and G-M colonies to be quantitated in the same culture dish. In these studies, we have corroborated and extended our previous work by examining the dose-response relationships, the temporal relationships, and the effects of cell number on competition between the erythroid and granulocytic lines of differentiation.

MATERIALS AND METHODS

Animals

Female BDF_1 mice, 12 to 16 weeks old, were used.

Preparation of Cell Suspensions

Marrow cells were flushed from the femora and tibiae of mice into α-medium. Nucleated cells were counted in duplicate by normal hemocytometry. We consistently recovered approximately 1.5×10^7 nucleated cells/femur or tibia from these mice (range, 1.3 to 1.7×10^7).

Epo and CSF Preparations

Pure epo from human urine, with a specific activity of 70,000 units/mg of protein, was used. This epo, which was purified in our laboratory, is homogeneous by several criteria (12).

A preparation of CSF from human embryonic kidney cell medium was used routinely in these studies. This CSF had a specific activity of 5,300 units/mg of protein; it was generously supplied by Mr. Otto Walasek of Abbott Laboratories, North Chicago, Ill. Recently, CSF from two sources has been purified. The specific activity of CSF obtained from L-cell culture medium is 1.6×10^8 utnits/mg of protein (16) and that of CSF from mouse-lung-conditioned medium is 7×10^7 units/mg of protein (1). Pure CSF from the latter source was used in one experiment (Table 10.3); it was generously provided by Dr. Antony Burgess of the Walter and Eliza Hall Institute of Medical Research, Parkville, Australia.

A unit of CSF is defined as the amount required to stimulate the formation of one G-M colony in a culture containing 7.5×10^4 mouse marrow cells (11). In preliminary experiments, we established that this definition is valid at higher cell concentrations as well. The plot of log G-M colony–log cell number has a slope of 1 at several CSF concentrations in the range of 120 to 360 units/ml.

Assay for Burst-Forming Units-erythroid (BFU-e)

The method of Iscove and Sieber (11) was used. Mouse marrow cells were cultured in 35×10 mm dishes (model 5221-R; Lux Scientific Corp., Thousand Oaks, Ca.) in 1 ml of medium consisting of 68% α-medium (Flow Laboratories, Rockville, Md.), 30% fetal calf serum (Flow Laboratories, Rockville, Md.), 0.8% methyl cellulose (4,000 cps, Dow Chemical Co., Midland, Mich.), 10^{-4} M β-mercaptoethanol (Eastman Kodak Co., Rochester, N.Y.), and 1% bovine serum albumin (Armour Pharmaceutical Co., Chicago, Ill., lot No. K382119). Cultures were incubated in a fully humidified atmosphere of 5% CO_2 in air for 8 days; erythroid colonies and bursts were then scored at 25 to 40 X under a dissecting microscope. The bursts were routinely stained in situ with benzidine according to the method of Ogawa et al. (13). Five replicates were used per point.

When erythroid bursts were quantified by measurement of hemoglobin synthesis, the following changes in the procedure were made (3): (a) Marrow cells were cultured in the medium described above, at a nucleated-cell density of 7.5×10^5/ml. (b) The size of the culture was reduced to 0.3 ml. which was added to the wells (16 mm in diameter) of Costar tissue culture plates (Model 3524; Costar Inc., Cambridge, Ma.). (c) Cultures were incubated in a fully humidified atmosphere of 3% CO_2 in air. (d) Transferrin-bound ^{59}Fe (0.4 μCi) was added to each culture in a volume of 20 μl 1 day prior to measurement of hemoglobin synthesis. Burst-associated hemoglobin synthesis was routinely measured on day 8. We modified the preparation of ^{59}Fe-labeled rat serum, described in detail previously (5), by substituting α-medium for NCTC-109. Incorporation of ^{59}Fe into hematin was measured according to a previously published method (5). Six replicates were used per point.

Assay for Colony-Forming Unit-Culture (CFU-c)

Granulocyte-macrophage (G-M) colonies, grown under the conditions described for the growth of erythroid bursts, were counted on day 8 at 25 X under a dissecting microscope. Morphologic characteristics were used for identification of the colonies.

Feeding Cultures with Fresh Medium

A modification of the method of Opitz et al. (14) was used. On day 3, 20 μl, and on day 6, 40 μl of double-strength α-medium were added to the 0.3 ml cultures.

Adding Indomethacin to Cultures

Indomethacin (Sigma Chemical Co., St. Louis, Mo.) was added to cultures on day 0 to give a final concentration of 1.4×10^{-7} M, and again on day 4 to give a concentration of 2.8×10^{-7} M, provided there was no loss of indomethacin between day 0 and day 4. This concentration has been shown to inhibit synthesis of prostaglandin E by macrophages (8).

TABLE 10.1 Effect of Cell Number on the Suppression of Granulocyte-Macrophage Colony Formation Caused by Erythropoietin

| NUCLEATED CELLS/ML | G-M COLONIES | | | | | % Suppression | p |
	No Additions	CSF[a]	Δ	CSF + epo[b]	Δ		
1×10^4	0	9	9	11	11	0	—
5×10^4	13	69	56	61	48	14	Not significant
1×10^5	42	217	175	163	121	31	$.02 > p > .01$
3×10^5	218	435	217	349	131	40	$.01 > p > .001$

[a] CFS = Colony stimulating factor (CSF), 180 units/ml.
[b] 1 unit/ml.

Adding Prostaglandin E_1 to Cultures

Synthetic prostaglandin E_1, a generous gift of Dr. John Pike, Upjohn Co., Kalamazoo, Mich., was added to cultures on day 0 to give a final concentration of 1×10^{-5} M.

RESULTS

We previously showed that epo caused a dose-dependent suppression of the number of CSF-stimulated granulocyte-macrophage (G-M) colonies formed in cultures of mouse marrow (18). Table 10.1 demonstrates that this relationship depends on the number of cells initially seeded in the culture dishes. Erythropoietin (1 unit/ml) caused a suppression of G-M colonies stimulated by 280 units/ml of CSF in cultures established with 1×10^5 and 3×10^5 mouse marrow cells/ml, but not in cultures at the two lower cell concentrations. Moreover, the level of suppression increased from 30 to 40% as the cell density was increased from 1×10^5/ml to 3×10^5/ml.

In agreement with our earlier findings (3), marrow cells that are grown in medium containing methyl cellulose and that are stimulated by epo have two hemoglobin synthesis peaks (Figure 10.1). The first corresponds to the development of CFU-e progeny after 2 days of culture; the second, larger peak corresponds to the development of erythroid bursts from BFU-e after 8 days of culture. When CSF was added simultaneously with epo to such cultures, the second peak, but not the first, was greatly reduced in size. In cultures with CSF and epo, hemoglobin synthesis peaked at day 7 and was reduced by 73% when compared with the 8-day peak in cultures without CSF. Moreover, the number of bursts was reduced by 60% (from 35 to 14) in the cultures containing CSF.

When we tested a range of CSF concentrations with several epo levels, we found that the degree of suppression of burst-associated hemoglobin synthesis due to CSF depended on the relative concentrations of CSF and epo (Table 10.2). When a low level

of CSF (11 units/ml) was combined with low concentrations of epo (0.1 and 0.5 units/ml), hemoglobin synthesis was augmented significantly. As the concentration of CSF was increased, however, the responses to increasing levels of epo were suppressed. The highest CSF concentration used (560

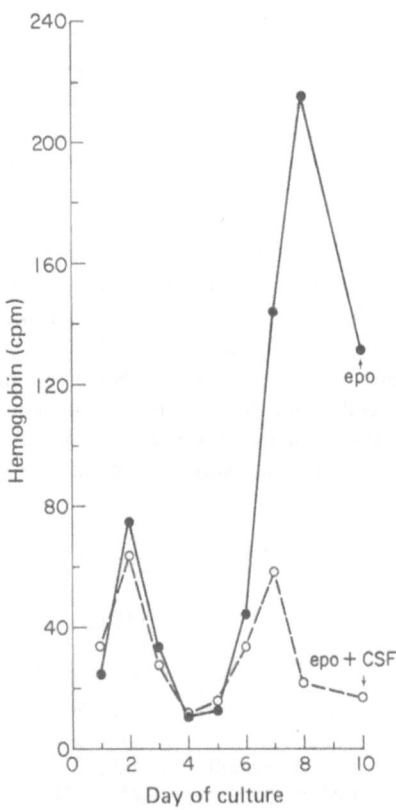

FIGURE 10.1. Effect of CSF on epo-stimulated hemoglobin synthesis. Each point represents ^{59}Fe cpm incorporation into hematin after a 24-hr pulse of 0.4 μCi ^{59}Fe; epo was used at 5 units/ml and CSF at 280 units/ml. Before the cells were harvested at 8 days, the number of bursts were counted; there was a mean of 35 in cultures with epo and 14 in cultures with epo and CSF. Each point represents the mean of six replicates.

TABLE 10.2 Effects of Increasing Colony-Stimulating Factor Concentration on Burst-Associated Hemoglobin Synthesis

CSF CONCENTRATION (UNITS/ML)	PERCENT OF HEMOGLOBIN SYNTHESIS (%)				
	Erythropoietin (Units/ml)				
	0.1	0.5	1.5	5.0	10.0
0	100	100	100	100	100
11	200	145	111	107	100
45	36	71	70	88	102
110	0	26	41	41	44
220	0	0	14	23	24
560	0	0	0	1.7	2.4

units/ml) completely suppressed burst-associated hemoglobin synthesis caused by epo, except at the two highest epo levels at which the effects were 98% suppressed. At a constant CSF level, the addition of increasing amounts of epo caused increased hemoglobin synthesis, suggesting that the relationship between the two inducers is competitive. Potentiation of the effects of low epo levels by low doses of CSF is a consistent finding that remains unexplained at present, but it does provide further evidence that the two inducers are acting at a common locus.

Colony-stimulating factor has recently been purified from mouse-lung-conditioned medium and was kindly made available to us by Dr. Antony Burgess (1). Low concentrations of this pure preparation (10 and 25 units/ml) augmented burst-associated hemoglobin synthesis by 80% (Table 10.3). One hundred units/ml and 500 units/ml of this CSF suppressed hemoglobin synthesis by 62% and 98%, respectively. A CSF concentration of 25 units/ml caused a small, but significant increase in hemoglobin synthesis in marrow cells alone, but none of the other concentrations, either higher or lower, had any effect. These data rule out the possibility that a contaminant in the

CSF preparation was responsible either for potentiation of the epo effect, when CSF was used at low concentrations, or suppression of the epo effect at higher levels.

In agreement with our findings that suppression of CSF-stimulated G-M colony formation caused by epo was cell density dependent (Table 10.1), we found that suppression of epo-stimulated hemoglobin synthesis caused by CSF was related to cell number (Table 10.4). None of the CSF concentrations caused suppression at 10^5 cells/ml; at 2×10^5 cells/ml, however, 560 units/ml of CSF caused a 33% suppression of hemoglobin synthesis. In cultures with twice as many cells (4×10^5/ml), the same level of CSF caused about 90% suppression. At the highest cell density used (1×10^6 cells/ml), the lowest CSF concentration used in this experiment, 56 units/ml, caused almost complete suppression of burst-associated hemoglobin synthesis.

To examine the possibility that cell crowding at densities between 10^5 and 10^6 cells/ml was responsible for the suppression of epo-stimulated hemoglobin synthesis, we determined the effect of epo and CSF on the number of nucleated cells in methyl cellulose

TABLE 10.3 Effect of Pure Lung-Cell Colony Stimulating Factor on Burst-Associated Hemoglobin Synthesis

GROUP	ADDITION	HEMOGLOBIN SYNTHESIS (^{59}Fe cpm ± SD)	INCREMENT DUE TO EPO (cpm)[a]	POTENTIATION OF EPO RESPONSE (%)	SUPPRESSION OF EPO RESPONSE (%)	p
1	None	15 ± 2	—	—	—	—
2	epo[b]	593 ± 249	578	—	—	—
3	CSF (5 units/ml)	15 ± 3	—	—	—	—
4	CSF + epo	833 ± 165	818	54	—	2 vs 4: N.S.
5	CSF (10 units/ml)	16 ± 3	—	—	—	—
6	CSF + epo	985 ± 221	969	82	—	2 vs 6: p < .05
7	CSF (25 units/ml)	39 ± 4	—	—	—	1 vs 7: p < .001
8	CSF + epo	996 ± 216	957	80	—	2 vs 8: p < .02
9	CSF (100 units/ml)	18 ± 2	—	—	—	—
10	CSF + epo	218 ± 26	200	—	62	2 vs 10: p < .01
11	CSF (500 units/ml)	16 ± 1	—	—	—	—
12	CSF + epo	28 ± 9	12	—	98	2 vs 12: p < .001

[a]Counts per minute.
[b]Erythropoietin (epo) was used at 5 units /ml.

TABLE 10.4 Effect of Cell Number on the Suppression of Burst-Associated Hemoglobin Synthesis Caused by Colony-Stimulating Factor

CELL NUMBER ($\times 10^5$/ml)	BURST-ASSOCIATED HEMOGLOBIN SYNTHESIS (%) WHEN CSF WAS ADDED WITH EPO (5 UNITS OF EPO = 100%)[a]		
	(units of CSF/ml)		
	56	220	560
1	120	99	104
2	95	117	67
4	128	61	8
6	84	3	2
8	15	0	0
10	3	0	0

[a]Erythropoietin = epo.

cultures. When we initially plated 7.5×10^5 marrow cells/ml, we observed the largest number of cells in cultures stimulated with epo alone (Figure 10.2). Cell numbers reached a peak of about 2×10^6 cells/ml on days 7 and 8. The addition of both CSF and epo resulted, on days 7 and 8, in a cell density intermediate to the number of cells found in cultures with either inducer added alone. The simultaneous effects of CSF and epo on cell number were not additive; thus, a high cell density did not seem to be responsible for suppressing hemoglobin synthesis in burst-associated erythroid cells.

The addition of extra medium to cultures on days 3 and 6 had no effect on the suppression of burst-associated hemoglobin synthesis caused by CSF (Table 10.5). Hemoglobin synthesis was more than 90% suppressed in both cases, which indicated that depletion of a medium constituent was not responsible for the observed effects.

To test whether G-M colonies growing in the same cultures with erythroid bursts interfered with erythroid burst development, we plated marrow cells at several cell concentrations with epo, but without CSF, and at 8 days, counted the numbers of bursts and G-M colonies (Figure 10.3). When the number of bursts and the number of cells plated were compared

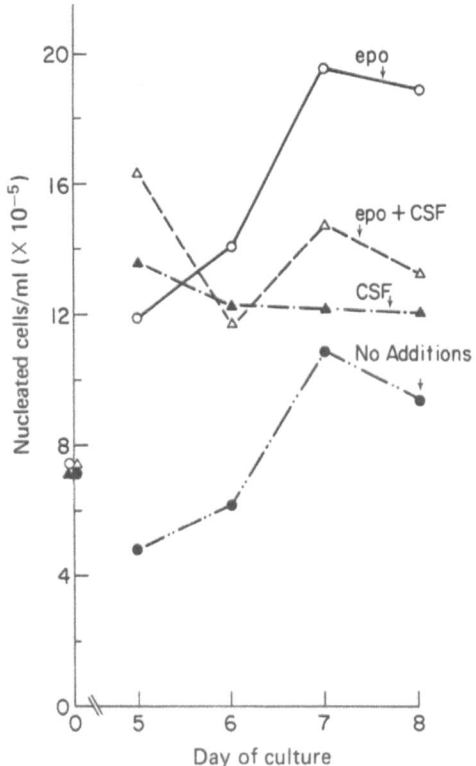

FIGURE 10.2. Effects of epo and CSF on cell numbers in culture. Mouse marrow cells (7.5×10^5/ml) were cultured with epo (5 units/ml), CSF (560 units/ml), epo and CSF, or neither. The cells were harvested, washed once with α-medium, and counted in duplicate with a hemocytometer. Each point represents the mean of six replicates.

on a log-log plot, a linear relationship was obtained from 5×10^4 to 1×10^6 cells/ml. A similar plot of the endogenously generated G-M colonies in these cultures also fits a straight line over the same cell density range. The salient finding in this experiment was that, at cell densities up to 1×10^6ml, the number of bursts increased independently of the increase in G-M colonies. At the 1×10^6/ml cells concentration, the presence of nearly 600 G-M colonies did not interfere with the growth of the 250

TABLE 10.5 Effect of Extra Medium on Suppression of Burst-Associated Hemoglobin Synthesis

	BURST-ASSOCIATED HEMOGLOBIN SYNTHESIS			
	(^{59}Fe cpm ± SD)[c]			
	No epo[a] or CSF[b]	2.5 units/ml epo	560 units/ml CSF	Epo + CSF
No extra medium	9 ± 2	827 ± 119	11 ± 2	12 ± 2
Extra medium added on days 3 and 6	9 ± 1	725 ± 176	15 ± 1	21 ± 4

[a]Erythropoietin.
[b]Colony-stimulating factor.
[c]Counts per minute (cpm).

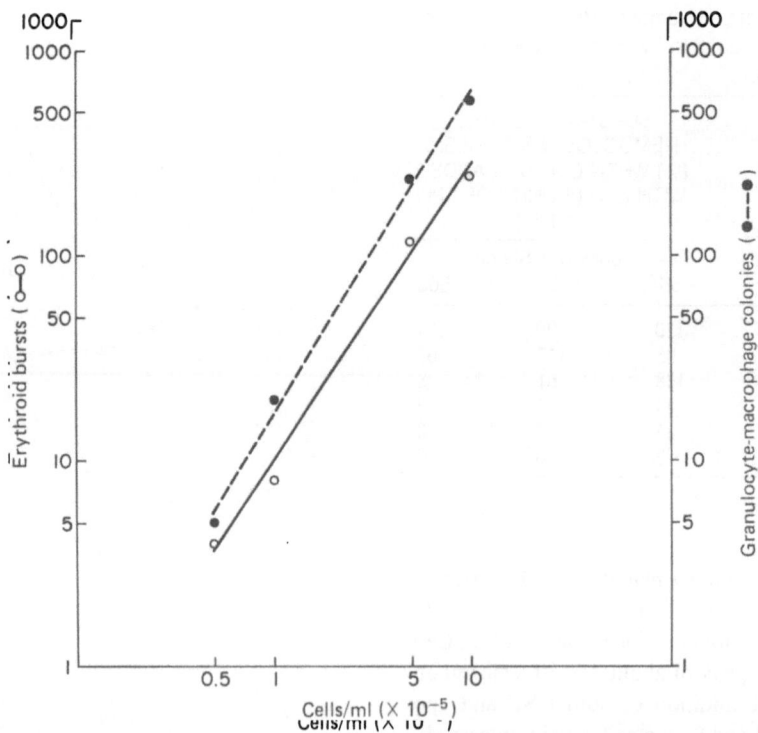

FIGURE 10.3. Effect of cell number on erythroid bursts and G-M colonies. Cells were cultured with epo (5 units/ml), and after 8 days, bursts and spontaneous G-M colonies were counted in the same dishes. Each point represents the mean of five replicates.

erythroid bursts expected from the number of bursts found at the three lower cell densities.

Prostaglandin E_1 has been shown to inhibit CSF action by suppressing proliferation of CFU-c (8,9). To test the possibility that prostaglandin E, synthesized by macrophages in culture plays a role in the suppression, by CSF, of burst-associated hemoglobin synthesis, we added indomethacin to cultures, with the results shown in Table 10.6. Indomethacin, an inhibitor of prostaglandin synthetase, had no effect on the action of CSF in suppression of the epo effect. Erythropoietin-stimulated hemoglobin synthesis was reduced by more than 90% in both the presence and absence of indomethacin.

When we added prostaglandin E, $(10^{-5}M$ final concentration) to methyl cellulose cultures with epo, we found that it had no significant effect on 2- or 8-day hemoglobin synthesis, nor on the number of bursts in these cultures (47 versus 44 for epo in the absence of CSF) (Figure 10.4). We found repeatedly that CSF suppressed burst-associated hemoglobin synthesis; prostaglandin E, however, abrogated this suppression, so that the burst number was restored from 7 to 42, a figure that does not differ significantly from the 44 bursts found with epo alone.

DISCUSSION

Our results show that, under our experimental conditions, both epo and CSF can modulate the response of marrow cells to each other. The level of suppression by CSF of erythroid function, and by epo of G-M colony formation, depends on the rela-

TABLE 10.6 Effect of Indomethacin on Suppression of Burst-Associated Hemoglobin Synthesis

| | BURST-ASSOCIATED HEMOGLOBIN SYNTHESIS | | | |
| | (^{59}Fe cpm $\pm$ SD)^c | | | |
	No epo^a or CSF^b	2.5 units/ml epo	560 units/ml CSF	Epo + CSF
No indomethacin	7 ± 1	610 ± 218	11 ± 2	28 ± 11
Indomethacin added	12 ± 4	732 ± 157	12 ± 3	31 ± 11

^{a}Erythropoietin.
^{b}Colony-stimulating factor.
^{c}Counts per minute (cpm).

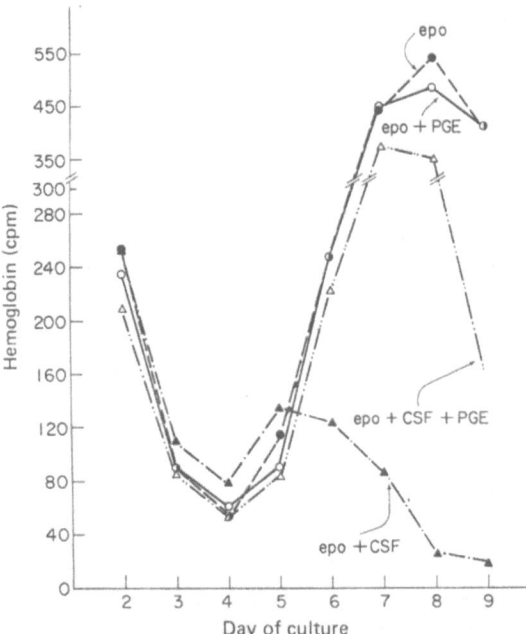

FIGURE 10.4. Effects of CSF and prostaglandin E, on epo-stimulated hemoglobin synthesis where epo (5 units/ml) was added to marrow cells with CSF (560 units/ml) or prostaglandin E, (10^{-5} M) or both. Each point represents ^{59}Fe cpm incorporation into hematin after a 24-hr pulse of 0.4 μCi ^{59}Fe. Before the cells were harvested on days 6 through 8, the number of bursts were counted: epo stimulated a peak of 44 bursts; epo + PGE, 47 bursts; epo + CSF + PGE, 42 bursts; and epo + CSF, 7 bursts. Each point represents the mean of six replicates.

tive concentrations of the two factors. Furthermore, in the case of suppression of erythroid function by CSF, the cellular locus is the most primitive erythroid progenitor, the BFU-e, and not the CFU-e, which is a relatively more differentiated precursor (Figure 10.1). Our results indicate that, when CSF stimulates G-M colony formation in the presence of epo, granulocytic differentiation occurs at the expense of erythroid burst formation. This suggests that CSF and epo compete for a common target cell that is capable of eventually forming either erythroid bursts or granulocytic colonies. This interpretation is strengthened by the finding that, in 8-day cultures, the suppressive effect of one inducer can be overcome by increased amounts of the other (Table 10.2).

We used pure human urinary epo in all the experiments reported here. This epo had a specific activity of 70,000 units/mg of protein and was homogeneous by several criteria (12). The CSF preparation used routinely was partially purified from the culture medium of human embryonic kidney cells and had a specific activity of 5,300 units/mg of protein. The fact

that qualitatively similar results were obtained with pure CSF from mouse-lung-conditioned medium (Table 10.3) strongly suggests that CSF, rather than an impurity, was responsible for the observed effects. The augmentation of epo action by low concentrations of CSF, seem most strikingly with pure CSF, is puzzling. It is possible that CSF, at low concentration, causes proliferation, but not differentiation, of the dipotent target cell it may share with epo. Thus in the presence of epo, erythroid function is potentiated due to an increased number of responsive cells. Regardless of the mechanism of potentiation, this effect indicates that CSF and epo can act on the same or on closely related cells.

Cell number was found to play an important role in the suppression of CSF action by epo and of erythroid function by CSF. It is noteworthy that suppression followed the same pattern in both cases; suppression was found at cell densities above 10^5/ml, but not below that density. The mechanism underlying the large effects of cell concentration, as seen, for example, in Table 10.4, is not yet understood. One possible explanation is that the high concentration of cells, especially granulocytes and macrophages formed as a result of CSF action, depletes the medium of other substances essential for erythroid cell differentiation. This possibility has been tested in several experiments reported here and is not compatible with the results reported in (a) Table 10.5, in which we show that feeding cultures with fresh medium had no effect on the suppression of epo action by CSF; (b) Table 10.2, in which we show that, at high cell concentration, increasing the amount of epo acted to diminish the suppressive effect; (c) Figure 10.2, in which high levels of both epo and CSF did not have additive effects on the number of cells in cultures at 8 days, but instead resulted in a cell concentration intermediate to the cell numbers obtained when epo and CSF were added individually; and (d) Figure 10.3, in which we show that approximately 600 G-M colonies did not prevent the growth of the 250 erythroid bursts predicted by extrapolation from three other points in a cell density experiment. Whereas these results tend to rule out cell crowding and depletion of medium nutrients as causes of the competitive interactions we observed, they do not provide a satisfactory explanation. Hemopoietic cell differentiation *in vivo* takes place in a milieu in which cells are closely associated within the bone marrow; one estimate of cell concentration is 1.8×10^9 cells/ml (2). Our findings suggest that relatively high cell concentrations may be required *in vitro* for studies of physiologic mechanisms regulating differentiation. Cellular interactions at higher cell densities may be important in physiologic responses of marrow cells, such as epo-stimulated hemoglobin synthesis, but not in influencing the number of clonogenic cells at limiting cell dilutions.

Macrophages have been shown to be a source of factors that have specific positive and negative effects on granulopoiesis; these factors are CSF and prostaglandin E, respectively (8,9). Prostaglandin E_1 suppresses the effect of CSF by acting at the level of the CFU-c to prevent proliferation of this primitive granulocyte precursor (8,9). We found that prostaglandin E abrogated the suppressive effect of CSF on epo action (Figure 10.4), which suggested that, when the action of one of the inducers is blocked, the competitive effect on the action of the other inducer is lost. This finding, like others we have reported, is compatible with the existence of a dipotent target cell that is responsive to both epo and CSF.

SUMMARY

We studied the simultaneous effect of erythropoietin (epo) and colony-stimulating factor (CSF) on mouse bone marrow cells *in vitro*. Pure human urinary epo decreased the number of granulocyte-macrophage colonies formed due to CSF; conversely, CSF reduced the number of erythroid bursts and caused a dose-dependent suppression in burst-associated hemoglobin synthesis induced by epo. When we blocked CSF action with prostaglandin E, the suppressive effect of that inducer on epo action was lost. These competitive effects did not occur in cultures with less than 10^5 marrow cells/ml; but between about 10^5 and 10^6 cells/ml, the level of suppression was directly related to cell density. Suppression of erythroid function at higher cell densities was not due to depletion of nutrients in the culture medium as a result of cell crowding.

In addition to a suppressive effect on erythroid differentiation at higher concentrations, pure mouse lung CSF at low levels augmented the effect of epo on burst-associated hemoglobin synthesis. Although we do not yet understand this phenomenon, it and the other findings reported here suggest that epo and CSF may act on a pluripotent stem cell capable of giving rise to erythroid bursts and granulocyte-macrophage colonies depending on the relative concentrations of the two inducers.

ACKNOWLEDGMENTS

This work was supported in part by Grants CA 18375 and CA 19265 from the National Cancer Institute. The Franklin McLean Memorial Research Institute is operated by The University of Chicago for the U.S. Department of Energy under Contract No. EY-76-C-02-0069.

We are grateful to Dr. Antony Burgess, Walter and Eliza Hall Institute of Medical Research, Melbourne, Australia, for the pure lung CSF; to Mr. Otto Walasek, Abbott Laboratories, North Chicago, Ill. for human embryonic kidney cell CSF; and to Dr. John Pike, Upjohn Co., Kalamazoo, Mich., for the synthetic prostaglandin E_1. We gratefully acknowledge Nancy Pech's superb technical assistance.

REFERENCES

1. Burgess, A. W., Camakaris, J., and Metcalf, D. Purification and properties of colony-stimulating factor from mouse lung-conditioned medium. *J. Biol. Chem.,* 252:1998, 1977.

2. Crafts, R. C., and Meineke, H. A. The anemia of hypophysectomized animals. *Ann. N.Y. Acad. Sci.,* 77:501, 1959.

3. Eliason, J. F., Van Zant, G., and Goldwasser, E. Unpublished results.

4. Goldwasser, E. Erythropoietin and the differentiation of red blood cells. *Fed. Proc., 34*:2285, 1975.

5. Goldwasser, E., Eliason, J. F., and Sikkema, D. An assay for erythropoietin *in vitro* at the milliunit level. *Endocrinol.,* 97:315, 1975.

6. Hellman, S., and Grate, H. E. Haematopoietic stem cells: Evidence for competing proliferative demands. *Nature, 216:*65, 1967.

7. Iscove, N. N., and Sieber, F. Erythroid progenitors in mouse bone marrow detected by macroscopic colony formation in culture. *Exp. Hematol., 3*:32, 1975.

8. Kurland, J. I., Broxmeyer, H. E., Pelus, L. M., Bockman, R. S., and Moore, M. A. S. Role for monocyte-macrophage derived colony stimulating factor and prostaglandin E in the positive and negative feedback control of myeloid stem cell proliferation. *Blood, 52*:388, 1978.

9. Kurland, J. I., Broxmeyer, H. E., and Moore, M. A. S. Limitation of excessive myelopoiesis by the intrinsic modulation of macrophage derived prostaglandin E. *Science, 199*:552, 1978.

10. Lajtha, L. G., and Schofield, R. On the problem of differentiation in haemopoiesis. *Differentiation, 2*:313, 1974.

11. Metcalf, D., and Moore, M. A. S. *Maemopoietic Cells.* Amsterdam: North-Holland, 1971, p. 409.

12. Miyake, T., C., Kung, K. -H., and Goldwasser, E. The purification of human erythropoietin. *J. Biol. Chem.,* 252:5558, 1977.

13. Ogawa, M., Parmley, R. T., Bank, H. L., and Spicer, S. S. Human marrow erythropoiesis in culture. I: Characterization of methylcellulose colony assay. *Blood, 48*:407, 1976.

14. Opitz, U., Seidel, H. -J., and Bertoncello, I. Erythroid stem cells in Friend-virus infected mice. *J. Cell Physiol., 96*:95, 1978.

15. Stanley, E. R., Hansen, G., Woodcock, J., and Metcalf, D. Colony stimulating factor and the regulation of granulopoiesis and macrophage production. *Fed. Proc., 34*:2272, 1975.

16. Stanley, E. R., and Heard, P. M. Factors regulating macrophage production and growth: Purification and some properties of the colony stimulating factor from medium conditioned by mouse L. cells. *J. Biol. Chem., 252*:4305, 1977.

17. Van Zant, G., and Goldwasser, E. The effects of erythropoietin *in vitro* on spleen colony forming cells. *J. Cell Physiol., 90*:241, 1977.

18. Van Zant, G., and Goldwasser, E. The simultaneous effects of erythropoietin and colony stimulating factor on bone marrow cells. *Science, 198*:733, 1977.

PART III

Erythropoiesis and Megakaryocytopoiesis

W. Fried

11

Progress in Erythropoiesis and Megakaryocytopoiesis

W. Fried

Toward the latter part of the nineteenth century, quantitative methods for measuring the numbers of erythrocytes in the blood had developed to the extent that research into the mechanisms that regulate erythrocyte production could be undertaken in earnest. By the turn of the century, it was reasonably well established that anemia accompanies various illnesses, including renal failure; that the red cell count recovers rapidly after hemorrhage; and that the red cell count increases in persons residing in the thin atmosphere of high altitudes. The hypothesis proposed to explain these phenomenon was that hypoxia directly effects erythrocyte cell precursors in the bone marrow to increase red cell production.

In 1906, Carnot and DeFlandre (3), influenced by the recent discovery of the first hormone, secretin, postulated that the effect of hypoxia on erythropoiesis is mediated by a hormone, "hemopoietine." Their experimental verification of this hypothesis was, however, not readily reproducible, and this concept was not widely accepted. Although interest in the humoral regulation of erythropoiesis was renewed by the report of Grant (5) that anemia accompanying marrow erythroid hyperplasia is not associated with marrow hypoxia, direct evidence for the existence of a humoral regulator of erythropoiesis was first convincingly provided by the experiments of Reissman (13) in parabiotic rats. In 1953, Erslev (4) showed that plasma from anemic rabbits contained a substance that increased the rate of erythrocyte production when injected into normal rats. Subsequently, numerous investigators confirmed the existence of a hormonal mediator for erythropoiesis (erythropoietin) and extensively studied its role in the regulation of erythropoiesis in normal and diseased states (8).

Because hemopoietic precursors less differentiated than the proerythroblast, myeloblast, and megakaryoblast were undistinguishable morphologically, studies of the effects of erythropoietin on specific target cells became possible only after the development of clonal assays for the various early hemopoietic precursor cells. The earliest such assay, the spleen colony assay introduced by Till and McCulloch in 1961 (16), permitted the study of the kinetics of multipotential hemopoietic stem cells (CFU-s). This assay, being an *in vivo* assay in mice, could not be applied to studies of CFU-s in any species other than the mouse or rat. But it did give rise to numerous investigations into the kinetics of these cells and the factors that cause them to differentiate. Although there is still controversy over the issue of whether or not erythropoietin can induce CFU-s to differentiate into the erythroid series (2,17), there is now considerable experimental support for the concept that differentiation is influenced by fixed microenvironmental elements. These as yet

75

unidentified elements in the hemopoietic stroma have been termed by Wolf and Trentin (18) the hemopoietic inductive microenvironment (HIM). In the mouse, the splenic HIM favors the growth of erythroid cells, whereas bone marrow HIM favors the growth of myeloid cells. In Chapter 15, this volume, Gurney et al. discuss some of the conditions that influence the spleen and its HIM, capitalizing on an animal model in which the marrow but not the spleen is destroyed by radiation.

In 1966, Bradley and Metcalf (1) and Pluznik and Sachs (12) described culture methods suitable for the clonal growth of granulocytic precursors, designated colony-forming unit-culture (CFU-c) *in vitro*. These methods were subsequently modified to permit the cloning of granulocytic precursors from various animals including humans (11). Shortly thereafter, techniques for cloning erythroid precursors were described by Stephenson et al. (15), and in ensuing years, techniques for clonal culture of erythroid precursors at varying levels of maturation were developed (6). Human colony-forming unit-erythroid (CFU-e) produce colonies of eight to 50 cells after 7 days in culture. They are exquisitely sensitive to changes in the titer of erythropoietin, both *in vivo* and *in vitro*. They are considered a relatively mature erythroid precursor and are present in human marrow, but not in the peripheral blood. The burst-forming unit-erythroid (BFU-e) are more similar than CFU-e in physical and kinetic properties to CFU-c and are considered the earliest detectable erythroid precursors. They produce large colonies of erythroid cells that consist, actually, of "bursts" of smaller colonies. These colonies appear after 10 to 15 days in culture; they are responsive to erythropoietin only when it is added to cultures in very high titers, yet are dependent on erythropoietin for *in vitro* growth. They are not influenced by transfusion- or hypoxia-induced polycythemia. These BFU-e are present in both marrow and peripheral blood. Recently, factors other than erythropoietin, have been shown to influence BFU-e growth *in vitro*, T-lymphocytes in particular (10). Addition of these cells, or of a substance liberated by them after lectin or tetanus toxoid stimulation (7), into the culture system enhance BFU-e growth. In Chapter 12, this volume, Nissen et al. report on the presence of a substance in the plasma of patients with aplastic anemia that stimulates BFU-e growth *in vitro* and that is distinct from erythropoietin. There is evidence, however, that T-lymphocytes influence erythropoiesis in congenitally anemic mice (14). This paper now provides support for the concept that human BFU-e growth *in vivo* is modulated by factors other than erythropoietin.

Although CFU-e growth; under normal circumstances is regulated by erythropoietin, that of patients with polycythemia rubra vera and of mice infected by Friend's virus are, in part, independent of erythropoietin; that is, marrow or spleen cells from these sources will produce erythroid colonies in cultures devoid of erythropoietin. (Yet such colonies will respond to addition of erythropoietin by an increase in colony formation.) For this reason, the Friend's virus infection of mice has been an important tool in studying factors that influence erythroid cell maturation. Hankins et al. in Chapter 13, this volume, report the effect of infecting mouse cells *in vitro* with Friend's virus and their propensity to produce BFU-e in culture. It thereby furthers our knowledge of the target cell for Friend's virus infection and provides new insights for the study of erythroid transformation in health and in disease.

Cloning of megakaryocytic precursor cells (CFU-m) has been a relatively recent accomplishment (9); this procedure is already playing an important role in deciphering the complex kinetics of platelet production. In Chapter 14, Nakeff et al., this volume, report on their use of this method to obtain information on the responses of CFU-m to various chemotherapeutic agents.

The papers presented at this year's Annual Scientific Meeting of the International Society of Experimental Hematology demonstrate our growing awareness of the complexities of mechanisms for regulating hemopoiesis since the turn of the century when scientists knew that variations in the oxygen supply to the marrow regulated erythrocyte production.

SUMMARY

The historical development of mechanisms for the study of erythropoiesis are described. Particular emphasis is given to recent advances using clonal growth *in vitro*. Specific chapters in Part III are introduced and their relevancy is indicated.

REFERENCES

1. Bradley, T. R., and Metcalf, D. The growth of mouse bone marrow cells *in vitro*. *Aust. Exp. Biol. Med. Sci.* 44:287, 1966.
2. Bruce, W. R., and McCulloch, E. A. The effect of erythropoietic stimulation on hemopoietic colony forming cells of mice. *Blood*, 23:246, 1964.
3. Carnot, P., and Deflandre, C. Sur l'activite hemopoietique des differents organes au cours de la regeneration du sang. *CR Acad Sci., 143*:432, 1906.
4. Erslev, A. J. Humoral regulation of red cell production. *Blood, 8*:349, 1953.
5. Grant, W. C. Oxygen saturation in bone marrow and in

arterial and venous blood during prolonged hemorrhagic erythropoiesis. *Am. J. Physiol., 153*:521, 1948.

6. Iscove, V. W., and Wieber, F. Erythroid progenitors in mouse bone marrow detected by macroscopic colony formation in culture. *Exp. Hematol., 3*:32, 1975.

7. Johnson, G. R., and Metcalf, D. Pure and mixed erythroid colony formation *in vitro* stimulated by spleen conditioned medium with no detectable erythropoietin. *Proc. Nat. Acad. Sci. U.S. 74*:3869, 1977.

8. Krantz, S. D., and Jacobson, L. O. *Erythropoietin and the Regulation of Erythropoiesis.* Chicago: University of Chicago Press, 1970.

9. Nakeff, A., and Dicke, K. A. Stem cell differentiation into megakaryocytes from mouse bone marrow cultured with the thin layer agar technique. *Exp. Hematol., 22*:58, 1972.

10. Nathan, D. G., Chess, L., Hillman, D. G. Human erythroid burst-forming unit: T cell requirement for proliferation *in vitro. J. Exp. Med., 147*:324, 1978.

11. Pike, B. L., and Robinson, W. A. Human Bone marrow colony growth in agar gel. *J. Cell Physiol., 76*:77, 1970.

12. Pluznik, D. H., and Sachs, L. The induction of clones of normal mast cells by a substance from conditioned medium. *Exp. Cell Res. 43*:553, 1966.

13. Reissman, K. R. Studies on the mechanism of erythropoietic stimulation in parabiotic rats during hypoxia. *Blood, 5*:372, 1950.

14. Sharkis, S. V., Ahmed, A., Sensenbrenner, L. L., Jedzrejczak, W. W. Goldstein, A. L., and Sell, K. W. The regulation of hemopoiesis: Effect of thymosin or thymocytes in a diffusion chamber. In Baum, S. V., and Ledney, G. D., eds., *Experimental Hematology Today 1978.* New York: Springer-Verlag, 1978.

15. Stephenson, J. R., Axelrod, A. A., McLeod, D. L., and Shreeve, M. M. Induction of colonies of hemoglobin synthesizing cells by erythropoietin *in vitro. Proc. Nat. Acad. Sci. U.S., 68*:1542, 1971.

16. Till, J. E., and McCulloch, E. A. A direct measurement of the radiation sensivity of normal mouse bone marrow cells. *Radiat Res., 14*:212, 1961.

17. Van Zant, G., and Goldwasser, E. Simultaneous effects of erythropoietin and colony stimulating factor on bone marrow cells. *Science, 198*:733, 1977.

18. Wolf, N. S., and Trentin, J. J. Hemopoietic colony studies. V. Effect hemopoietic organ stroma on differentiation of pluripotential stem cells. *J. Exp. Med., 127*:205, 1968.

12

High Burst-Promoting Activity (BPA) in Serum of Patients with Acquired Aplastic Anemia

C. Nissen, N. N. Iscove, and B. Speck

The precursors of erythrocytes, granulocytes, and macrophages can be induced to proliferate and mature under appropriate tissue culture conditions. The role of specific glycoproteins—erythropoietin and macrophage and granulocyte colony-stimulating factors—in these processes is now widely recognized.

Recently, an additional class of factors, which is distinct from those mentioned, has been identified. Their discovery depended on the ability to distinguish very primitive (BFU-e) from relatively mature (CFU-e) erythrocyte precursors on the basis of their differing cloning properties in either semi-solid (1,9) or viscous culture medium (11). Once this was possible, it became apparent that the very earliest stages of erythropoiesis depended in culture not on erythropoietin, but rather on glycoproteins having burst-promoting activity (BPA) (2,13,14,18,19,27).

In this study, the sera of patients with a variety of anemias were tested for their effects on colony formation by human hemopoietic precursors in methyl cellulose cultures. The cultures were set up so as to be limiting in, and therefore sensitive to, BPA and granulocyte and macrophage colony-stimulating activity. This was accomplished by eliminating incidental sources of these activities, i.e., by removing plastic-adherent cells from the marrow populations (16,2) and by using a batch of fetal calf serum selected for its low colony-stimulating and burst-promoting activity. Conversely, the cultures were made insensitive to variations either in erythropoietin or in the nonspecific growth-supporting constituents of serum. This was accomplished by the routine inclusion of erythropoietin and nonlimiting amounts of fetal calf serum and the relevant active serum constituents albumin and transferrin (8).

We report here the presence of high levels of macrophage colony-stimulating and burst-promoting activities in the sera of many patients with aplastic anemia. These activities were not high in healthy control sera nor in sera from patients with comparable degrees of anemia of other origins.

This report places the major emphasis on serum BPA. The elevation of serum BPA in aplastic anemia is of considerable interest on fundamental grounds alone, but the most striking aspect of the study was the high frequency of autologous remission in the patient group with high serum BPA (idiopathic acquired and chloramphenicol-associated aplastic anemia) as contrasted with the low frequency of remission in the group associated with low serum BPA (post-hepatitic and congenital hypoplasia).

MATERIALS AND METHODS

Patients

The 27 patients with aplastic anemia are listed in Table 12.1. All met the criteria for severe aplastic

TABLE 12.1 PATIENTS WITH APLASTIC ANEMIA

CLASSIFICATION OF ANEMIA	PATIENT	AGE	SEX	TRANSFUSION HISTORY	DRUGS WHEN SERUM SAMPLED		TREATMENT
					ANDROGEN	GLUCOCORTICOID	
Idiopathic	L.M.	15	M	−	+	−	ALG + BM
	S.C.	15	M	−	+	+	ALG + BM
	E.M.	20	F	−	−	−	ALG + BM
	S.E.	20	F	−	+	+	ALG + BM
	E.K.	22	M	+	−	−	ALG + BM
	A.B.	14	M	+	+	+	ALG + BM
	A.Ch.	18	M	−	+	+	ALG + BM
	J.L.C.	10	M	+	+	+	ALG + BM
	M.R.	10	M	+	+	+	ALG + BM
	M.P.	55	F	+	+	+	ALG + BM
	J.K.	7	F	+	+	+	ALG
	B.S.	28	M	+	+	+	ALG
	R.T.	35	M	+	+	+	ALG
	I.St.	28	F	−	−	−	Cy + BM
	R.L.	16	M	−	−	−	Cy + BM
	I.B.	17	F	+	−	−	Cy + BM
	A.M.	45	F	+	+	+	Cy + BM
	K.K.	26	F	+	−	−	Cy + BM
Chloramphenicol	I.O.	17	F	+	+	+	ALG
	H.M.	28	M	+	−	−	ALG + BM
	P.J.	6	F	+	+	+	Cy + BM
Post-hepatitic	A.R.	14	M	+	+	+	ALG + BM
	M.M.	15	M	−	−	−	—
	G.S.	22	M	−	−	−	Cy + BM
Fanconi's	A.E.	10	F	+	−	−	ALG + BM
Congenital	A.W.	6	F	+	+	+	ALG
	M.B.	5	M	+	+	+	ALG + BM

anemia formulated by Camitta et al. (4). Aplasia was attributed to chloramphenicol in three patients who had taken the drug 3 and 6 months before onset. Three patients who had had severe hepatitis within 3 months of hospital admission were classified as post-hepatitic. Of these, G. S. was positive and A. R. and M. M. negative for Australia antigen. Only A. E. met the classical criteria (including chromosomal anomalies) for Fanconi's anemia (3). Two other children with mental retardation and long histories of slowly progressive pancytopenia were presumed to have congenital disease.

Patients with HLA-matched siblings were given at least 3×10^8 nucleated donor marrow cells/kg body weight after preparation with cyclophosphamide (Cy) (4×50 mg/kg). Patients without matched siblings received at least 2×10^8 single haplotype-matched donor marrow cells/kg after preparation with horse anti-human lymphocyte globulin (40 mg/kg on each of 4 successive days) (25).

Ammonium Sulfate Precipitation and Gel Permeation Chromatography

Solid ammonium sulfate was added to 10 ml of serum from patient S.C. to 25% of saturation and stirred at room temperature for 30 min. The sample was centrifuged at 10,000 *g* for 30 min at room temperature. The procedure was repeated on successive supernatants to yield precipitates at 50 and 75% saturation. These were dissolved in 0.16 M NaCl, dialyzed thoroughly against a 0.16 M NaCl solution containing 2 mM Na.EDTA pH 7.4, and then filter-sterilized. They were stored at 4°C.

After determining that most of the BPA was in the 25 to 50% fraction, this fraction was chromatographed in a 3 ml volume on a calibrated 2.5 × 95-cm column of Sephadex G-150 in 160 mM NaCl, 4 mM HEPES (pH 7.3), and 0.05% polyethylene glycol (MW 6,000) at 4°C. The fractions were concentrated seven times over an Amicon UM-10 membrane, filter-sterilized, and tested directly in culture for erythropoietin and BPA activity.

Methyl Cellulose Cultures

Human bone marrow cells were suspended in culture medium. Where indicated, adherent cells were removed by a single, 30-min incubation in plastic tissue culture dishes, essentially as described (16). They were plated in a substantially modified Dulbecco's medium ("IMDM," GIBCO formula 78-5220) (12) to which were added 0.8% methyl cellulose, 10^{-4} M α-thioglycerol, 360 μg/ml human transferrin (one-third iron saturated), 1% delipidated and deionized (12) bovine serum albumin, 20% fetal calf serum (GIBCO batch K971101 D), and 1 erythropoietin unit/ml, all expressed in final concentrations. Plates (35 mm Petri, Greiner), always containing 10^5 nucleated cells in 1 ml, were incubated at 37°C in a fully humidified, 5% CO_2-air mixture. The CO_2 concentration was regularly checked with a Fyrite CO_2 indicator and strictly maintained.

Orange-to-red colonies of at least eight cells were counted on the 7th day of culture. These were considered to have arisen from CFU-e. Orange-to-red bursts of at least three subclusters, or a single cluster if macroscopic, were scored on the 14th day of culture. These were considered to have arisen from BFU-e. Granulocytic colonies (neutrophilic and eosinophilic) and mononuclear cell colonies, distinguished by their characteristic appearance (10,17), were scored on the 14th day of culture.

Erythropoietin was obtained from the urine of patients with aplastic anemia and partially purified by passage over DEAE-cellulose, concanavalin A-Sepharose, and Sephadex G-150. The preparations were devoid of BPA and granulocyte or mononuclear cell colony-stimulating activity. Their erythropoietin content was measured by comparing their capacity to stimulate colony formation by mouse CFU-e to that of a commercial standard (Connaught sheep erythropoietin step III).

Human leukocyte-conditioned medium was prepared in IMDM, containing 10% fetal calf serum and 10^{-4} M ü-thiogylcerol, over blood cells immobilized in agar as described previously (10).

Mixed Leukocyte Cultures

Human leukocytes were isolated from blood by centrifugation over Ficoll-Hypaque (22). Responder cells, 5×10^5, were cultured with 1×10^6 irradiated (1,500 R x-rays) stimulating cells/ml IMDM containing 10% fetal calf serum. Microcultures (0.2 ml) were inoculated with 2 μCi [³H]TdR/well on the 5th day. After 16 hr of labeling, the incorporated label was counted. The supernatant medium from macrocultures was harvested after 5 days of incubation.

Results

The requirement of human BFU-e for BPA is demonstrated in Table 12.2. Non-plastic-adherent human marrow cells (10^5/ml) were plated in methyl

TABLE 12.2 Requirement of BFU-e for BPA

CULTURE CONDITIONS		BURSTS/PLATE
N.A.[a]	+ 1 U[2b]	1.3 ± 1.0^{8c}
Unseparated	+ 1 U	6.9 ± 3.3
N.A.	+ 1 U + LCM[4d]	22 ± 4.3
N.A.	+ 20 U	6.2 ± 2.7
N.A.	+ 20 U + LCM	22 ± 5.1

Data are pooled from six experiments

[a]Non-adherent cells.
[b]Erythropoietin units.
[c]95% confidence limits expected on the basis of Poisson variation.
[d]20% human leukocyte-conditioned medium.

cellulose with 20% fetal calf serum and 1 erythropoietin unit/ml. Few large erythroid colonies (bursts) grew unless either adherent marrow cells or human leukocyte-conditioned medium were present. A 20-fold increase in concentration of erythropoietin had little effect alone. This culture system, always with non-adherent normal human marrow cells, 1 erythropoietin unit/ml, and 20% fetal calf serum, formed the basis for all the following assays for BPA.

Burst promotion by leukocyte-conditioned medium was concentration dependent, as shown in Figure 12.1. The conditioned medium also stimulated granulocytic and mononuclear cell colonies. In addition, dose-dependent inhibition of colony formation by CFU-e was consistently observed.

When normal human sera were tested, a low level of BPA and a slight inhibition of CFU-e growth were observed [Figure 12.2(A)]. These sera also weakly promoted granulocyte and mononuclear cell colony formation with mononuclear cell colonies predominating [Figure 12.3(B)]. They did not inhibit burst formation upon addition to cultures stimulated with leukocyte-conditioned medium.

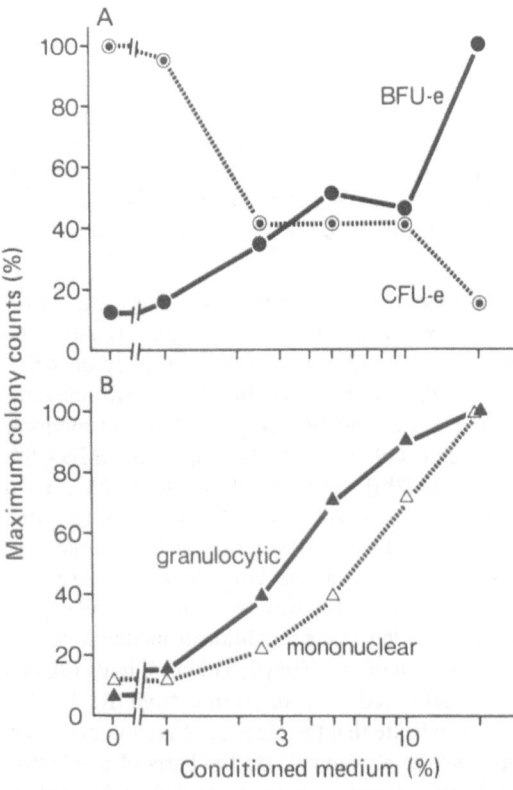

FIGURE 12.1. Effect of leukocyte-conditioned medium on growth of hemopoietic colonies. Colony counts are normalized to per cent of maximum (18 BFU-e, 27 CFU-e, 18 granulocyte, and 28 mononuclear colonies per plate). **(A)** Erythroid colony data represent three experiments; **(B)** myeloid data, one experiment.

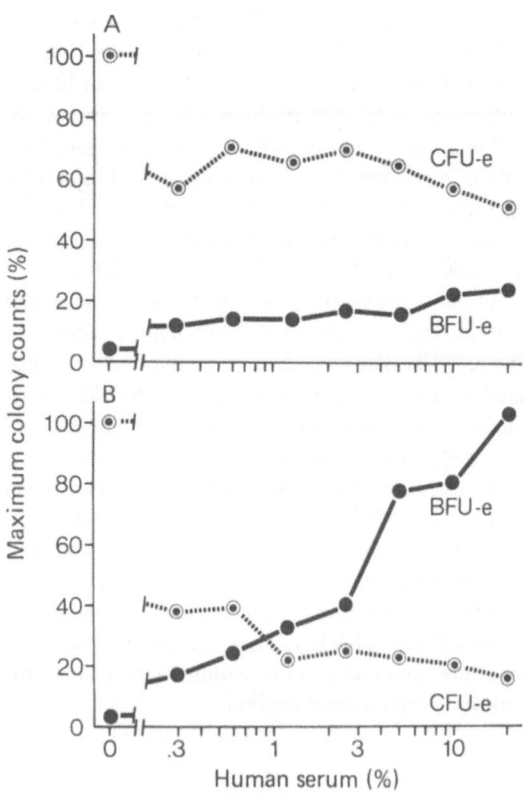

FIGURE 12.2. Effect of serum from (**A**) healthy controls (data pooled from five experiments) and (**B**) patients with aplastic anemia (data pooled from nine experiments) on colony formation by CFU-e and BFU-e. Colony counts are normalized to per cent of mean pooled maxima for both experiment groups—31 BFU-e and 59 CFU-e plate.

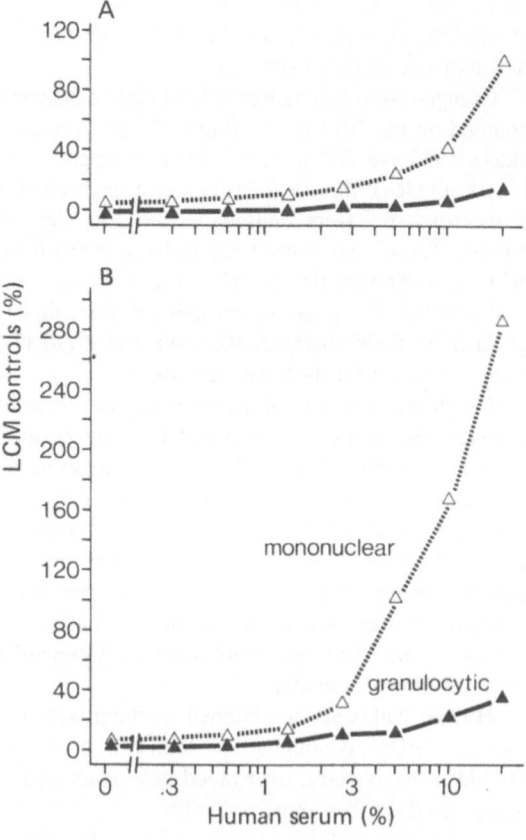

FIGURE 12.3. Effect of serum from (**A**) healthy controls (five experiments) and (**B**) patients with aplastic anemia (nine experiments) on growth of granulocytic and mononuclear cell colonies. Colony counts are normalized to per cent of the mean counts per plate (44 granulocytic and 12 mononuclear) of a series of 20 independent determinations made with 20% leukocyte-conditioned medium (LCM).

Markedly stronger BPA was found in sera from patients with aplastic anemia [Figure 12.2(b)]. All sera in this study were tested exclusively on ABO-compatible marrow cells. In all sera with elevated BPA, CFU-e inhibition (Figure 12.2) and mononuclear cell colony-stimulating activity (Mφ-CSA) [Figure 12.2(b)] were also above the levels seen in normal controls. Also Mφ-CSA was stronger than that in 20% LCM, consistently yielding larger and more numerous colonies. Granulocytic (neutrophilic or eosinophilic) CSA was very low. Mixing experiments with leukocyte-conditioned medium, to test for inhibition of granulocytic colonies, have thus far been performed only on serum from R. L. The results indicate that the absence of granulocyte colonies may be explainable on the basis of a selective inhibitory activity toward granulocytic colony growth.

All the serum activities survived 56°C for 30 min, and did not decline significantly during 6 months' storage at 20°C, even with repeated freezing and thawing.

Because it was likely that erythropoietin was ele-vated in the aplastic anemia sera, control experiments were performed to rule out erythropoietin as the basis for the burst promotion. Additional eryth-ropoietin in concentrations up to 20 units/ml consis-tently failed to provide significant burst formation alone in this culture system (Table 12.2). More direct evidence was provided when an active serum (from patient S. C.) was chemically separated by gel per-meation chromatography. As measured by capacity to induce colonies from CFU-e in erythropoietin-free cultures, the serum erythropoietin eluted with the expected apparent molecular weight of 60,000. In contrast, BPA eluted mainly in the apparent molecu-lar weight range of 15,000 to 20,000.

Prior to treatment, sera from 27 aplastic anemia patients were tested at a concentration of 20% in BPA-depleted cultures of ABO-compatible marrow. The results, expressed as bursts per plate, are shown in Table 12.3. The distribution of values is shown in

TABLE 12.3 Serum BPA in Patients with Aplastic Anemia

CLASSIFICATION OF ANEMIA	PATIENT	INITIAL Hb (μ%)	BPA[a] BEFORE TREATMENT	AFTER	HEMOPOIETIC RECOVERY	PLATELETS/mm³	NEED FOR SUPPORT[2]
Idiopathic	L.M.	5.0	85	14	Autologous	21,000	—
	S.C.	10.8	85	30	Autologous	8,000R	R
	E.M.	10.7	43	2	Autologous	218,000	—
	S.E.	6.4	39	9	Autologous	68,000	—
	E.K.	4.9	26	0	Autologous	48,000	—
	A.B.	6.9	24	0	Autologous	80,000	—
	A.Ch.	6.9	22	2	Autologous	300,000	—
	J.L.C.	9.1	22	n.d.	Autologous	90,000	—
	M.R.	7.5	19	2	Autologous	18,000	—
	M.P.	9.0	8	n.d.	—	1,000	R,G,Pl
	J.K.	10.0	47	10	Autologous	46,000	—
	B.S.	8.4	51	6	Autologous	105,000	—
	R.T.	8.0	28	24	—	1,000	R,Pl[d]
	I.St.	13.0	81	16	Autologous	52,000	—
	R.L.	4.0	68	n.d.	—[c]		
	I.B.	5.5	42	n.d.	Autologous	20,000	R
	A.M.	7.7	17	4	Autologous	55,000	—
	K.K.	5.8	5	11	Graft		—
Chloramphenicol	I.O.	8.1	34	n.d.	Autologous[c]	1,000	Pl
	H.M.	9.3	23	n.d.	Autologous	10,000	R
	P.J.	12.0	26	n.d.	—[c]		
Posthepatitic	A.R.	14.0	6	2	—	1,000	R,G,Pl
	M.M.	9.0	6	n.d.	Autologous	n.d.	—
	G.S.	9.0	6	n.d.	—[c]		
Fanconi's	A.E.	8.8	16	n.d.			
Congenital	A.W.	8.9	2	1	—	6,000	R,Pl
	M.B.	6.5	6	n.d.	—	6,000	R,G,Pl

[a]Expressed as bursts per plate (normal range, 0–7).
[b]R, red cells; G, granulocytes; Pl, platelets.
[c]Deceased.
[d]Later achieved autologous recovery after azathioprine and prednisone with no further requirement for support.

Figure 12.4. Since the values are not normalized, much of the scatter reflects variation in the number of precursor cells in the various test marrows. Twenty healthy control sera were tested on the same marrows and gave a range of zero to seven bursts per plate. Sixteen of 18 patients with acquired idiopathic anemia, three of three patients with chloramphenicol-associated aplastic anemia, and one patient with classical Fanconi's anemia had serum BPA clearly above the normal range. Sera from the two idiopathic patients without elevated BPA have not yet been tested for inhibition of burst formation stimulated by leukocyte-conditioned medium. In three of three patients with post-hepatitic aplasia, and two of two patients with presumptive congenital hypoplasia, BPA was in the normal range. Serum from one of the latter (A. W.) was tested for inhibition of burst promotion by 10% leukocyte-conditioned medium and none was observed.

Seventeen of the aplastic anemia patients were given anti-lymphocyte globulin (ALG) with or without infusion of single HLA haplotype-matched bone marrow in an attempt to provoke autologous recovery. Ten who achieved stable marrow reconstitution and two with incomplete responses had high serum BPA before treatment (Table 12.3). Five failed to improve and four of these five had low serum BPA. In three other patients, a permanent marrow graft was attempted after preparation with cyclophosphamide. All achieved autologous recovery (complete in two, incomplete in one) after rejection of the grafts, and all had high BPA. In the entire series, 15 of 16 patients with high pre-treatment serum BPA achieved autologous recovery, whereas none of the five patients without elevated BPA improved. The one patient (R. T.) with high BPA who did not respond to ALG now appears to be entering a remission, 3 years after ALG, following a recent course of prednisone and azathioprine. His serum BPA fell to the normal range before clinical evidence of remission.

After treatment, BPA fell to normal in those who achieved complete hemopoietic recovery, while remaining high in patients with residual deficits such as mild thrombocytopenia (Table 12.3) . The coefficient of correlation between the post-treatment BPA level and the platelet concentration at which the patients stabilized was −0.49, which is just significant at the 5% level.

Sera from patients with other disorders were also

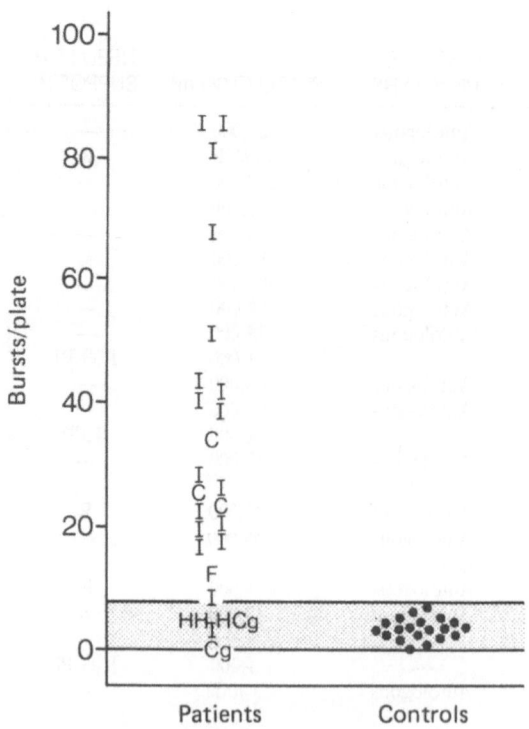

FIGURE 12.4. Distribution of BPA levels in sera from patients with idiopathic acquired anemia (I), chloramphenicol-associated anemia (C), post-hepatitic anemia (H), Fanconi's (F) and congenital (Cg) aplasia, and healthy controls (•——•).

tested (Table 12.4). The results are again expressed simply as bursts induced per plate by 20% test serum. Abnormal levels of BPA, CFU-e inhibition, and macrophage colony-stimulating activity were not seen in five patients with anemia (Hb, 7.0 to 8.8 g %) of different origins, in one patient with aplasia following cyclophosphamide and irradiation, in one patient with graft-versus-host disease, and in one patient with an autoimmune disorder. Interestingly, above-normal serum BPA was found in the hematologically normal father of one of the aplastic anemia patients (L. M.) and also in a patient with acute myelogenous leukemia in early relapse. Five additional related donors for other patients were also tested and none had elevated serum BPA.

Because of the possibility that immune processes may be involved in aplastic anemia, we asked whether increased amounts of BPA might be released as a consequence of cellular immune activiation. An obvious test was to examine the supernatant medium of mixed human leukocyte cultures. The results are shown in Figure 12.5 and indicate that additional amounts of BPA were indeed released when peripheral leukocytes reacted to unrelated leukocytes.

DISCUSSION

The essential observation in this study is the triad of mononuclear cell colony-stimulating activity, CFU-e inhibiting activity, and BPA at above normal levels in the serum of 19 of 21 patients with idiopathic

TABLE 12.4 Serum BPA in Patients with Anemias of Diverse Origin

DIAGNOSIS	AGE	SEX	DRUGS	HEMOGLOBIN (gm %)	BPA
Anemia:					
Iron deficiency	40	F	Iron	8.0	4
Blackfan-Diamond	17	F	Glucocorticoid	7.1	6
Pure red cell aplasia	16	F	Glucocorticoid	7.0	0
Chronic infection	70	F	Antibiotics	8.1	3
Congenital hemolytic	38	F	—	8.8	3
Acute granulocytic Leukemia; pancytopenia after chemotherapy and irradiation	16	M	Cy 3 days before (1,000 R 1 day before)	9.5 (transfused)	3
Graft versus host Disease, severe	6	F	Methotrexate 2 days before	10.5 (transfused)	5
Autoimmune Hypoparathyroidism + pernicious anemia	55	M	B_{12} + fo,ate	7.7	6
Healthy donor	37	M	—	Normal	18
Acute granulocytic leukemia, early relapse	23	F	—	12.0	25
Acute granulocytic leukemia, refractory	16	M	—	7.4 (transfused)	3

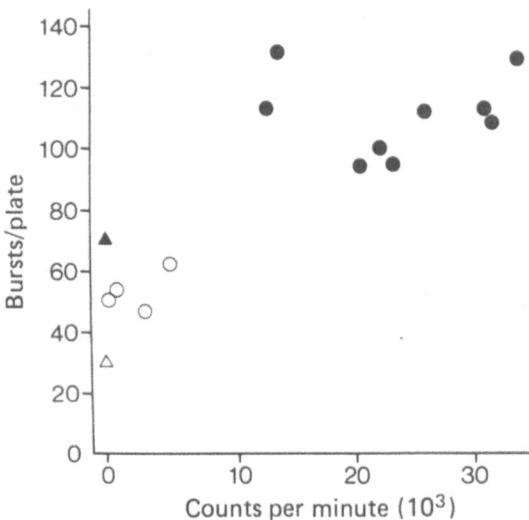

FIGURE 12.5. BPA in the supernatant medium of mixed leukocyte cultures, tested at 20% in methyl cellulose cultures. Supernatants of unirradiated responder cells alone at 1.5×10^6/ml (o——o); supernatants of mixed leukocyte reactions of the four donors in various combinations (●——●); no conditioned medium (△——△); 20% leukocyte-conditioned medium (▲——▲).

acquired or chloramphenicol-associated aplastic anemia. These serum activities were not elevated in three of three patients with post-hepatitic aplasia and in two of three patients with congenital disorders. The differentiation of post-hepatitic aplasic patients from those with idiopathic acquired disease on the basis of serum activity in culture was described earlier by Karp et al. (15). In that study, thymidine uptake by human marrow cells was measured after incubation in patient serum. Sera from patients with idiopathic acquired aplasia stimulated thymidine uptake relative to healthy control sera, whereas sera from patients with post-hepatitic aplasia did not. The use of colony assays in the present study allows a more specific assessment of the nature of the activities involved.

The mononuclear cell colony-stimulating activity in active aplastic anemia sera was quite striking. The numbers of mononuclear cell colonies obtained with 20% patient serum regularly exceeded those obtained with 20% leukocyte-conditioned medium. Despite the high level of mononuclear cell colony-stimulating activity, the numbers of granulocyte colonies in these cultures were very low. In agreement with Gordon (6), patient serum inhibited granulocytic colony growth in cultures containing leukocyte-conditioned medium in the single test for inhibition done so far.

The active sera also inhibited colony formation by CFU-e. It is not difficult to imagine that antibodies could be responsible, in some instances, and particularly in transfused patients. But, equivalent inhibition was also observed with sera from the untransfused patients in the series. Inhibition was also found with medium conditioned by normal leukocytes, in roughly the same proportion to BPA as was found in the active sera. Sources of mouse BPA, including medium conditioned by spleen cells stimulated by either concanavalin A or pokeweed mitogen, also inhibit colony formation by CFU-e, but the inhibitor(s) can be separated chemically from the BPA (13). It is conceivable that interferon is responsible, since it is known to be present in supernatants of activated lymphocytes (26) and has already been implicated as an inhibitor of hemopoietic cell growth (5,7,20,23).

The abbreviation BPA is an operational designation for recently discovered glycoproteins (2,13,14,-18,19,27) required for growth during the earliest phases of erythroid maturation. Erythropoietin alone is sufficient for colony formation in culture by the more mature cell stages represented by CFU-e. On the other hand, both erythropoietin and BPA are required for burst formation by BFU-e, and BPA is required only during the first few days of burst development. Erythropoietin, on the other hand, is needed only during the final days in order to allow terminal growth and the hemoglobin formation that render the bursts recognizable as being erythroid (13).

In addition to promoting purely erythroid bursts, BPA preparations also permit growth of mixed colonies from mouse bone marrow and fetal liver that contain megakaryocytes, granulocytes, and macrophage/monocytes in addition to erythroblasts (13,-14,18). The molecules promoting formation of mixed colonies are similar to BPA in size, isoelectric focusing pattern, and affinity for concanavalin A-Sepharose (13, Iscove, in progress). Until this activity can be separated from BPA, the possibility remains that the targets of BPA include pluripotential cells in addition to progenitors committed to erythropoiesis.

It is known that BPA is present in varying concentrations in fetal calf serum (13), a fact probably essential to the original achievements of burst formation in culture (1,11). This report indicates that it is also present at low concentrations in normal human serum. The known cellular sources of BPA are human fetal kidney (2); WEHI-3 cells (a mouse monocytic cell line (21) (Iscove and N. Williams, unpublished); adherent human marrow cells (2); high density mouse marrow cells (27); human peripheral leukocytes stimulated by PHA (2), agar extract (Hoang, Nissen, and Iscove, in progress), tetanus toxoid (19), or allogeneic cells; and mouse spleen cells stimulated by pokeweed mitogen (13,14,18), concanavalin A (13), or antigen to which the cells have previously been primed (Schreier and Iscove, in progress). In none of these instances is it yet clear whether the active molecules interact directly with primitive hemopoietic precursors or whether they exert their effect by causing factor release by other marrow cells.

Similarly, the tetanus toxoid (19) and mixed leukocyte experiments with human cells, and experiments with mouse spleen (18) implicate T-lymphocytes in the process of BPA release, after stimulation by either lectins or antigens. It is not clear, however, whether stimulated T-cells actually release BPA themselves or whether they cause some other type of cell to release it. The secretion of BPA by adherent human marrow cells and by the WEHI-3 mouse monocytic cell line is compatible with the latter possibility. If monocytes are the source of BPA, then T-lymphocyte activation may be only one of several signals for BPA release.

With so little known at the present time, the significance of the elevated serum levels of BPA in aplastic anemia remains purely speculative. Its apparent presence at a low level in normal serum, and its elevation in a disorder characterized by inadequate marrow function, are certainly compatible with a regulatory role. Alternatively, hypoplastic marrow might simply take up BPA from the circulation at a decreased rate (decreased "utilization"). But, the observations of apparent normal levels in radiation-induced, post-hepatitic, and congenital hypoplasias fail to add support to either concept.

The release of BPA in culture upon T-lymphocyte activation merits serious consideration in the light of evidence that idiopathic aplastic anemia may have an immune basis (24). The elevated level of circulating BPA could reflect cellular immune hyperactivity. If so, assay of serum BPA might distinguish aplasias of immune origin from those of non-immune origin and would, therefore, be predictive of response to immunosuppressive treatment. Convincing support, however, will require a much larger data base on serum BPA in aplastic anemia as well as in other disorders of established immune origin.

SUMMARY

Colony formation by primitive erythrocyte progenitors, burst-forming unit-erythropoietic (BFU-e) in culture is dependent on burst-promoting activity (BPA) in addition to erythropoietin. In this study, BPA in human serum was measured in cultures of normal human bone marrow from which incidental sources of BPA had been removed. Serum from healthy volunteers and from patients with anemias of diverse origin had very low levels of BPA or granulocyte or mononuclear cell colony-stimulating activity (G- or Mφ-CSA). In contrast, serum from 19 or 21 patients with idiopathic acquired anemia or chloramphenicol-associated aplastic anemia had high levels of BPA and Mφ-CSA; and it also inhibited colony formation by more mature erythroid precursors (CFU-e). These effects could not be reproduced by increasing erythropoietin concentrations in the cultures. The activities returned to low levels in patients who achieved complete autologous remission after treatment, but not in patients who achieved only partial remission. In the group characterized by high serum BPA, 16 of 20 patients achieved partial or complete remission. In contrast, only one of five patients achieved remission in the group associated with low BPA (post-hepatitic and congenital aplastic anemia).

ACKNOWLEDGMENTS

The authors are grateful to M. Nobile and F. Nobile for performing the mixed leukocyte cultures, Marianne Schweizer for excellent technical assistance, C. Peschle for supplying crude urinary erythropoietin, and H. -P. Stahlberger for the artwork.

C. Nissen is supported by grant numbers FOR 080.AK.75 and FOR 101.AK.77(2) from the Swiss Cancer League, and 3.3320.74 from the Swiss Science Foundation.

REFERENCES

1. Axelrad, A. A., McLeod, D. L., Shreeve, M. M., and Heath, D. S. Properties of cells that produce erythrocytic colonies in vitro. In Robinson, W., ed., *Hemopoiesis in Culture*. Washington, D.C.: U.S. Government Printing Office, 1974, p. 226.
2. Aye, M. T. Erythroid colony formation in cultures of human marrow: Effect of leukocyte conditioned medium. *J. cell Physiol.*, *91*:69, 1977.
3. Beard, M. E. J. Fanconi anemia. In *Congenital Disorders of Erythropoiesis*, Ciba Foundation Symposium 37 (new series). Amsterdam: Elsevier, 1976, p. 103.
4. Camitta, B. M., Thomas, E. D., Nathan, D. G., Santos, G. W., Gordon-Smith, E. C., Gale, R. P., Rapaport, J. M., and Storb, R. Severe aplastic anemia⁺ a prospective study of the effect of early marrow transplantation on acute mortality. *Blood*, *48*:63, 1976.
5. Fleming, W. A., McNeill, T. A., and Killen, M. The effects of an inhibiting factor (interferon) on the *in vitro* growth of granulocyte-macrophage colonies. *Immunology*, *23*:429, 1972.
6. Gordon, M. Y. Circulating inhibitors of granulopoiesis in patients with aplastic anemia. *Br. J. Haematol.*, *39*:491, 1978.
7. Greenberg, P. L., and Mosny, S. A. Cytotoxic effects of interferon *in vitro* on granulocytic progenitor cells. *Cancer Res.*, *37*:1794, 1977.
8. Guilbert, L. J., and Iscove, N. N. Partial replacement

of serum by selenite, transferrin, albumin and lecithin in haemopoietic cell cultures. *Nature, 263*:594, 1976.

9. Heath, D. S., Axelrad, A. A., McLeon, D. L., and Shreeve, M. M. Separation of the erythropoietin-responsive progenitors BFU-E and CFU-E in mouse bone marrow by unit gravity sedimentation. *Blood, 47*:777, 1976.

10. Iscove, N. N., Senn, J. S., Till, J. E., and McCulloch, E. A. Colony formation by normal and leukemic human marrow cells in culture: effect of conditioned medium. *Blood, 37*:1, 1971.

11. Iscove, N. N., and Sieber, F. Erythroid progenitors in mouse bone marrow detected by macroscopic colony formation in culture. *Exp. Hematol., 3*:32, 1975.

12. Iscove, N. N., and Melchers, F. Complete replacement of serum by albumin, transferrin, and soybean lipid in cultures of lipopolysaccharide-reactive B lymphocytes. *J. Exp. Med., 147*:923, 1978.

13. Iscove, N. N. Erythropoietin-independent stimulation of early erythropoiesis in adult marrow cultures by conditioned media from lectin-stimulated mouse spleen cells. In *Proceedings, ICN-UCLA Symposium on Hemopoietic Cell Differentiation* (in press) 1978.

14. Johnson, G. R., and Metcalf, D. Pure and mixed erythroid colony formation in vitro stimulated by spleen conditioned medium with no detectable erythropoietin. *Proc. Nat. Acad. Sci.* (U.S.), *74*:3897, 1977.

15. Karp, J. E., Schacter, L. P., and Burke, P. J. Humoral factors in aplastic anemia: Relationship of liver dysfunction to lack of serum stimulation of bone marrow growth *in vitro*. *Blood, 51*:397, 1978.

16. Messner, H. A., Till, J. E., and McCulloch, E. A. Interacting cell populations affecting granulopoietic colony formation by normal and leukemic human marrow cells. *Blood, 42*:707, 1973.

17. Messner, H., Till, J. E., and McCulloch, E. A. Density distributions of marrow cells from mouse and man. *Series Haematologica, 5*(2):22, 1972.

18. Metcalf, D., and Johnson, G. R. Production by spleen and lymph node cells of conditioned medium with erythroid and other hemopoietic colony-stimulating activity. *J. Cell. Physiol. 96*:31, 1978.

19. Nathan, D. G., Chess, L., Hillman, D. G., Clarke, B., Breard, J., Merler, E., and Housman, D. E. Human erythroid burst-forming unit: T-cell requirement for proliferation *in vitro*. *J. Exp. Med., 147*:324, 1978.

20. Nissen, C., Speck, B. Emödi, G., and Iscove, N. N. Toxicity of human leucocyte interferon preparations in human bone-marrow cultures. *Lancet, I*:203, 1977.

21. Ralph, P., Moore, M. A. S., and Nilsson, K. Lysozyme synthesis by established human and murine histiocytic lymphoma cell lines. *J. Exp. Med., 143*:1528, 1976.

22. Rubin, S. H., and Cowan, D. H. Assay of granulocytic progenitor cells in human peripheral blood. *Exp. Hematol., 1*:127, 1973.

23. Smith, K. A., Frederickson, T. N., Mobraaten, L. E., and DeMaeyer, E. The interaction of erythropoietin with fetal liver cells. II. Inhibition of the erythropoietin effect by interferon. *Exp. Hematol., 5*:333, 1977.

24. Speck, B., Cornu, P., Nissen, C., Groff, P., Weber, W., and Jeannet, M. On the pathogenesis and treatment of aplastic anemia. In Ledney, G. D., and Baum, S. J., eds., *Experimental Hematology Today*. New York: Springer-Verlag, 1978, p. 143.

25. Speck, B., Gluckmann, E., Haak, H. L., and van Rood, J. J. Treatment of aplastic anemia by anti-lymphocyte globulin with and without allogeneic bone marrow infusion. *Lancet, II*:1145, 1977.

26. Stobo, J., Green, I., Jackson, L., and Baron, S. Identification of a subpopulation of mouse lymphoid cells required for interferon production after stimulation with mitogens. *J. Immunol., 112*:1589, 1974.

27. Wagemaker, G. Cellular and soluble factors influencing the differentiation of primitive erythroid progenitor cells (BFU-E) *in vitro*. In Murphy, M. J. Jr., ed., *In vitro Aspects of Erythropoiesis*. New York: Springer-Verlag, 1978.

13

An *In Vitro* System for Erythroid Transformation by Friend Leukemia Virus

W. D. Hankins, T. A. Kost,
M. J. Koury, and S. B. Krantz

In 1957, Dr. Charlotte Friend discovered a leukemia virus that produced hepatosplenomegaly in mice 3 to 4 weeks after inoculation (7). Mirand (13) later demonstrated that certain preparations of Friend virus (FV) led to a striking increase in erythropoiesis and polycythemia. In 1964, Axelrad and Steeves (1) published an *in vivo* assay founded on the observation that injection of Friend virus into susceptible strains of mice led to the appearance of erythroid colonies under the splenic capsule 9 days later. The number of colonies were shown to be directly proportional to the amount of virus solution injected and a unit of virus activity (focus-forming unit, or FFU) was defined as the amount of virus necessary to produce an average of one focus per spleen 9 days after infection. The study of focus formation in susceptible and resistant strains of mice led to the finding that the Friend virus is a complex of at least two biologic entities (6,17): the spleen focus-forming virus (SFFV), which is thought to be responsible for the increase in erythropoiesis, and a murine leukemia virus (MuLV), which can be isolated free of SFFV and which produces a leukemia not accompanied by an increase in erythropoiesis.

In addition to spleen focus formation, increased splenic ^{59}Fe uptake (10), and hemoglobin synthesis have been employed to characterize the Friend virus-induced increase in erythropoiesis. We previously reported that spleen cells from plethoric mice, cultured *in vitro* soon after FV infection, exhibit increased hemoglobin synthesis (9). Spleen cells from uninfected mice do not share this property. This increased ^{59}Fe incorporation has now been shown to be a part of a well-defined wave of hemoglobin synthesis that occurs 4 to 5 days post-infection (11). The magnitude of this wave of hemoglobin synthesis varies with the dose of virus and with the length of time the cells are allowed to reside in the animal following the infection. On the other hand, the time-course of hemoglobin synthesis is not altered by the dose or the time *in vivo* (or a variety of other experimental manipulations (10). The time required between infection *in vivo* and the peak of hemoglobin synthesis *in vitro* was consistently 4 to 5 days.

To more exactly define the target cell(s), the transforming component(s) of the Friend complex, and the events that mediate transformation, a complete *in vitro* system allowing for infection, transformation, and expression of the transformed phenotype is needed. In 1975, Clarke et al.[5] (5) reported that mixing murine marrow cells with tissue culture fluids from cultures of Friend virus-infected fibroblasts led to an increase in the number of erythroid colonies observed in plasma clot cultures 2 days after infection. The erythroid colonies produced by the Friend virus complex occurred only at 2 days and were single randomly distributed colonies, similar to those produced by erythropoietin (2,12) (EP) at 2 days. Although culture of marrow cells with EP for 3 to 8 days in this system gives rise to clusters of erythroid colonies, termed bursts (1,8), no erythroid bursts were observed after longer periods of culture with FV. The number of virus-induced single erythroid colonies was proportional to the number of focus-forming units used for infection, but the induction of EP-independent colonies was limited to the tissue culture fluids from a particular cell culture line and was not observed with plasma from FV-infected mice.

Since we observed a peak of hemoglobin synthesis *in vitro* 4 to 5 days after infection *in vivo*, we looked for erythroid colonies at a similar time after infection of hemopoietic cells with the Friend leukemia virus *in vitro*. We now report that incubation of mouse marrow cells *in vitro* with infectious plasma from Friend virus-treated mice does lead to production of erythroid colonies in plasma clot cultures 5 days later. The erythroid colonies appear in clusters resembling EP-induced bursts and are, therefore, referred to as virus-induced erythroid bursts. Substantial numbers of virus-induced bursts, each containing multiple erythroid colonies, are observed and thus represent a very large increase of erythropoiesis *in vitro*. For this reason, it is hoped that this virus-induced burst formation will provide the much needed system in which to study the effects of Friend virus on mouse hemopoietic cells.

METHODS AND MATERIALS
Virus

The polycythemia-producing strain of Friend leukemia virus was originally obtained from Dr. Robert Holdenreid at the National Institutes of Health and has been maintained by serial passage in BALB/c mice. The particular preparation of infectious plasma used in most of these experiments (designated "M19") was prepared by bleeding animals that had received 2,000 spleen focus-forming units intravenously in an injection volume of 0.2 ml 18 days prior to bleeding. Blood was collected from the brachial artery and placed in heparinized tubes. Small aliquots of plasma from this blood were stored at $-70°$ C until used. Virus titers determined in the spleen focus-forming assay (2) and XC plaque assay (15) were 5×10^5 FFU/ml and 1×10^7 plaque-forming units/ml, respectively.

Animals

Our BALB/c mice (8 to 12 weeks old) were obtained from the National Cancer Institute, Bethesda, Md. In all the experiments reported here, the mice were rendered anemic by treatment with phenylhydrazine. Phenylhydrazine hydrochloride (Fisher Scientific, St. Louis, Mo.) was administered by three intraperitoneal injections at a dose of 40 μg/ g body weight. Injections were given at 40, 48, and 62 hr prior to sacrifice (16). Marrow from phenylhydrazine-treated mice consistently produced more bursts upon FV infection than did marrow from untreated mice. For the preparation of bone marrow suspensions, tibias and femurs were removed from the mice, and the bone marrow cells were flushed into Petri dishes by forcing Hanks' balanced salt solution (HBSS) through the bones using a syringe attached to a 27-gauge needle. Suspensions of separated cells were prepared and washed in 50 volumes of HBSS. A cell pellet was collected by centrifugation and resuspended in α-medium containing 2% fetal calf serum. Nucleated cell counts were determined with a hemocytometer after being appropriately diluted with a fresh aqueous solution of 0.2% methylene blue.

Incubation of Bone Marrow Cells with Virus

A solution containing 4×10^6 nucleated bone marrow cells was dispensed into 12×75-mm sterile plastic tubes (Falcon Plastics, Oxford, Ca.). After centrifugation at $1,000 \times g$ for 5 min, the supernatant was discarded and the cell pellet resuspended in 300 μl of either α-medium of the virus test solution diluted in α-medium. This mixture was incubated at 4°C for 1 to 2 hr. The virus-cell mixture was then diluted to 2 ml to give the following final concentrations of reagents: bovine serum albumin (prepared as described in ref. 14), 1% v/v; penicillin, 100 units/ml; streptomycin, 100 μg/ml; β-mercaptoethanol, 10^{-4} M; fetal calf serum, 30% v/v; beef embryo extract, 0.85% v/v; bovine citrated plasma, 10% v/v. This mixture was dispensed (0.5 ml/well) into triplicate tissue culture wells (Disposo-Trays, Model 96-SC, Linbro Scientific, Inc., Hamden, Conn.) and incubated at 37°C in humidified atmosphere of 5% CO_2— 95% air. After a 5-day incubation, the clots in the wells were placed on glass slides, fixed with 5% glutaraldehyde, stained in a 1% 3,3'-dimethoxybenzidine solution, and placed in 2.5% hydrogen peroxide in ethanol. The slides were counterstained with Harris hematoxylin and washed in tapwater. The data depicted in Figure 13.5 were obtained from clots removed daily.

RESULTS
Morphology of Erythroid Bursts

An erythroid burst formed 5 days after treatment of BALB/c marrow cells with a 1:20 dilution of infectious plasma is presented at three different magnifications in Figure 13.1. The burst is composed of multiple colonies of benzidie-positive cells exhibiting

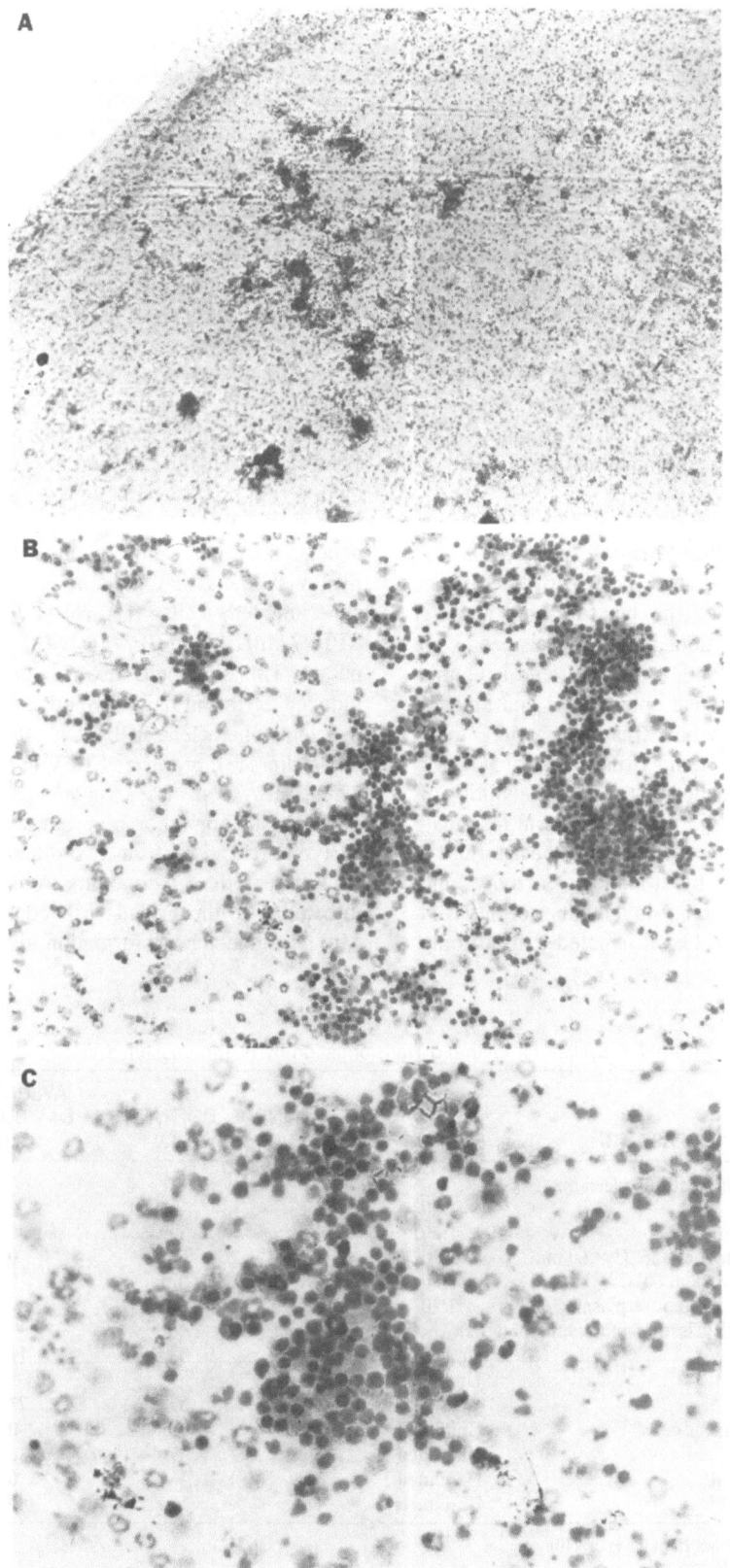

FIGURE 13.1. Morphology of the virus-induced erythroid burst: A cell pellet of 4×10^6 marrow cells was resuspended in a 1:20 dilution of infectious plasma (5×10^5 FFU/ml) and assayed for burst formation as described in Methods and Materials. (**A**) A Friend virus-induced erythroid burst is shown in the lower one-half of the photograph (40 X). Some colonies of the erythroid burst in (A) are shown at 200 X (**B**) and 400 X (**C**).

typical erythroid morphology. Within a single clot, there is considerable variation in the size and number of colonies present in different bursts. It is possible that growth-promoting cells are randomly distributed, which might provide a better microenvironment in some portions of the plate. Alternatively, bursts of different size and cell number may reflect the expression of virus-induced burst formation from target cells with different proliferative potentials. We have been unable to detect any differences in the appearance of untreated cultures and those incubated with virus except for the absence of erythroid bursts in the former. A background of randomly distributed single hemopoietic cells with many myeloid colonies are the predominant feature in slides of nontreated cultures.

Viral Nature of Burst-Forming Activity

Several experiments have been performed in an attempt to characterize the burst-forming activity (BFA) present in infectious plasma. It has been well documented that the ability of the Friend leukemia virus complex to produce splenic foci and splenomegaly is abolished by heat treatment at 56°C for 1 hr. Heat treatment of Friend virus plasma at 56°C for 1 hr or at 80°C for 3 min led to a complete abolition of its BFA (Table 13.1). If SFFV is the agent responsible for burst formation, one would expect to find a clear association of burst-inducing activity and spleen focus-forming ability in a number of different preparations. In Table 13.1 it is noted that not only

did a separate preparation of infectious plasma induce burst formation, but tissue culture supernatants that contained SFFV also exhibited BFA. This BFA was not detected in either normal mouse plasma or tissue culture supernatants from uninfected cells.

The tissue culture preparations containing BFA were harvested from cell lines infected with Friend virus in three different laboratories. Thus, SC-1 (M-16) denotes SC-1 cells infected with Friend virus plasma in our laboratory (W. D. Hankins, unpublished) that produced SFFV, MuLV, and BFA (Table 13.1). Cell lines 1902 and EY-26 were obtained from Dr. David Troxler at the National Institutes of Health, Bethesda, Md. (18). The 1902 cells do not produce infectious SFFV or MuLV, but have the SFFV genome integrated into cellular genetic material. Upon infection of 1902 cells with MuLV-F, the SFFV genome was rescued and the superinfected cells (EY-26) released infectious SFFV, MuLV, and BFA (Table 13.1) into the supernatant. The FP52 cells also produced an entity that induced erythroid burst formation (Table 13.1). This cell line, characterized by Dr. Alan Bernstein and co-workers (3), produces SFFV in excess of MuLV. Thus, BFA was not a unique property of SFFV passaged in our laboratory but was also found in SFFV preparations from two other laboratories.

Whereas the supernatants from all the cell lines infected with the Friend virus complex (SFFV and MuLV) induced burst formation, tissue culture fluids

TABLE 13.1 An *In Vitro* System for Erythroid Transformation by Friend Leukemia Virus[a]

TEST PREPARATION	DILUTION	CONTAINS SFFV	CONTAINS MuLV-F[b]	AVERAGE 5-DAY BURSTS/ CLOT
Experiment I(infectious plasma):				
M19 plasma	1:20	Yes	Yes	23
M18 plasma	1:20	Yes	Yes	30
M19 plasma treated, 56° C × 60 min	1:20	No	n.d.	0
M18 plasma treated, 80° C × 3 min	1:20	No	n.d.	0
Normal BALB/c mouse plasma	1:20	No	n.d.	0
Experiment II (Tissue culture supernatants):				
SC-1 (M16)	Undiluted Undiluted	Yes	Yes	11
FP52[c]		Yes	Yes	23
1902 nonproducer cells[d]	Undiluted	No	No	0
EY26[d]	Undiluted	Yes	Yes	36
SC-1 MuLV-M[e]	Undiluted	No	Yes	0
SC-1 MuLV-F[f]	Undiluted	No	Yes	0

[a]Modified from Hankins et al. (10).
[b]n.d., not determined.
[c]A gift from Dr. Alan Bernstein, described in ref. 30.
[d]A gift from Dr. David Troxler, described in ref. 18.
[e]SC-1 cells were infected with Moloney leukemia virus obtained from the American Type Culture Collection.
[f]SC-1 cells were infected with FV plasma that had been diluted sufficiently so that no SFFV could be detected in the spleen focus-forming assay. A supernatant from these cells was absorbed to SC-1 cells at a dilution of 10^{-8}. The resulting cells produced 10^4 to 10^5 PFU/ml and no splenic foci could be detected up to 2 months after injection of undiluted supernatant into BALB/c mice.

from SC-1 cells infected with Moloney leukemia virus or purified Friend MuLV had no BFA, although these preparations were active in the XC plaque assay.

The sedimentation properties of RNA tumor viruses are well known. To determine the sedimentation properties of the BFA, infectious plasma was layered over a 15 to 60% sucrose density gradient, which was then centrifuged at 4°C for 16 hr at 45,000 rpm in an SW-50.1 rotor. The gradient was fractionated, and the percent sucrose was determined by measuring the refractive index of selected fractions. Each fraction was assayed for BFA and for activity in the XC and spleen focus-forming assays. The results are shown in Figure 13.2. The BFA was found to co-sediment with the SFFV and MuLV activity on this linear sucrose equilibrium gradient sedimentation. As expected for most RNA tumor viruses, all three activities were recovered between 1.14 and 1.16 g/cc.

Characteristics of Burst Formation

To examine the relation between virus dose and erythroid burst formation, a constant number of cells was incubated with increasing dilutions of infectious plasma. Figure 13.3 shows that the number of erythroid bursts increased with increasing concentrations (decreasing dilutions) of infectious plasma. The linear portion of the dose-response curve can be extrapolated back through the origin. Similar dose-response curves are observed if one infects with increasing dilutions of tissue culture-derived virus instead of infectious plasma (data not shown). Increasing the plasma concentration or tissue culture fluid in the inoculum ultimately led to a plateau or, in some experiments, an inhibition of burst formation.

Figure 13.4 depicts an experiment in which increasing numbers of cells were incubated with a constant volume of diluted infectious plasma and then seeded in plasma clot cultures. As expected, the number of erythroid bursts were increased as the number of cells increased. The observed sigmoid relationship suggests that burst formation is more efficient when higher cell numbers were used. Whether this increased efficiency was produced by better conditions for virus-cell interactions or cell-cell interaction is not known. On the basis of these results, we chose to incubate routinely 4×10^6 cells

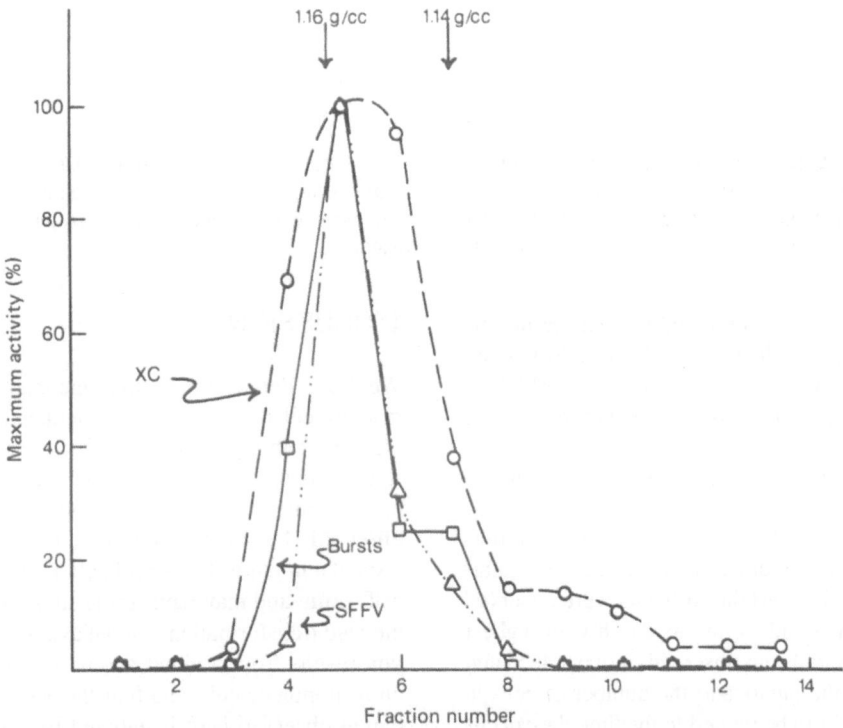

FIGURE 13.2. Equilibrium sedimentation of burst-forming activity: 0.5 ml of infectious plasma (5×10^5 FFU/ml) was layered over a 5-ml linear sucrose gradient 15 to 60% w/v and centrifuged for 16 hr at 45,000 rpm at 4°C in a Beckman SW50.1 rotor; 0.4 ml fractions were collected from the bottom and after 1:50 dilution in αMEM, assayed for BFA (□——□) (see Methods and Materials), spleen focus-forming activity (Δ–··–··Δ) (2), or XC plaque-forming activity (16) (o–––o). Densities were determined by refractive index measurements on selected fractions.

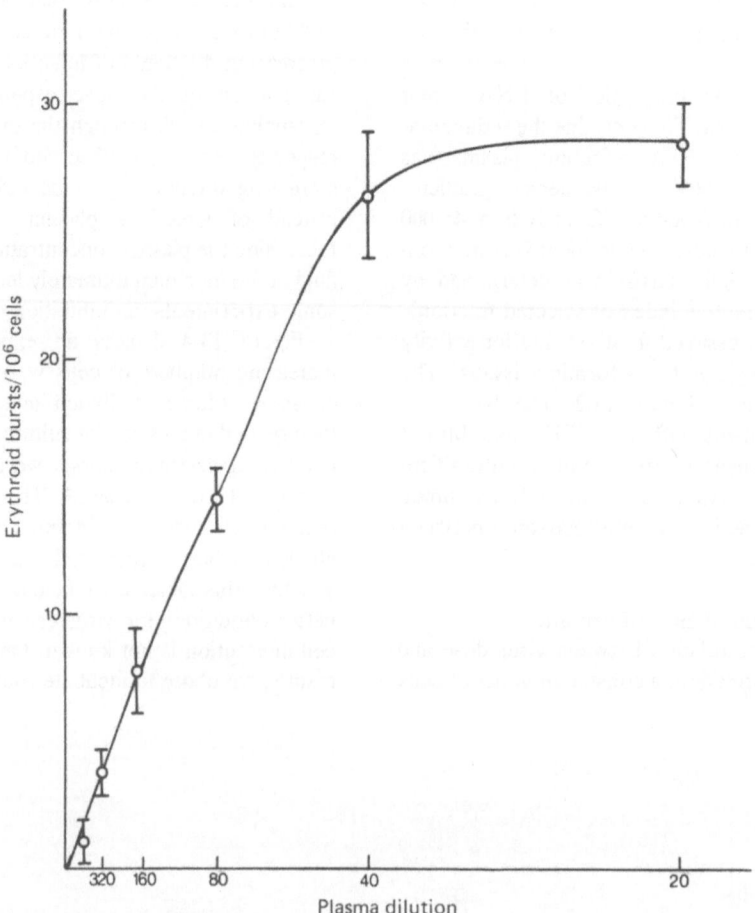

FIGURE 13.3. Virus dose and erythroid burst formation: A cell pellet containing 4×10^6 cells was resuspended in 200 μl of the infectious plasma (5×10^5 FFU/ml) diluted in α-medium as indicated. After incubation for 1 hr, the other plasma clot components were added, and the resulting cell suspensions cultured as described in Methods and Materials.

with 300 μl of virus solution and to seed the plasma clot cultures at a cell density of 10^6 cells/0.5 ml. Although slightly higher numbers were obtained above this density, the erythroid bursts formed were less well defined and more difficult to count.

Figure 13.5 shows a time-course of virus-induced burst formation. Whereas, in most of the experiments described in this report the erythroid bursts have been scored on day 5 after incubation of marrow cells with virus, erythroid bursts were observed as early as day 3 and as late as day 8 with highest numbers of bursts being observed on days 4, 5, and 6. The size of the bursts and the number of cells in the bursts appear to be related to the time they occur post-infection. In general, the bursts observed on day 3 are smaller than those on days 4 to 7, with the bursts observed on day 8 being the largest. The distribution of burst size could represent a continuum of stages in burst formation or infection of target cells with different proliferative potentials.

DISCUSSION

We have demonstrated that incubation of mouse marrow cells with the Friend leukemia virus complex and subsequent culture in the plasma clot culture system leads to the production of erythroid colony bursts. Maximum numbers of bursts are observed, 4 to 6 days after virus infection. Although a small amount of EP is probably present in the fetal calf serum and may have a role subsequent to virus-induced transformation, we believe that burst formation results from a direct action of the Friend virus on hemopoietic cells and that the erythroid differentiation observed here is initiated by a virus and not EP.

Three experimental results support this. First, BFA is found not only in infectious plasma from Friend virus-infected mice, but also in the tissue culture supernatants of several different cell lines infected with Friend virus that release SFFV into the

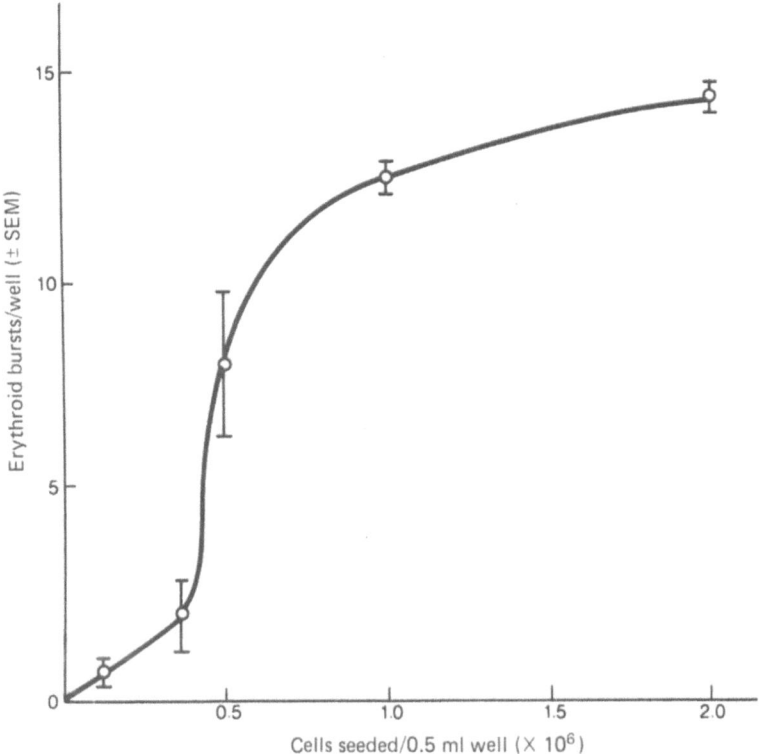

FIGURE 13.4.   Relation of cell density to burst formation: 300 μl of a 1:20 dilution of infectious plasma (5 × 10⁵ FFU/ml) was used to resuspend various numbers of bone marrow cells. Sufficient cells were added during the incubation to yield the final number of cells (0.5 ml/well) indicated on the X-axis after all the plasma clot components had been added. Cultures were treated as described in Methods and Materials.

medium. Normal plasma and tissue culture supernatants from control, uninfected cell lines had no BFA. Although it could be argued that infectious plasma simply contained more EP than normal plasma, it is very unlikely that three different lines of fibroblasts all produced EP. Second, the BFA was demonstrated to be heat labile and it has been previously shown that EP is heat stable even after incubation at 100°C (4). Third, after equilibrium density centrifugation, the BFA was recovered at the same density as SFFV and MuLV. When centrifuged under identical conditions, EP was recovered near the top of a 15 to 60% sucrose gradient (W. D. Hankins, unpublished data).

Virus-induced burst formation was dose dependent, with the number of bursts formed in plasma clots increasing as a linear function of the proportion of infectious plasma included in the cell suspension during the virus absorption period. The reason for the plateau at low plasma dilutions is not clear. An exhaustion of target cells is one possible explanation. But, since a decrease in burst formation occurs in some experiments as the concentration of virus is increased beyond the dose at the plateau, it is more

likely that a contaminant may be present in the virus preparation and may, through general toxicity, decrease burst formation at higher concentrations. Another contributing factor may be a direct toxic effect of the increased virus concentration.

In the dose-response experiment presented in Figure 13.3, 40 FFU produced an average of 1 burst/ 10⁶ cells. In other experiments, we have observed burst formation with as few as 3 FFU/10⁶ cells (unpublished data, authors). Since the percentage of target cells, contaminating inhibitors in the test preparation, and direct toxicity of the virus may affect the efficiency of burst formation, the ratio of bursts to FFU may vary among experiments. Studies are now underway in our laboratory to establish a standard set of conditions for virus burst formation in order to determine if an *in vitro* assay for the spleen focus-forming virus is feasible.

The time-course of virus-induced burst formation is interesting. We have observed erythroid bursts as early as day 3 and as late as day 8 with the highest number of bursts appearing on days 4, 5, and 6. Burst formation then defines a wave of erythropoiesis that occurs *in vitro* with a time-course that is

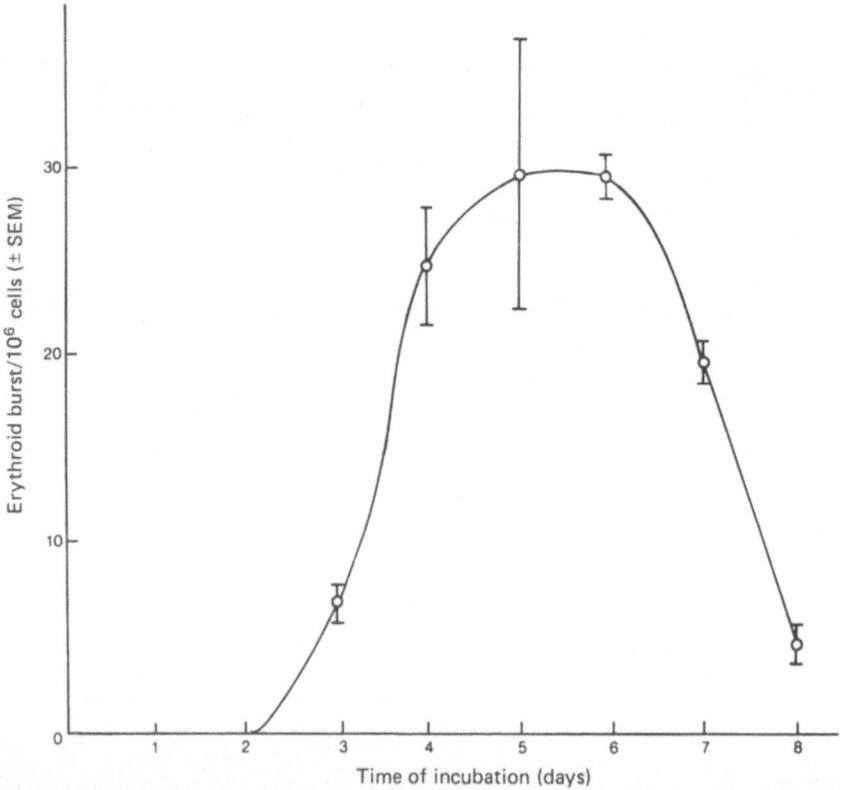

FIGURE 13.5. Time of appearance of virus-induced erythroid bursts: 4×10^6 bone marrow cells were incubated for 2 hr with 300 μl of a 1:20 dilution of infectious plasma (5×10^5 FFU/ml) and subsequently cultured in plasma clots as described in Methods and Materials. Clots were removed at the indicated times.

similar to the wave of hemoglobin synthesis we previously observed in an -in vivo-in vitro culture system (9,11). In those studies, spleen cells were removed from mice previously infected with Friend virus and a wave of hemoglobin synthesis was observed when these cells were cultured *in vitro*. The time required between infection *in vivo* and and the peak of hemoglobin synthesis *in vitro* was 4 to 5 days. Since these two aspects of Friend virus-induced erythropoiesis are monitored in two different culture systems, we cannot be sure that both represent the same biologic effect. Our recent unpublished studies using the methyl cellulose culture system, however, suggests that this may be the case. With this system, we find that both burst formation and incorporation of ^{59}Fe into hemoglobin follow a similar time-course with peaks occurring 4 to 6 days after virus infection.

The data taken together suggest that it is possible to define at least one aspect of transformation by the Friend leukemia virus complex. Transformation would include the early interaction of Friend virus with marrow cells that initiates a sequence of events leading to a wave of erythropoiesis, with a peak 4 to 6 days subsequent to virus infection. In contrast to uninfected cells, Friend virus-transformed cells can proliferate and differentiate to the point of synthesizing hemoglobin *in vitro*. Since exogenous EP is not added to these cultures, the transformed cells either differentiate without EP or they are hypersensitive to the minute quantities that may be present in the fetal calf serum or other medium components. With the use of burst formation as an end point for virus-induced erythroid transformation, the earlier events of this transformation can now be studied *in vitro*.

ACKNOWLEDGMENTS

We thank Paul Matthai and Anne Brockman for expert technical assistance. This work was supported by the National Institutes of Health (AM-15555 and T32 AM07186) and by Veterans Administration Research Funds.

REFERENCES

1. Axelrad, A. A., and Steeves, R. A. Assay for Friend leukemia virus: Rapid quantitative method based on enumeration of macroscopic spleen foci in mice. *Virology, 24*:513, 1964.
2. Axelrad, A. A., McLeod, D. L., Shreeve, M. M., and Heath, D. S. Properties of cells that produce erythrocytic colonies *in vitro*. In Robinson, W. A., ed., *Second International Workshop in Hemopoiesis in Culture*. p. 226.
3. Bernstein, A., Mak, T. W., and Stephenson, J. R. The Friend virus genome: Evidence for the stable association of MuLV sequences and sequences involved in erythroleukemic transformation. *Cell, 12*:287, 1977.
4. Borsook, H. A discussion of humoral erythropoietic factors. *Ann. N.Y. Acad. Sci., 77*:725, 1959.
5. Clarke, B. J., Axelrad, A. A., Shreeve, M. M., and McLeod, D. L. Erythroid colony induction without erythropoietin by Friend leukemia virus *in vitro*. *Proc. Nat. Acad. Sci., 72*:3556, 1975.
6. Dawson, P. J., Rose, W. M., and Fieldsteel, A. H. Lymphatic leukaemia in rats and mice inoculated with Friend virus. *Brit. J. Cancer, 20*:114, 1966.
7. Friend, C. Cell-free transmission in adult Swiss mice of a disease having the character of a leukemia. *J. Exp. Med., 105*:307, 1957.
8. Gregory, C. J. Erythropoietin sensitivity as a differentiation marker in the hemopoietic system: Studies of three erythropoietic colony responses in culture. *J. Cell Physiol., 89*:289, 1976.
9. Hankins, W. D., and Krantz, S. B. *In vitro* expression of erythroid differentiation induced by Friend polycythaemia virus. *Nature, 253*:731, 1975.
10. Hankins, W. D., Vessell, D., and Krantz, S. B. The effect of Friend polycythemia virus on splenic uptake of ^{59}Fe in fasted mice. *Proc. Soc. Exp. Biol. Med., 142*:829, 1973.
11. Hankins, W. D., Rosenblatt, P., and Krantz, S. B. Splenic erythroid response to Friend polycythemia virus: Time course *in vitro* after infection *in vivo*. *J. Nat. Cancer Inst., 59*:107, 1977.
12. Iscove, N. N., and Sieber, F. Erythroid progenitors in mouse bone marrow detected by macroscopic colony formation in culture. *Exp. Hemat., 3*:32, 1975.
13. Mirand, E. A. Murine viral-induced polycythemia. *Ann. N.Y. Acad. Sci. 149*:486, 1968.
14. McLeod, D. L., Shreeve, M. M., and Axelrad, A. A. Improved plasma culture system for production of erythrocytic colonies *in vitro*: Quantitative assay method for CFU-E. *Blood,4 -4*:577, 1974.
15. Rowe, W. P., Pugh, W. E., and Hartley, J. W. Plaque assay techniques for murine leukemia viruses. *Virology, 42*:1136, 1970.
16. Spivak, J. L., Marmor, J., and Dickerman, H. W. Studies on splenic erythropoiesis in the mouse. I Ribosomal ribonucleic acid metabolism. *J. Lab. Clin. Med., 79*:526, 1972.
17. Steeves, R. A., Eckner, R. J., Bennett, M., Mirand, E. A., and Trudel, P. J. Isolation and characterization of a lymphatic leukemia virus in the Friend virus complex. *J. Nat. Cancer Inst., 46*:1209, 1971.
18. Troxler, D. H., Parks, W. P., Vass, W. C., and Scolnick, E. M. Isolation of a fibroblast nonproducer cell line containing the Friend strain of the spleen focus-forming virus. *Virology, 76*:602, 1977.

14

Response of Megakaryocyte, Erythroid, and Granulocyte-Macrophage Progenitor Cells in Mouse Bone Marrow to Gamma-Irradiation and Cyclophosphamide

A. Nakeff, W. L. McLellan,
J. Bryan, and F. A. Valeriote

In vitro culture systems have recently been established for supporting the growth of megakaryocyte colonies from mouse bone marrow in either semisolid plasma or agar (7,8,12,19). This advance has permitted the first quantitative assay for a class of megakaryocyte progenitors entitled the colony-forming unit, megakaryocyte, or CFU-m (12). Although CFU-m are obviously important because they are the only megakaryocyte progenitors capable of proliferation (i.e., mitosis), their physiologic role in new negakaryocyte production *in vivo* remains to be more fully defined (10,11).

This report presents data defining new properties characteristic of marrow CFU-m from studies of their physiologic response to perturbations induced by a series of agents, including whole-body irradiation and specific chemotherapeutic agents. Responses elicited in the pluripotential stem cell (CFU-s) and clonogenic progenitors committed to granulocyte-macrophage (CFU-gm) and late erythroid (CFU-e3) differentiation are also compared in order to better understand the role of CFU-m in the overall scheme of hemopoietic cell development.

METHODS AND RESULTS

Radiation Dose Response

Groups of male B6D2F$_1$ mice 8 to 10 weeks of age were exposed to increasing doses of wholebody cesium-137 gamma rays delivered at a dose rate of 112 rad/min in a cesium irradiator (2). Twenty-four hours later, groups of three mice each were killed at each dose and the marrow from a single femur pooled and assayed for CFU-s by the spleen colony technique (16) and CFU-m, CFU-gm, and CFU-e3 by the plasma culture technique as described previously (12,13). Total nucleated cell counts were also obtained. The number of colony-forming units was calculated for each femur and expressed as a fraction of control (unirradiated) values.

The dose-survival curve for CFU-m presented in Figure 14.1 showed an exponential decrease in colony-forming ability with increasing dose and no clear evidence of a shoulder. The average D_0 value obtained by linear regression analysis was 128 rad, making CFU-m somewhat more radioresistant than CFU-s with a D_0 of 86 rad and CFU-e3, which were the most radiosensitive of the four clonogenic progenitors, with a $D_0 = 66$ rad. Interestingly, CFU-gm displayed the same radiation sensitivity as CFU-m with a D_0 of 128 rad.

Radiation Recovery

To determine the recovery pattern of the different progenitors as a function of time after irradiation, mice were exposed to a single dose of 500 rad. Then groups of three mice each in each experiment were

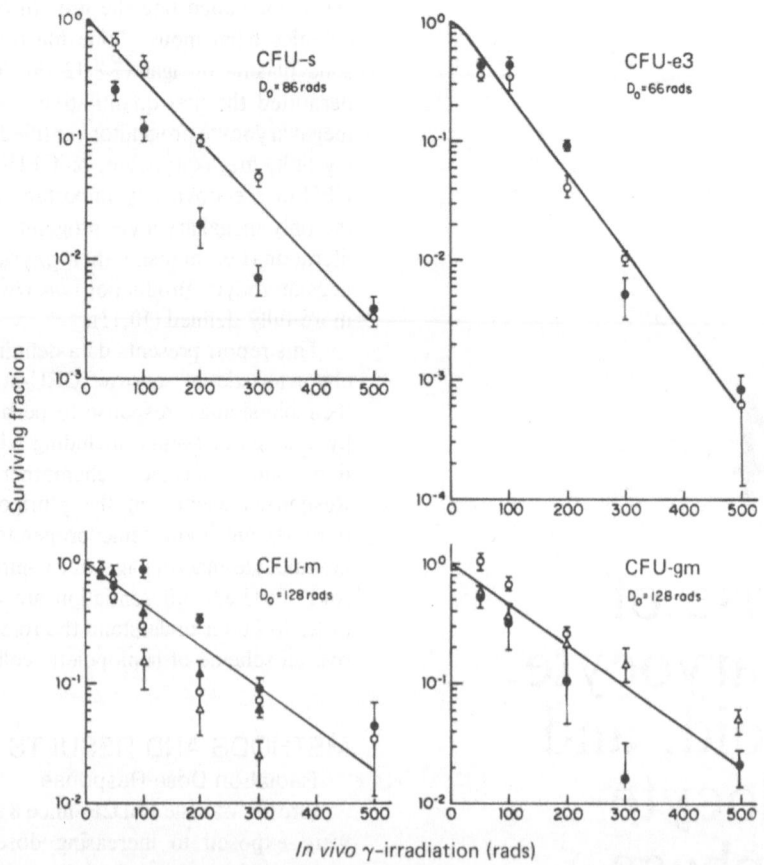

FIGURE 14.1. Dose-response curves of colony-forming cells in mouse femoral marrow to gamma irradiation *in vivo*. Assays were performed 24 hr after irradiation. The different symbols denote different experiments and errors shown ±1 SE. Marrow for each point was pooled from three mice. Each value obtained in culture was the mean of four to six dishes and each CFU-s value was the mean of 10 to 15 mice.

killed at various times thereafter, for up to 14 days. Their marrows were pooled and the total femoral content of each of the different progenitors determined as described previously and normalized to control values in unirradiated mice killed at the same time.

As shown in Figure 14.2, the nadir of the CFU-m was about 2% of control at day 2, followed by a delayed recovery that was still not complete by day 14. This pattern of recovery was similar to that observed for CFU-s and CFU-gm but different from that observed for CFU-e3. Although CFU-e initially decreased almost four decades, recovery was extremely rapid, reaching preirradiation levels by day 7.

Cyclophosphamide Dose Response

To determine the specificities of the previous responses of the clonogenic progenitors to ionizing radiation, we studied the same parameters in mice

after injection of the alkylating agent cyclophosphamide.

Mice were injected via the tail vein with graded doses of cyclophosphamide (Mead Johnson) in a volume of 0.5 ml saline. Dose-response relationships were determined 24 hr later for each of the progenitors as described previously. As presented in Figure 14.3, the dose-survival curve for CFU-m was similar to that for CFU-s and CFU-gm with a dose of 4 mg/ mouse (approximately 200 mg/kg body weight) resulting in a decrease of about a decade. The CFU-e3 were somewhat more sensitive.

Recovery Following Cyclophosphamide

The pattern of recovery for each of the progenitors is shown in Figure 14.4 following the intravenous injection of 2 mg/mouse of cyclophosphamide. Recovery was much more rapid for all the progenitors as compared to that following gamma-irradiation although the initial decrease was less than that

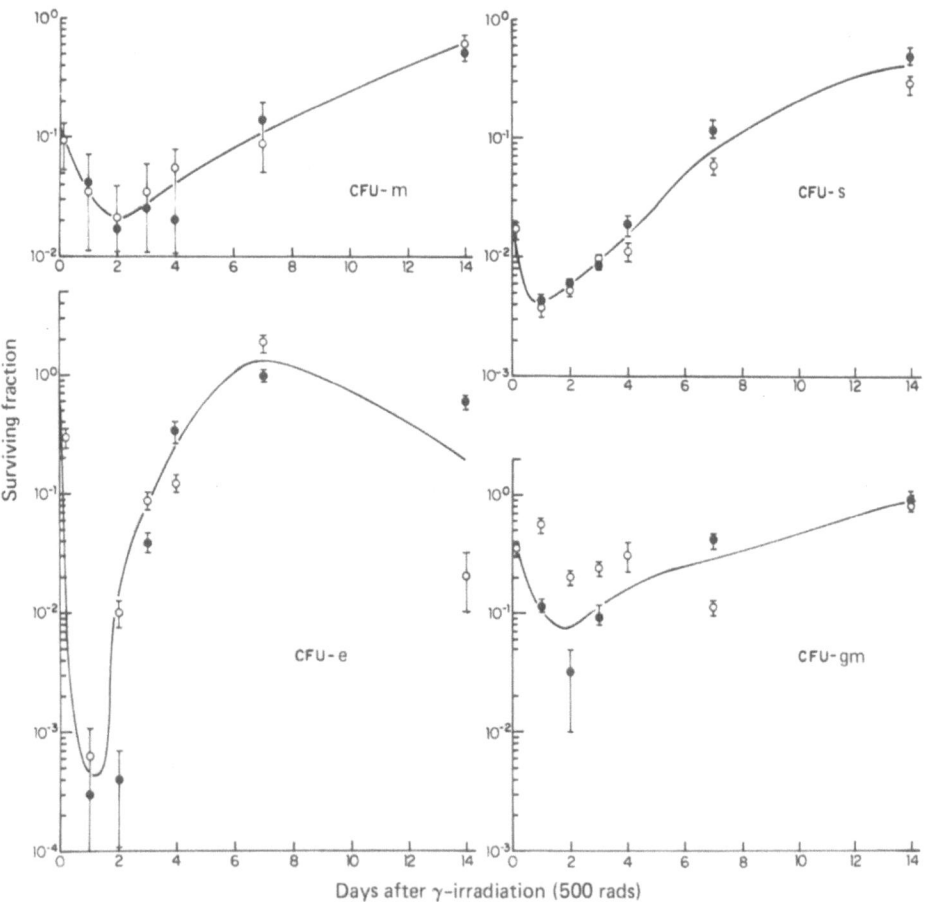

FIGURE 14.2. Response of colony-forming cells in mouse femoral marrow as a function of time after 500 rad wholebody gamma irradiation. Other information as in Figure 14.1.

observed following 500 rad. All three committed progenitors reached pretreatment levels within 2 to 3 days, with CFU-s exhibiting an aborted recovery on day 4.

DISCUSSION

These studies provide the first radiation dose-survival curve for mouse femoral marrow CFU-m. They demonstrate that CFU-m exhibit a delayed recovery from radiation damage similar to that seen for CFU-s and CFU-gm, although it was not observed following treatment with cyclophosphamide.

When the differences in irradiation and culture conditions and assay times and procedures are taken into consideration, the D_0 values obtained in this study for mouse marrow CFU-s, CFU-gm, and CFU-e3 generally agree with those in the literature. The previously reported D_0 value for *in vivo* irradiation of mouse marrow CFU-s was 95 rad by

McCulloch and Till (6). Reported D_0 values for *in vitro* irradiation of CFU-gm were 95 rad by Chen and Schooley (1) and 160 rad by Senn and McCulloch (14). Lin (4) reported a D_0 for mouse marrow CFU-gm of 135 rad and 124 rad irradiated *in vitro* and *in vivo*, respectively. Senn and McCulloch (14) also showed that human CFU-gm were slightly less radiation sensitive than mouse CFU-gm under the same conditions, with a D_0 of 137 rad. The CFU-e3 in mouse marrow have been reported by Monette et al. (9) to be radiation sensitive (D_0 70 rad). Tepperman et al. (15) measured a D_0 of 113 rad for human CFU-e, although it was not clear whether this assay may have included erythroid progenitors more analogous to mouse BFU-e; the latter may be more radiation resistant than the more differentiated CFU-e3, resulting in a correspondingly higher D_0. The agreement between our measurements of the radiation response of CFU-s, CFU-gm, and CFU-e3 and those previously published provides a measure of confidence in the D_0 value of 128 rad obtained for CFU-m in this study. Although human CFU-m cannot pres-

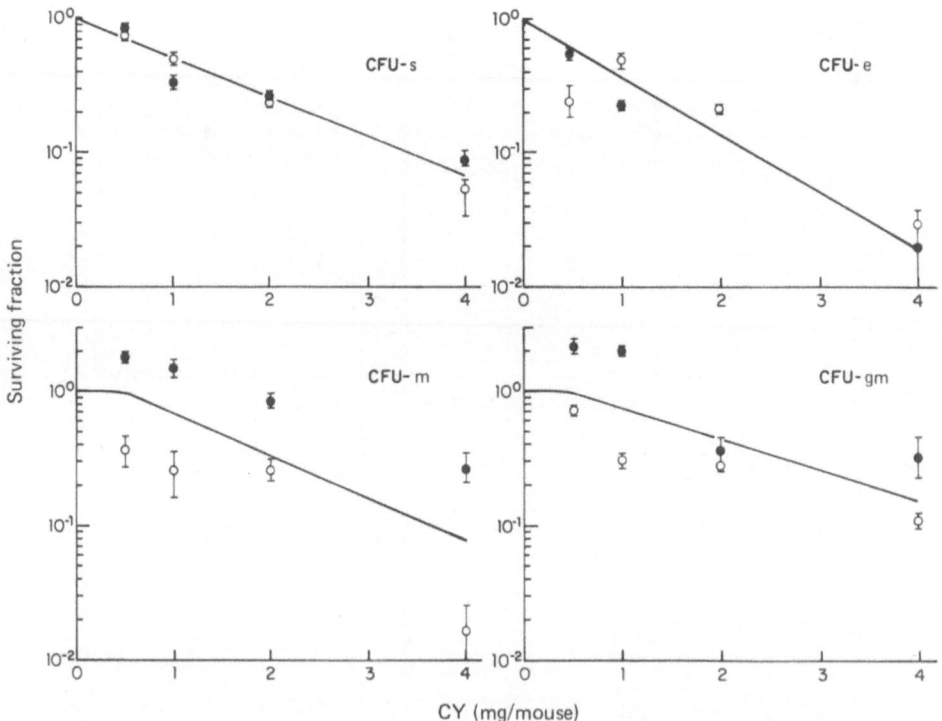

FIGURE 14.3. Dose-response curves for colony-forming cells in mouse femoral marrow to cyclophosphamide injected intravenously. Assays were performed 24 hr after injection. Control animals received saline. Other information as in Figure 14.1.

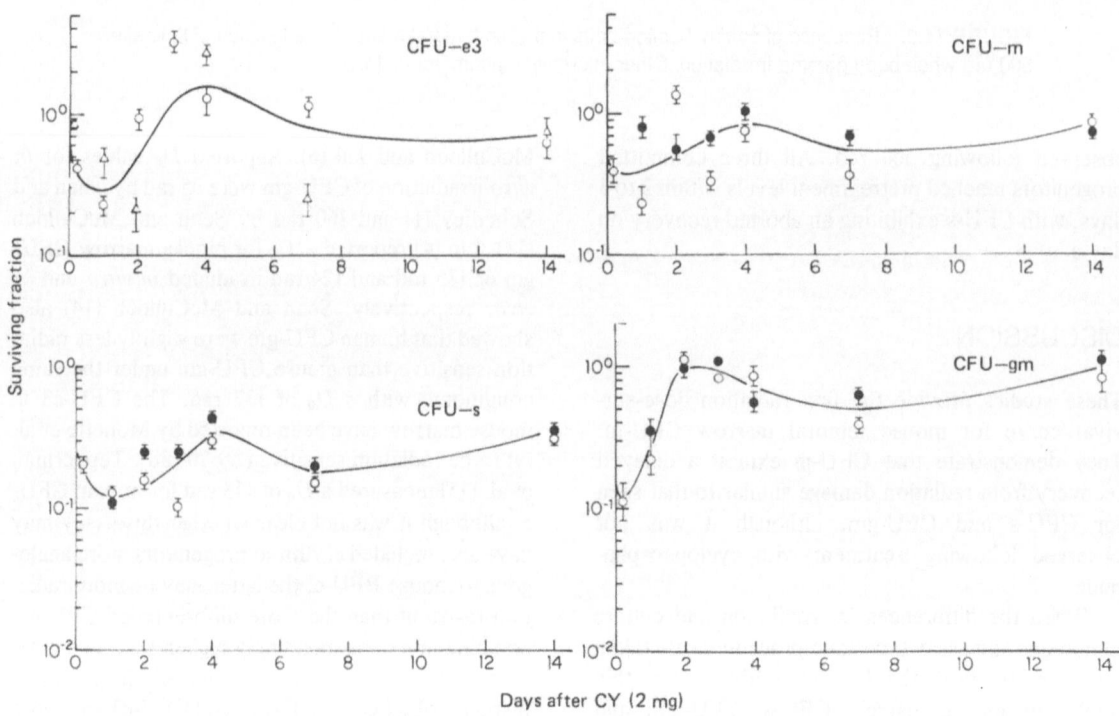

FIGURE 14.4. Response of colony-forming cells in mouse femoral marrow as a function of time after a single dose of 2mg of cyclophosphamide. Other information as in Figure 14.1.

ently be cultured, an analogy with CFU-gm and the similarity between D_0 values for this progenitor in mouse as compared to human marrow (14) indicates that the D_0 value for human marrow CFU-m may not be very different from that reported here for mouse marrow.

The similarity in the shape and D_0 of the radiation dose-survival curves for CFU-m and considered with the similarity in the radiation recovery between these progenitors and CFU-s, provides evidence that CFU-m are relatively immature progenitors that are probably closely linked developmentally to CFU-s. The delayed recovery of CFU-m indicates that these progenitors may be somewhat restricted in their proliferative response, with maintenance of their normal cellularity depending on an influx of new progenitors from the CFU-s compartment. To determine to what extent the recovery of CFU-m involves recruitment of CFU-m into the cell cycle, we plan to use hydroxyurea and cytosine arabinoside as proliferative probes to measure the fraction of CFU-m in S phase at different times throughout the recovery period. These data, together with a more extensive analysis including changes in the megakaryocyte and platelet compartments, will be the subject of a subsequent publication.

At present, we do not know why the CFU-e3 were the most radiosensitive of all the progenitors since they recovered so rapidly. It is known that CFU-e3 are the most differentiated of the four progenitors in this study and are preceded by at least two erythroid progenitor populations (3). To what extent the latter contribute to the recovery of CFU-e3 is not clear, since data are not available on their response to ionizing radiation. These compartments could conceivably provide a large influx of CFU-e3 following irradiation, which initially may be independent of erythroid differentiation from CFU-s.

The dose- and time-response relationships of CFU-s, CFU-gm, and CFU-e3 to cyclophosphamide were similiar to those described previously (5,17,18) when differences in mouse strain, culture conditions, and route of administration of the drug were taken into account. The CFU-m were not remarkably different in these respects from the other progenitor cells.

The reason for the difference in the rates of recovery of progenitors following gamma irradiation or cyclophosphamide remains to be determined. The delayed recovery of CFU-s following irradiation and the rapid recovery following cyclophosphamide are similiar to those reported previously (17). This difference in recovery rate has now been shown to be a property of not only the CFU-s, but also of progenitors committed to megakaryocyte and granulocyte-macrophage differentiation. The latter are assayed under conditions that are very different than that for CFU-s and respond uniquely to different environmental factors and regulation. We still do not know if this difference reflects a different mode of action of the agents on the progenitors or a difference in their environment following treatment. We believe that the difference in CFU-m response reported here is not simply due to a difference in the degree to which the CFU-m initially decreased, since when mice were given 6 mg of cyclophosphamide (data not shown), which reduced their CFU-m at 24 hr to a level approaching that observed after 500 rad, recovery was still complete by day 4. Also, the difference in recovery of CFU-s following radiation and cyclophosphamide could not be explained in this manner, since the difference persisted even when the initial starting values of CFU-s were the same (17).

This study, together with others (5,17,18), shows that not only do hemopoietic progenitors respond differently to the same agent but they also respond differently to different agents, for reasons that are not clearly understood. Nevertheless, it is important to consider the consequences of these differences in the scheduling of anticancer chemotherapy.

SUMMARY

Dose- and time-survival curves were compared for different progenitor cells (CFU-s, CFU-m, and CFU-e3) in mouse femoral marrow following either wholebody gamma irradiation or treatment with cyclophosphamide. The first radiation dose-survival curve for CFU-m obtained was characterized by a D_0 of 128 rad, which was the same as that for CFU-gm but higher than that for CFU-s (D_0 86 rad) and CFU-e3 (D_0 66 rad). A delayed recovery that was not complete by 14 days after 500 rad of gamma radiation was observed that was similar for CFU-m, CFU-s, and CFU-gm in contrast to that for CFU-e3, which was substantially greater and more rapid. Recovery following the administration of 2 mg of cyclophosphamide was similar for all four progenitors and much more rapid than that following gamma irradiation, being complete by 4 days. These data define characteristics specific to each progenitor class and have important consequences for the scheduling of anticancer chemotherapy.

ACKNOWLEDGMENTS

We wish to thank Mrs. Donna Troeckler for typing this manuscript.

This work was supported by Grant CA 13053 from the National Cancer Institute and Grants HL

20826 and Research Career Development Award HL 00440 (A.N.) from the National Heart, Lung and Blood Institute, U.S. Public Health Service.

REFERENCES

1. Chen, M. G., and Schooley, J. C. Recovery of proliferative capacity of agar-colony-forming cells and spleen colony-forming cells following ionizing radiation or vinblastine. *J. Cell. Physiol.* 75:89, 1970.

2. Cunningham, J. R., Bruce, W. R., and Webb, H. P. A convenient ^{137}Cs unit for irradiating cell suspensions and small animals. *Phys. Med. Biol., 10*:381, 1965.

3. Gregory, C. J. Erythropoietin sensitivity as a differentiation marker in the hemopoietic system: Studies of three erythropoietic colony responses in culture. *J. Cell Physiol., 89*:289, 1976.

4. Lin, H. Peritoneal exudate cells. III Effect of gamma-irradiation on mouse peritoneal colony-forming cells. *Radiat. Res., 63*:560, 1975.

5. Marsh, J. C. The effects of cancer chemotherapeutic agents on normal hematopoietic precursor cells: A review. *Cancer Res., 36*:1853, 1976.

6. McCulloch, E. A., and Till, J. E. The sensitivity of cells from normal mouse bone marrow to gamma-radiation *in vitro* and *in vivo. Radiat. Res., 16*:822, 1962.

7. McLeod, D. L., Shreeve, M. M., and Axelrad, A. A. Induction of megakaryocyte colonies with platelet formation *in vitro. Nature, 261*:492, 1976.

8. Metcalf, D., MacDonald, H. R., Odartchenko, N., and Sordat, B. Growth of mouse megakaryocyte colonies *in vitro. Proc. Nat. Acad. Sci. U.S., 72*:1755, 1975.

9. Monette, F. C., Kent, R. B., Weiner, E. J., Lavanis, E. C., and Lydon, P. J. Cell cycle properties and proliferation kinetics of late erythroid progenitor cells. (Abst.) *Exp. Hematol. 6*(Suppl. 3):12, 1978.

10. Nakeff, A. Colony-forming unit, megakaryocyte (CFU-M): Its use in elucidating the kinetics and humoral control of the megakaryocytic committed progenitor cell compartment. In Baum, S. J., and Ledney, G. D., eds. *Experimental Hematology Today 1977.* New York: Springer-Verlag, 1977, p. 111.

11. Nakeff, A., and Bryan, J. E. Megakaryocyte proliferation and its regulation as revealed by CFU-M analysis. In Golde, D. W., Cline, M. J., Metcalf, D. and Fox, C. F., eds. *Hematopoietic Cell Differentiation* (ICN-UCLA Symposia on Molecular and Cellular Biology Vol. X). New York: Academic Press, 1978.

12. Nakeff, A., and Daniels-McQueen, S. *In vitro* colony assay for a new class of megakaryocyte precursor: Colony-forming unit, megakaryocyte (CFU-M). *Proc. Soc. Exp. Biol. Med., 151*:587, 1976.

13. Nakeff, A., and Valeriote, F. A. Use of *in vitro* cell colony assays for measuring the cytotoxicity of chemotherapeutic agents to hematopoietic progenitor cells committed to myeloid, erythroid and megakaryocytoid differentiation. In Action, R., ed., *Cell Culture and its Applications.* New York: Academic Press, 1977, p. 433.

14. Senn, J. S., and McCulloch, E. A. Radiation sensitivity of human marrow cells measured by a cell culture method. *Blood, 35*:56, 1970.

15. Tepperman, A. D., Curtis, J. E., and McCulloch, E. A. Erythropoietic colonies in cultures of human marrow. *Blood, 44*:659, 1974.

16. Till, J. E., and McCulloch, E. A. A direct measurement of the radiation sensitivity of normal mouse bone marrow cells. *Radiat. Res., 14*:213, 1961.

17. Valeriote, F. A., Collins, D. C., and Bruce, W. R. Hematological recovery in the mouse following single doses of gamma radiation and cyclophosphamide. *Radiat. Res., 33*:501, 1968.

18. Valeriote, F. A., and Nakeff, A. Effects of anti-cancer agents on hematopoietic progenitor cells. In Silber, R. D., Gordon, A. S., and LoBue, J., eds. *The Year in Hematology, 1978,* Vol. II, New York: Plenum, in Press.).

19. Williams, N., Jackson, H., Sheridan, A. P. C., Murphy, M. J. Jr., Elste, A., and Moore, M. A. S. Regulation of megakaryopoiesis in long-term murine bone marrow cultures. *Blood, 51*:245, 1978.

15

Treatment of Strontium-89-Induced Hypoplastic Anemia in Mice by Intraperitoneal Implants of Spleen Slices

C. W. Gurney, J. W. Lyon,
E. L. Simmons, E. O. Gaston,
and B. A. Malcolm

Strontium-89 (^{89}Sr), a bone-seeking, beta-emitting isotope, causes marrow damage in proportion to the dose administered. With progressively severe marrow aplasia as the ^{89}Sr dose is increased, extramedullary hemopoiesis in the spleen is increased, and essentially normal erythrocyte and thrombocyte levels are maintained (4). Following splenectomy in such animals, however, hypoplastic anemia (pancytopenia) ensues, and if the ^{89}Sr dosage exceeds 2.5 μCi/g, death from hemopoietic failure is common. The mild anemia produced by a small dose of ^{89}Sr and splenectomy can be corrected by weekly injections of testosterone enanthate (3). The hematocrits will also rise if such animals are placed in a hypoxia chamber (3), but the anemia cannot be reversed by injections of syngeneic bone marrow. Marrow cells injected into mice may initially fail to reverse the anemia because colonizing cells that lodge in the femur are killed, or their growth is suppressed by the radiation emitted by the ^{89}Sr absorbed in the bone. Since the anemia remains refractory long after most of the isotope has decayed, however, we suspect that the lining of the bone marrow spaces, areas normally rich in primitive hemopoietic precursor cells (2), and the bone marrow stroma may have been altered irreversibly by the ^{89}Sr and thus are no longer a physiologically attractive site for the growth of colonizing cells.

In the present experiments, we performed intraperitoneal implantations of normal spleen slices, or slices x-irradiated *in vitro* immediately before implantation, in order to determine whether these slices would serve as fresh hemopoietic sites under the stress that follows administration of ^{89}Sr and splenectomy. We also studied the hemopoietic consequences of irradiation of the slices immediately before implantation.

MATERIALS AND METHODS

In each experiment, there were at least 10 mice in every group; these were used to provide values from which each average hematocrit or platelet levels were obtained. For statistical analysis of the data, the Students t test and p values were used (5).

Mice
The mice used were BDF$_1$ Cox females, obtained from Laboratory Supply Company, Indianapolis, Ind.

Strontium-89
The ^{89}Sr chloride was obtained from the Oak Ridge National Laboratory, Oak Ridge, Tenn. The stock solution is carried in 0.86 N HCl and is neutralized before intraperitoneal injection.

X-ray Procedure

The spleens to be irradiated were placed in cold Hanks' solution in a Petri dish. A General Electric Maxitron 250 was the source of x-rays. The desired dose was delivered at the rate of 640 R/min at 250 kVp, 30 ma, with a filter of 0.25-mm Cu and 1.00-mm Al. The distance from the source to target was 26.5 cm.

Anesthesia

Surgical removal of the spleen and implantation of spleen slices was performed under Nembutal anesthesia. Pentobarbital Sodium (Abbott Laboratories, North Chicago, Ill.) was diluted with isotonic saline and the dose (70 mg/kg body weight) was administered intraperitoneally.

Hematologic Procedures

Peripheral blood was obtained from the retroorbital venus sinus. The hematocrit reading was determined by the microcapillary method, and the heparinized microhematocrit tubes (Clay-Adams) containing blood were spun in an International Micro-capillary centrifuge, Model MB. The platelet count in a blood suspension was counted with a Coulter Counter Model F, as previously described (1).

RESULTS

In the first experiment, young adult BDF$_1$ female mice were given 2.0 μCi/g body weight of ^{89}Sr intra-peritoneally. Eleven days later, the spleens were removed surgically under Nembutal anesthesia. Before each incision was closed, thin slices from an entire BDF$_1$ spleen were inserted into the peritoneal cavity. These slices were either normal or irradiated with 800 R x-rays *in vitro* in Hanks' solution just prior to implantation. A third group of mice received no implants; these animals served as splenectomy-only controls.

The depression of the hematocrit level 3 weeks after splenectomy was least severe in mice that had received normal spleen slices (Figure 15.1). Hematocrit values in this group rose from 44 to 47% during the next 4 weeks, and this level was subsequently maintained. Hematocrit values in mice that received 800 R slices had fallen to 39% in 3 weeks, but recovered dramatically during the next 3 weeks, to 45%. This level was maintained (being slightly below the hematocrit values of mice that had received non-irradiated slices) for a total of 17 weeks. The fall in hematocrit values was most severe in splenectomized mice that had received no spleen implants, with the lowest point being 34% in 6 weeks, after which there was gradual recovery to 44% at 12 weeks and 47% at 16 weeks.

Hematocrit levels in all groups were essentially identical 5 months after the original splenectomy and surgery. At this time, we subjected the mice to a new hematologic stress by administering a second dose of 2.0 μCi/g ^{89}Sr. In this way, we attempted to determine whether the spleen slices that had been transplanted were still functional and capable of blood

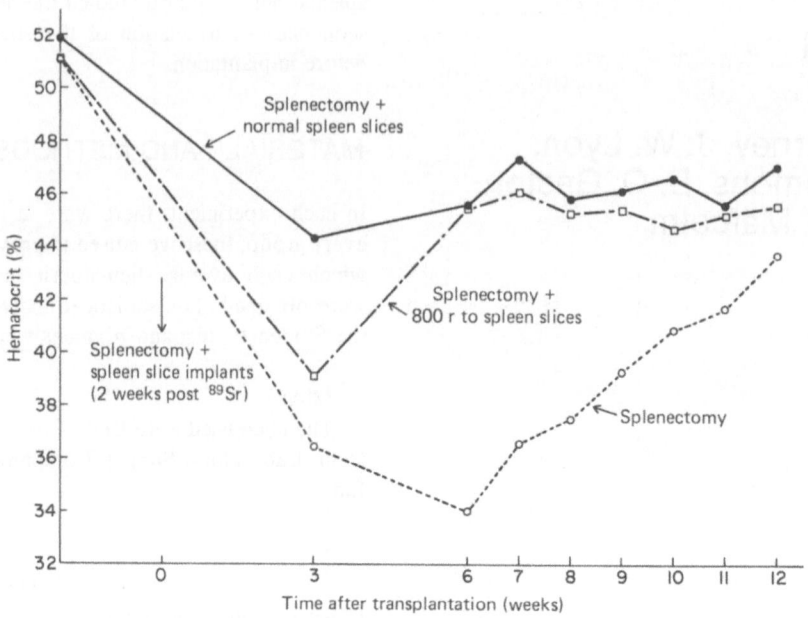

FIGURE 15.1. Changes in hematocrit values of ^{89}Sr-treated mice following splenectomy with and without implantation of spleen slices (BDF$_1$, ♀).

formation. The control splenectomy-only mice were severely affected by this new insult (Figure 15.2). Their hematocrit values fell precipitously to a level of 19% by 5 weeks, and all the mice died. The hematocrit response of the mice with spleen slice implants was reminiscent of the initial phase of the experiment. The fall was less severe and recovery was faster in animals bearing non-irradiated slices, whereas a more severe hematocrit depression from 46 to 38 in 2 weeks followed in mice bearing irradiated spleen slices. Platelet levels were also followed in this phase of the experiment (Figure 15.3), and they were seen to follow the same rank order observed for the hematocrit levels.

In our next experiment, also conducted in BDF_1 female mice, we studied the regenerative ability of spleen slices when the time of their implantation was delayed. The mice were injected with 2.25 ± 0.25 $\mu Ci/g$ of ^{89}Sr and were splenectomized 2 weeks later. Six weeks later, at which time the hematocrit and platelet levels had fallen dramatically below the comparable values of the non-radiostrontium-treated splenectomized controls, two groups of mice were implanted either with normal spleen slices or with slices that had been given 1,200 R x-irradiation.

The hematocrits of mice receiving living spleen slices recovered rapidly from 25% to nearly normal levels of 44% within 3 weeks (Figure 15.4). The dose of 1,200 R prior to implantation apparently prevented support of erythropoietic function by the irradiated spleen slices, since the hematocrits of recipients of irradiated slices were indistinguishable from the values recorded in the sham-operated controls, in which slices were not implanted.

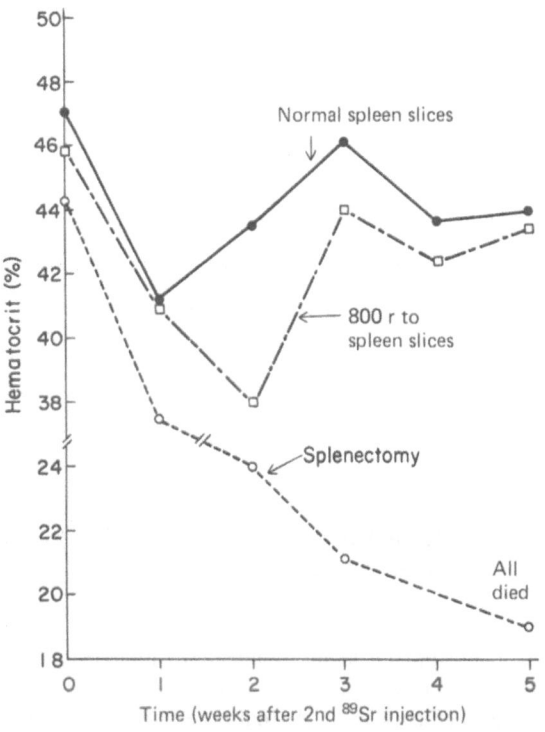

FIGURE 15.2. Effect of intraperitoneal spleen slice implants on the changes in hematocrit values following a second dose of ^{89}Sr (2.0 $\mu ci/g$ 5 months after implantation of spleen slices) in splenectomized mice (BDF_1, ♀).

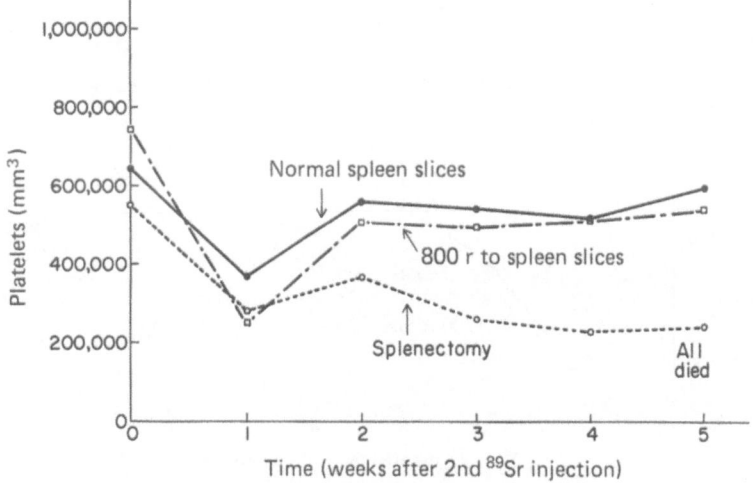

FIGURE 15.3. Effect of intraperitoneal spleen slice implants on the changes in the platelet count following a second dose of ^{89}Sr (2.0 $\mu ci/g$ 5 months after implantation of spleen slices) in splenectomized mice (BDF_1, ♀).

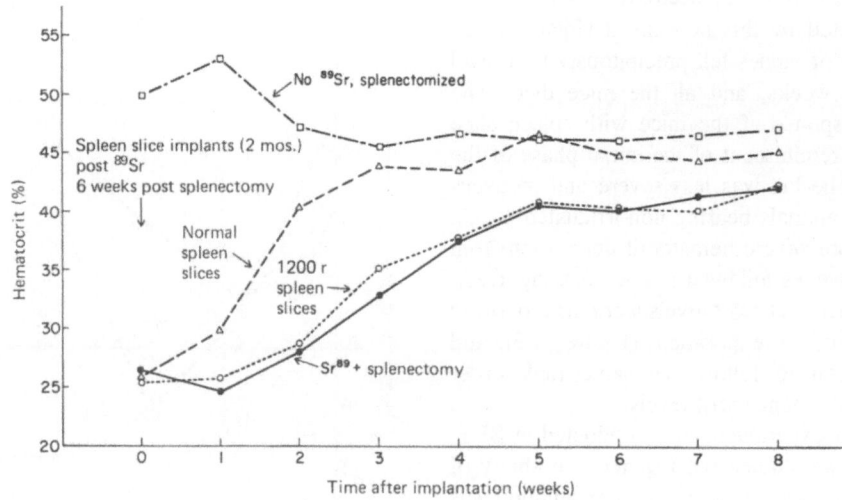

FIGURE 15.4. Effect of implantation of spleen slices on hematocrit values of splenectomized, ^{89}Sr-treated mice (BDF₁, ♀).

Platelet levels in mice implanted with living slices rose quickly within 1 week to between 700,000 and 800,000 (Figure 15.5). At that time, however, a plateau was reached, and the platelet count did not attain the 1-million level characteristic of normal mice and of the non-radiostrontium-treated splenectomized mice. The sham-operated group that received 1,200 R irradiated slices remained depressed in the 400,000 range for the duration of the experiment.

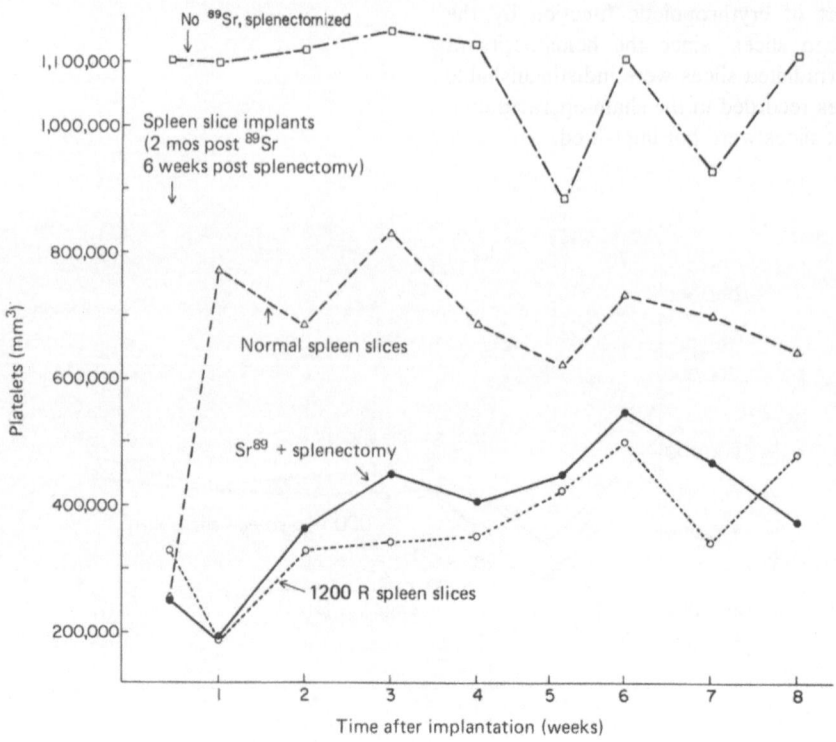

FIGURE 15.5 Effect of implantation of spleen slices on the platelet count of splenectomized, ^{89}Sr-treated miie (BDF₁, ♀).

DISCUSSION

In previous studies, it was demonstrated that the anemia produced in DBA/2 female mice by splenectomy and [89]Sr (a bone-seeking, beta-emitting isotope that produces a long-lasting pancytopenia) could not be improved by the injection of marrow cells (3), the present finding of hematologic improvement following implantation of spleen slices suggests that viable stem cells are not as important for the establishment of a normal rate of hemopoiesis in this model as is the reintroduction of a normal matrix. It is not surprising that spleen slices will serve this function, since erythropoiesis and thrombocytopoiesis continue at nearly normal rates in intact spleens following total destruction of the bone marrow by administration of a large amount of [89]Sr (4).

The present experiments demonstrate conclusively that spleen slices can be implanted in the peritoneal cavity and that they help restore a erythropoiesis and thrombocytopoiesis to normal levels. The implants remain viable for several months and maintain blood production following further marrow damage severe enough to be fatal in comparably damaged, untreated animals.

In the first experiment, fresh spleen slices supplied both "soil" (splenic hemopoietic matrix) and "seeds" (multi potential stem cells, CFU-s), and the fall in hematocrit values observed after splenectomy was dampened substantially if such normal spleen slices were implanted. Irradiation of the spleen slices had a deleterious effect on erythropoiesis, which was most marked at 3 weeks; but the irradiated spleen slices were then able to sustain erythropoiesis at a rate such that hematocrit levels approximated those of recipients of normal spleen slices by the 6th week.

Following the second dose of radiostrontium, severe thrombocytopenia developed. In animals lacking spleen tissue, a rapid fall, followed by stability of the platelet count in the region of 250,000/ml (normal value, approximately 1,100,000/ml), was observed. This rapid fall and quick establishment of a new stable level is to be expected for a cell compartment with a rapid turnover. The slower fall in hematocrit values over a 5-week span, from an average of 44% before the second dose of ^{89}Sr to a pre-terminal average of 19%, is consistent with the longer lifetime and slower replacement rate of erythrocytes. We suspect, but do not have direct evidence to prove, that death was a consequence of infection, coincident to leukopenia and perhaps abetted by anemia. We know from other studies in our laboratory that granulocytopenia and lymphopenia are also severe in these animals.

In the second experiment, 1,200 R-irradiated spleen slices did not support erythropoiesis and

thrombocytopoiesis upon implantation. Perhaps the stem cells that escaped destruction at 800 R in the first experiment were eliminated by the larger radiation dose in the second experiment. Alternatively, implantation may have been impaired. Finally, some stromal function might have been irreversibly damaged by the irradiation, such that viable stem cells (CFU-s) from the mouse could not successfully colonize the implanted slices. Further studies will be directed at a better understanding of the mechanism underlying this observation.

We are particularly interested in the discrepancy between the support of erythropoiesis and of thrombopoiesis afforded by normal spleen slices. Whereas erythropoiesis returned nearly to the level characteristic of control splenectomized mice as a consequence of the implantation of normal spleen slices, the production of platelets rose quickly from 250,000 to approximately 800,000, but then plateaued at a value only 60 to 70% of that in the normal splenectomized control animals. Perhaps the amount of transplanted stroma established in the peritoneum of recipients was saturated by erythropoietic cells in response to physiologic demand, and further hypertrophy to supply space and nutrient function necessary for complete restoration of thrombocytopoiesis was not possible. Alternatively, perhaps thrombocytopoiesis is sluggish and coarsely regulated and can meet the animals' needs satisfactorily at 70% of the normal level. When order is finally established from the diversity of opinion regarding the interrelationship and responsiveness of the various primitive hemopoietic cell compartments *in vitro,* we will still have a fascinating series of problems of differential expression of hemopoietic activity, rate, and regulation to explore in the whole animal.

Occasionally, patients are encountered whose idiopathic pancytopenia persists for a long time after a possbly pancytopenia-inducing medication has been discontinued or in the absence of continued exposure to any known hematoxin. If the injury occurred in a short period of time and if its target were the hemopoietic stem cells, if the noxious agent were removed, and if enough stem cells remained to sustain survival, one would expect replication of stem cells and recovery that is prompt and total. This is not always the case, however, and pancytopenia may persist for months or years. It is possible that such patients have profound and irreversible depletion of CFU-s and that they are able to sustain life only on the basis of continued, but suboptimal blood formation from compartments of commited stem cells that cannot be completely reconstituted. Persistent and irreversible damage to the marrow stroma or the hemopoiesis-inductive microenvironment, however, also remains a theoretical possibility, the consideration of which initiated the series of experiments reported here.

SUMMARY

Administration of ^{89}Sr, a bone-seeking, beta-emitting isotope, causes marrow damage in proportion to the dose administered. Following splenectomy in mice, hypoplastic anemia (pancytopenia) ensues. In the present experiments, we implanted, intraperitoneally, normal spleen slices from an intact spleen, or spleen slices x-irradiated *in vitro* with 800 R or 1,200 R immediately before implantation. This was done to determine if such slices would become hemopoietic environments in mice subjected to 2.0 μCi/gm of ^{89}Sr and splenectomy. Living spleen slices and slices irradiated with 800 R restored hematocrit values to near-normal levels, but left a large platelet deficit (750,000 versus 1.1 million). But these platelet levels were substantially higher than those in sham-operated controls or in mice receiving spleen slices irradiated with 1,200 R, with a platelet count of only 400,000. That the implanted spleen fragments remain functional and capable of producing erythrocytes and platelets was demonstrated by the administration of an additional ^{89}Sr dose of 2.0 μCi/g 5 months after the implants had been introduced. Following this additional stress, the blood counts in control animals lacking such fragments fell drastically and all animals died. The blood counts in mice containing implanted spleen slices recovered within 1 month.

ACKNOWLEDGMENTS

These studies were supported by National Cancer Institute Grant CA-22039. The Franklin McLean Memorial Research Institute is operated by The University of Chi ago for the U.S. Department of Energy under Contract EY-76-C-02-0069.

REFERENCES

1. Birks, J., Klassen, L., and Gurney, C. W. Hypoxia-induced thrombocytopenia in mice. *J. Lab. Clin. Med.,* 86:230, 1975.
2. Gong, J. K. Endosteal marrow: A rich source of hematopoietic stem cells. *Science, 199*:1443, 1978.
3. Gurney, C. W., and Simmons, E. L. Unpublished results.
4. Klassen, L. W., Birks, J., Allen, E., and Gurney, C. W. Experimental Medullary Aplasia. *J. Lab. Clin. Med.,* 80:8, 1972.
5. Snedecor, G. W. *Statistical Methods,* 6th ed. Ames: Iowa State University Press, 1967.

PART IV
Transplantation Immunology

P. J. Tutschka

16

Suppressor Cells as Possible Mediators of Immune Function after Marrow Transplantation

Peter J. Tutschka,
Grover M. Hutchins,
and George W. Santos

The complex immunologic situation that evolves after allogeneic bone marrow transplantation is characterized by three key phenomena: graft-versus-host (GvH) reactions (17), an operational immunodeficiency lasting in man for up to 1 year post-transplantation (9,55), and finally, the development of a specific operational tolerance (4), at which a stable situation is seemingly neither perturbed by GvH or host-versus-graft (HvG) reactions. A considerable number of clinical data has been accumulated describing these situations without the mechanisms involved being fully elucidated or the possible causal interrelations between mechanisms identified.

Of the *in vitro* correlates of a clinical transplant situation, the mixed lymphocyte reaction is probably the closest one (37), allowing the study, in a well-defined, though artificial system, of the afferent as well as efferent limbs of the immune response. The testing of lymphocytes from the chimera in this system should allow some insight into the immunologic reactivity of the chimera and permit preliminary conclusions about the nature of the immune defect, as well as the nature of transplantation tolerance, to be drawn.

In the study described here, the reactivity of chimeric rat lymphocytes is examined in a mixed lymphocyte culture (MLC) at various times after transplantation and compared to the ability of the chimeras to mount an immune response against sheep red blood cells (SRBC) and dinitrochlorobenzene (DNCB).

MATERIALS AND METHODS
Animals

Female Lewis (Ag-B1), ACI (Ag-B4), and BN (Ag-B3) rats, 10 to 12 weeks old were obtained from Microbiological Associates (Bethesday, Md.). The animals were housed in polycarbonate cages, four to a cage. They were provided with tapwater and purina chow *ad libitum*.

Drugs and Irradiation

Cyclophosphamide (CY) was prepared in saline and injected intraperitoneally in a volume of 10 mg/kg body weight. It was given at the dose of 200 mg/kg as preparation for engraftment and at a dose of 7.5 mg/kg to Lewis rats grafted with ACI bone marrow as a means of reducing the incidence and severity of graft-versus-host disease (GvHD). A dual source ^{137}Cs small animal irradiator delivering 136 R/min was used to irradiate test cells.

Contact Sensitization to 2-4 Dinitrochlorobenzene

Rats were anesthetized with chloral hydrate, the hair on the abdomen was removed with an electric hair cutter, the abdominal skin was cleansed with acetone, and the DNCB solution was applied to the abdomen with a micropipette; the DNCB was spread over an area of approximately 1 cm². Fourteen days after application of the sensitizing dose (660 μg, right side of abdomen) a challenging dose was applied (16 μg, left side of abdomen), and the site of the challenging dose was biopsied after another 48 hr (6). The histologic reaction was graded on a scale from I to IV as described previously (41).

Immunization with Sheep Red Blood Cells

Immunization with sheep red blood cells (SRBC) and determination of saline agglutination and 2-ME resistant antibody titers were performed as described previously (40). Briefly, groups of Lewis rats were injected with 1 ml of 10% SRBC intravenously on days 21, 40, 60, or 250 after grafts of syngeneic or allogeneic (ACI) bone marrow. Agglutination titers were determined weekly for 3 weeks after immunization.

Preparation of Marrow Cell Inocula and Assessment of Chimerism

The marrow cell inocula were prepared as described previously (47). Antisera specific for ACI immunoglobulin allotype was used in a gel-diffusion test to assess lymphoid chimerism indirectly, as described before (23). The peripheral blood cells of ACI-Lewis (donor-host) chimeras were typed with strain-specific cytotoxic isoantisera (48) to assess the completeness of the chimerism.

Preparation of Long-term Chimeras

Lewis rats were given 200 mg/kg of CY followed in 24 hr by the infusion of 60×10^6 nucleated ACI bone marrow cells. The CY was given every other day for five doses to prevent severe GVHD. Posttransplantation (day 1 to 12), polymyxin and neomycin were given in the drinking water. The immune status of these animals was evaluated at 21, 40, 60, and 250 days post-transplantation and compared to the reactivity of Lewis rats given syngeneic marrow or untreated control Lewis rats.

Mixed Lymphocyte Culture Technique

Depletion of Macrophages Since the spleen cells from normal rats give erratic results in an MLC, probably due to naturally occurring suppressor cells of presumed macrophage origin (53), we used a method modified after Fitch and Nordin to remove macrophages (21,53). Briefly, all the spleen cells used in the MLC are incubated with carbonyl iron, and the iron-containing cells are removed with a powerful magnet. This procedure allows recovery of 65 to 75% of the original cell preparation.

Mixed Lymphocyte Culture The spleen cells, after depletion of iron-containing cells, were suspended in RMPI 1640 supplemented with 5% heat inactivated fresh BN serum, 5×10^{-5} M 2-mercaptoethanol (2-ME), 0.25 μg/ml glutamine, penicillin (50 μ/m), and streptomycin (50 mcg/ml). Aliquots of 4×10^6/ml responding cells were cultured with 6×10^6/ml stimulating cells in U-bottom microtiter plates (Cooke) for 5 days at 37°C in a humidified atmosphere of 5% CO_2 in air. Then 1 μCi of [³H]-thymidine was added and the culture was incubated for another 18 hr before counting in a liquid scintillation counter. Results are expressed as mean counts per minute (cpm ± SD) from triplicate cultures. To assess for suppression of the MLC, macrophage-depleted splenocytes from chimeras or normal ACI donors are added at day 0 of the culture.

RESULTS
Operational Immunodeficiency of ACI-Lewis Chimeras

Response to Immunization with Sheep Red Blood Cells Groups of Lewis rats, 10 per group, were either not treated, given 200 mg/kg of CY and syngeneic Lewis marrow, or 200 mg/kg of CY and allogeneic ACI marrow cells. At day 21, 40, 60, and 250 post-grafting they were immunized with SRBC, and the saline and 2-ME resistant agglutination titers were determined weekly for 3 weeks. Figure 16.1 shows the results of this experiment. When immunization was on day 21, the saline agglutination titers to SRBC of the allogeneic chimeras did not differ from that of the control groups, although the 2-ME resistant titers were slightly lower. Compared to the controls, a slight reduction in the titers of the allogeneic group was seen after immunization on day 40 and a marked reduction of the titers on day 60. Animals immunized on day 250 showed no difference in the agglutination titers whether they were untreated or syngeneic or allogeneic chimeras.

Delayed Cutaneous Hypersensitivity to 2-4 Dinitrochlorobenzene Groups of Lewis rats were either not treated or given a syngeneic or allogeneic (ACI) bone marrow graft. Table 16.1 shows their delayed cutaneous hypersensitivity (DCH) reactivity to DNCB on days 21, 40, 60, and 250 post-transplantation. On day 21, the allogeneic chimeras showed a heightened response compared to the controls, but lost their reactivity between 40 and 60 days. By 250 days, these animals had regained their ability to show DCH.

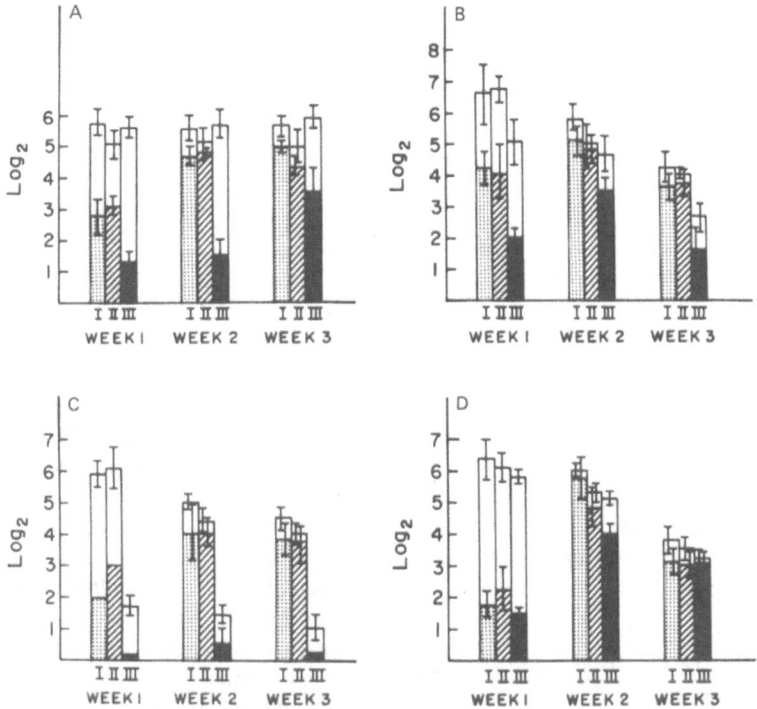

FIGURE 16.1. Animals were immunized at (**A**) 21, (**B**) 40, (**C**) 60, and (**D**) 250 days and bled 1, 2, and 3 weeks after immunization. Normal Lewis rats (I); Lewis rats grafted with Lewis marrow (II); Lewis rats grafted with ACI marrow (III); saline agglutination titers (open bars); 2-ME resistant titers (shaded bars).

Ability of Lymphocytes from Allogeneic Chimeras to Respond in Mixed Lymphocyte Culture Splenic lymphocytes, pooled from three allogeneic chimeras, were obtained 21, 40, 60, and 250 days post-bone marrow transplantation and used as responders in a MLC against host (Lewis) and third-party (BN) lymphocytes. Splenic lymphocytes from allogeneic chimeras responded well to host-type lymphocytes on day 21, but poorly on day 40. They lost their reactivity against host lymphocytes completely by day 60 and were still unreactive on day 250. Similarly, these chimeric lymphocytes responded well to third-party (BN) lymphocytes on day 21, poorly on day 40, and quite poorly on day 60. On day 250, post-bone marrow transplantation, however, they regained their reactivity to BN lymphocytes (Table 16.2).

Suppressive Effect of Splenic Lymphocytes from Allogeneic Chimeras on Mixed Lymphocyte Cultures

Suppressive Effect on Mixed Lymphocyte Cultures between Donor and Host Lymphocytes The spleens of three chimeric animals were removed 250 days after bone marrow grafting, the spleen cells were pooled, and various numbers of these pooled spleen cells were added to the MLC between donor (ACI) and host (Lewis) lymphocytes. To describe further the suppressive effect of chimeric lymphocytes on the MLC, the net cpm can be calculated: The net cpm is defined as the cpm of [(Responder (resp) + additional (add'l) cells) versus (vs) Stimulator (stim)] − [(Responder (resp) + irradiated (irrad) responder (resp)) versus (vs) Stimulator (stim)]. This formula allows us to determine whether the chimeric cells actually have a suppressive effect or whether

TABLE 16.1 Delayed Cutaneous Hypersensitivity to 2-4 Dinitrochlorobenzene in Normal Rats and Syngeneic or Allogeneic Marrow Recipients at Various Times After Transplantation

DAYS AFTER TRANSPLANTATION	DONOR-RECIPIENT COMBINATION	BIOPSY RESULT
21	ACI-Lewis	4+
	Lewis-Lewis	1–2+
	Untreated Lewis	3+
40	ACI-Lewis	Neg.
	Lewis-Lewis	2+
	Untreated Lewis	2+
60	ACI-Lewis	Neg.
	Lewis-Lewis	2+
	Untreated Lewis	2+
250	ACI-Lewis	2+
	Lewis-Lewis	2+
	Untreated Lewis	2–3+

TABLE 16.2 Response of Splenocytes from ACI-Lewis Chimeras (Chim) to Donor (ACI), Host (Lewis) and Third-Party Splenocytes (BN)

CELLS USED IN MLC		CPM ± SD (STIMULATION INDEX)							
Responder	Stimulator	Day 21		Day 40		Day 60		Day 250	
Chimera	Lewis	20,810 ± 1403	(34.0)	3,641 ± 474	(7.43)	583 ± 68	(1.16)	487 ± 101	(2.20)
Chimera	ACI	471 ± 68	(0.76)	733 ± 65	(1.49)	343 ± 107	(0.68)	193 ± 184	(0.87)
Chimera	BN	12,711 ± 988	(20.77)	8,916 ± 1074	(18.19)	4,712 ± 444	(9.41)	9,413 ± 676	(42.54)
ACI	Chimera	774 ± 64	(2.14)	413 ± 70	(0.72)	378 ± 44	(1.33)	279 ± 112	(0.67)
Lewis	Chimera	9,712 ± 814	(31.53)	10,070 ± 1171	(22.08)	12,114 ± 697	(18.13)	10,913 ± 877	(14.99)
BN	Chimera	5,712 ± 680	(14.28)	4,843 ± 712	(9.35)	6,661 ± 569	(10.88)	5,443 ± 999	(5.77)

they are behaving like a tolerant but inert cell population similar to a irradiated responder cell. A negative net cpm, therefore, would indicate active suppression, whereas a positive net cpm would indicate the participation of the chimeric lymphocytes in the MLC. The suppressive effect can further be described by calculating a relative suppressive index (RSI), which is defined as

$$\frac{[(Resp + Add'l\ Cells)\ vs\ Stim] - [(resp + Irrad\ Resp)\ vs\ Stim]}{[(Resp + Add'l\ Resp)\ vs\ Stim] - [(Resp + Irrad\ Resp)\ vs\ Stim]}$$

This index, in essence, describes the ratio of the net experimental count and the net count of normal, nontolerant responders. Theoretically, the ratio would be zero if the tolerant cells behaved as an inert cell population, one if the tolerant cells behaved like normal responder cells, and greater than one if the tolerant cells contributed to the MLC to a greater extent than normal responder cells (secondary

MLC); it would be negative if suppression indeed occurred. A clear suppressive effect is seen on the response of donor cells to host lymphocytes, which depends on the number of suppressor cells added (Table 16.3).

Suppressive Effect of Chimeric Splenocytes on Mixed Lymphocyte Culture between Donor and Host Lymphocytes at Various Times after Bone Marrow Transplantation The spleens of four groups of three chimeric animals each were removed at 21, 40, 60, and 250 days after marrow grafting. The splenocytes of these animals were pooled, and a fixed number of these cells (10^5) were added to MLC between donor (ACI) and host (Lewis) lymphocytes. Table 16.4 shows the result of this experiment. At 21 days, the addition of chimeric cells led to a heightened response (RSI, 2.03) compared to the addition of normal or irradiated ACI cells. At 40 days postbone marrow transplant, however, a possible suppressive effect was seen, which was clearly demon-

TABLE 16.3 Addition of Chimeric or ACI Splenocytes in Various Numbers to an ACI-Lewis MLC (ACI Responder-Lewis Stimulator). Age of Chimeras: 250 Days after Transplantation

NATURE OF ADDITIONAL CELLS	NUMBER OF ADDITIONAL CELLS		
	10×10^5	4×10^5	4×10^4
ACI, Live			
(CPM ± SD)	22,813 ± 780	14,437 ± 1,212	12,004 ± 989
(Net CPM)[a]	+13,028	+3,724	+984
(RSI)[b]	1	1	1
Chimeric, Live			
(CPM ± SD)	3,312 ± 676	6,742 ± 901	10,419 ± 1,205
(Net CPM)[a]	−6,473	−3,971	−597
(RSI)[b]	−1.26	−1.06	−0.61
ACI, Irradiated			
(CPM ± SD)	9,785 ± 912	10,713 ± 868	11,016 ± 1,104
(Net CPM)[a]	0	0	0
(RSI)[b]	N.A.	N.A.	N.A.
Chimeric, Irradiated			
(CPM ± SD)	10,612 ± 1,480	11,944 ± 1,029	11,699 ± 880
(Net CPM)[a]	+827	+1,231	+683
(RSI)[b]	0.06	0.33	+ 0.69

[a]Net CPM = CPM of [(Resp + Add'l Cells) vs Stim] − [(Resp + Irrad Resp) vs Stim]
[b]RSI = CPM of [(Resp + Add'l Cells) vs Stim] − [(Resp + Irrad Resp) vs Stim]/CPM of [(Resp + Live Resp Cells) vs Stim] − [(Resp + Irrad Resp) vs Stim]

TABLE 16.4 Addition of Chimeric or ACI Splenocytes at a Fixed Cell Number (10) to an ACI-Lewis MLC Various Days After Transplantation

CELLS USED IN MLC				TIME POST-TRANSPLANTATION				
Re-sponder	Stimu-lator	Additional Cells, Live	Additional Cells Irradiated	Day 21	Day 40	Day 60	Day 250	
ACI	Lewis	ACI		16,512 ± 1,407 +6,032 1	15,004 ± 971 +3,983 1	15,112 ± 1,101 +4,804 1	22,813 ± 780 +13,028 1	(CPM ± SD) (Net CPM)[a] (RSI)[b]
ACI	Lewis	Chimeric		22,731 ± 1,804 +12,251 +2.03	9,339 ± 610 −1,682 −0.42	2,110 ± 481 −8,198 −1.71	3,312 ± 676 −6,473 −1.20	(CPM ± SD) (Net CPM)[a] (RSI)[b]
ACI	Lewis		ACI	10,480 ± 843 0 N.A.	11,021 ± 1,260 0 N.A.	10,308 ± 870 0 N.A.	9,785 ± 912 0 N.A.	(CPM ± SD) (Net CPM)[a] (RSI)[a]
ACI	Lewis		Chimeric	11,812 ± 1,004 +1,332 +0.22	11,200 ± 1,341 +179 +0.04	11,430 ± 610 +1,122 +0.23	10,612 ± 140 +827 +0.06	(CPM ± SD) (Net CPM)[a] (RSI)[b]

[a]Net CPM = CPM of [(Resp + Add'l Cells) vs Stim] − [(Resp + Irrad Resp) vs Stim]

[b]RSI = CPM of [(Resp + Add'l Cells) vs Stim] − [(Resp + Irrad Resp) vs Stim]/CPM of [(Resp + Live Resp Cells) vs Stim] − [(Resp + Irrad Resp) vs Stim]

strable at 60 and 250 days, post-bone marrow transplant.

Suppressive Effect of Chimeric Splenocytes on Mixed Lymphocyte Cultures between Donor and Third-Party (BN) Lymphocytes at Various Times after Bone Marrow Transplantation The spleens of four groups of three chimeric animals each were removed at 21, 40, 70, and 250 days after marrow grafting, and 10⁵ spleen cells were added to a MLC between donor (ACI) and third-party (BN) lympho-

cytes. Table 16.5 shows the result of this experiment in time. A suppressive effect of the chimeric spleen cells was only demonstrated at 60 days post-transplantation, but is absent at 21, 40, or 250 days.

DISCUSSION

Immunodeficiency, post-allogeneic bone marrow transplantation is a well-described phenomenon in animals (13,30,49) and man (9,18,43,55) and at pres-

TABLE 16.5 Addition of Chimeric or ACI Splenocytes at a Fixed Cell Number (10) to an ACI-BN MLC (ACI Responder-BN Stimulator) at Various Days after Transplantation

CELLS USED IN MLC				TIME POST TRANSPLANTATION				
Re-sponder	Stimu-lator	Additional Cells, Live	Additional Cells, Irradiated	Day 21	Day 40	Day 60	Day 250	
ACI	BN	ACI		11,907 ± 508 +1,706 1	16,101 ± 1,471 +4,027 1	17,713 ± 1,200 +6,042 1	17,002 ± 1,400 +4,090 1	(CPM ± SD) (Net CPM)[a] (RSI)[b]
ACI	BN	Chimeric		13,402 ± 1,032 +3,201 +1.88	15,403 ± 980 +3,329 +0.83	6,812 ± 470 −4,859 −0.80	15,712 ± 973 +2,800 +0.68	(CPM ± SD) (Net CPM)[a] (RSI)[b]
ACI	BN		ACI	10,201 ± 1,503 0 N.A.	12,074 ± 1,070 0 N.A.	11,671 ± 801 0 N.A.	12,912 ± 1,071 0 N.A.	(CPM ± SD) (Net CPM)[a] (RSI)[b]
ACI	BN		Chimeric	9,992 ± 477 −209 −0.12	13,701 ± 1,864 +1,627 +0.40	10,913 ± 512 −758 −0.12	11,702 ± 864 −1,210 −0.29	(CPM ± SDO) (Net CPM)[a] (RSI)[b]

[a]Net CPM = CPM of [(Resp + Add'l Cells) vs Stim] − [(Resp + Irrad Resp) vs Stim]

[b]RSI = CPM of [(Resp + Add'l Cells) vs Stim] − [(Resp + Irrad Resp) vs Stim]/CPM of [(Resp + Live Resp Cells) vs Stim] − [(Resp + Irrad Resp) vs Stim]

ent is being extensively evaluated. The operational immune defect ranges from a global immunologic nonreactivity of the chimera to a defect primarily of the efferent T-cell limb (16,43,51) and is considered a primary reason for the susceptibility of transplant recipients to opportunistic infections (28). This immunoincompetence furthermore correlates with a decreased *in vitro* proliferative lymphocyte response to antigenic or mitogenic stimuli (29), which suggests that the immune defect is caused by an absence or a functional impairment of mature T-lymphocytes (6,27) that lack the proper inductive environment (44,54) or certain humoral factors presumed to be of thymic origin (24,25). Of interest is the fact that, despite the impaired *in vivo* and *in vitro* reactivity of the chimeras, certain *in vitro* parameters of immunity post-grafting rapidly return to normal; these include total lymphocyte counts and absolute numbers of peripheral T- and B-cells (11,24). Our data presented here agree with these results, showing a decreased antibody response to SRBC, a markedly impaired DCH against DNCB, and a functional proliferative defect of chimeric lymphocytes against host as well as third-party alloantigens. On day 20, however, when the recipient grafted with strongly histoincompatible marrow showed the peak of the GvHD, the chimeras behaved like fully immunocompetent animals, losing their immunocompetence between day 40 and day 60 and regaining it by day 250. This response on day 20 can be explained by a phenomenon similar to the allogeneic effect (9,39) followed later by a loss of immunocompetent cells; on the other hand, it might indicate the presence of an active suppressor mechanism rather than passive deletion.

The role of suppressor cells in the immune phenomena seen after transplantation remains unclear. In the classical theory of clonal selection (2,5), transplantation tolerance is regarded as a specific deletion or a specific and nonreversible inactivation of reactive cells, a view that is supported by many *in vivo* and *in vitro* data (10). A number of workers, however, have identified donor-type cells, in tolerant individuals, capable of reacting against host-type cells in mixed lymphocyte reaction (MLR) and local GvH reaction assays (42,57). These workers and others (34,45), furthermore, were not able to confirm the initial results, that certain humoral factors were blocking the reactivity of these donor cells (19,20,-33), and showed rather convincingly that serum factors were not a necessary prerequisite for the maintenance of a stable donor-to-host tolerance (7,38). Circumstantial evidence for a cellular suppressor system operative in transplantation tolerance is suggested by the experiments of Weiden et al. who were not able to affect graft-host tolerance *in vivo* by infusion of large numbers of donor cells into stable, long-term canine chimeras (52). More direct evidence for this concept comes from studies by Gengozian et al. (13) and studies in our laboratory (46), which indicate that a cortisone, acetate-resistant but CY-sensitive cellular suppressor system is present in long-term rat chimeras and that the suppressor cells specifically suppressed the development of GvHD in new hosts when mixed with original donor cells (50). The data presented here extend these *in vivo* observations: Lymphocytes from strongly histoincompatible chimeras obtained more than 60 days after transplantation suppressed the MLR of original donor cells against host alloantigens. This phenomenon depended upon the presence of live chimeric lymphocytes and further upon the number of chimeric lymphocytes added to the cultures.

Suppression of MLR has been reported in other experimental systems (1). The initial observation that concanavalin A-activated T-cells suppress proliferative responses of syngeneic responder cells in MLR (31,35) and alloantigen-induced generation of cytotoxic lymphocytes (CTL) (32) was followed by reports that T-cells activated with allogeneic cells *in vitro* or *in vivo* suppress both alloantigen-induced T-cell proliferation and CTL generation (26). Rich et al., following other studies (3,22), described a soluble suppressor factor (36) as the basis of the cellular MLR suppressor system. Although the factor suppressed proliferative responses in the MLR in an antigen-nonspecific manner, it was effective only if the cells producing the factor and responding to it shared the same I-C subregion of the H-2 complex. Similar findings were reported by Wonigeit and Pichlmayr in a rat system (56), but Gershon et al. (12) identified suppressor cells that may be syngeneic with donor or recipient. The suppressor cells that we found to be present in radiation chimeras 250 days after transplantation showed specificity for the stimulating alloantigen; at 60 days, however, the suppressor action was antigen nonspecific and suppressed the response of donor-type cells against third-party alloantigens.

It is unclear if we are dealing with two different populations of suppressor cells or with one population that matures across time and develops characteristics that allows a specific action. Of interest is the fact that the specific suppressor cell is present at a time when the chimera has regained full immunocompetence, whereas the nonspecific suppressor cell was identified at the height of immunodeficiency. It is tempting to speculate that both phenomena are related and that the immunodeficiency is actually caused, or at least augmented by the action of this suppressor system.

Clearly, further studies are needed to determine whether this preliminary observation holds true and can be confirmed in other animal systems (12). The immuno-manipulative measures that have been proposed to improve the post-grafting immunodeficiency (12) might then be seen in a different light: They might not necessarily supplement the chimera with functioning immunocompetent cells but rather turn the nonspecific suppressor system into a more specific (15) one, thus creating the tolerant, immunocompetent chimera. Studies are in progress in our laboratory to test this attractive possibility.

SUMMARY

Lewis rats, conditioned with cyclophosphamide (CY) and transplanted with AgB-incompatible ACI marrow cells, are immunocompetent shortly after transplantation. At 60 days after transplantation, when the graft-versus-host disease (GvHD) has subsided, they develop an immunodeficiency as demonstrated by an absent dinitrochlorobenzene (DNCB) response and a markedly reduced humoral response to sheep red blood cells (SRBC). This immune defect resolves by 250 days post-bone marrow transplantation (BMT). At 60 to 80 days post-BMT, spleen cells from chimeric animals do not respond to stimulation with ACI, Lewis, or third-party (BN) spleen cells in a mixed lymphocyte culture (MLC). The addition of chimeric spleen cells to responding ACI spleen cells in a MLC will suppress the reaction of these ACI cells to stimulating Lewis or BN cells. At 250 days post-BMT, spleen cells from chimeric animals respond normally to third-party BN-stimulating lymphocytes in a MLC, but do not respond to ACI- or Lewis-stimulating lymphocytes. Mixing chimeric spleen cells with responding ACI lymphocytes specifically suppresses the MLC to stimulating Lewis, but not to stimulating BN lymphocytes. These studies suggest that 60 to 80 days post-BMT, when the GvHD has subsided, suppressor cells that suppress the reactivity of donor-type lymphocytes against host and third-party antigens are present. Later on, this suppressive activity becomes highly specific and suppresses only the reactivity of donor cells for host antigens. It is conceivable that the immunodeficiency seen in the post-transplantation period is caused by an active suppression, rather than the absence, of immunocompetent T-cells.

ACKNOWLEDGMENTS

Supported by grant numbers CA-06973 and CA-15396 awarded by the National Cancer Institute, DHEW and ACS grant number CH-113.

REFERENCES

1. Asherson, G. L., and Zembala, M. Suppressor T-cells in cell-mediated immunity. *Br. Med. Bull., 32*(2):158, 1976.
2. Billingham, R. E., Brent, L., and Medawar, P. B. Quantitative studies on tissue transplantation immunity. II. Actively acquired tolerance. *Philos. Trans. R. Soc., London (Ser. B), 239*:357, 1956.
3. Blomgren, H., and Jacobsson H. Inhibition of erythroid cell growth by allogeneic murine lymphocytes. Evidence for a synergism between lymph node cells and thymocytes. *Cell. Immunol., 13*:288, 1974.
4. Brent, L., Brooks, C. B., Medawar, P. B., and Simpson, E. Transplantation tolerance. *Br. Med. Bull., 32*:101, 1976.
5. Burnet, F. M. The clonal selection therapy of acquired immunity. The 1958 Abraham Flexner Lectures. Vanderbilt Univ. Press, Nashville, Tenn., 1959.
6. Cross, A. M., Leuchon, E., and Miller, J. F. Studies on the recovery of the immune response in irradiated mice thymectomized in adult life. *J. Exp. Med., 119*:837, 1964.
7. Dorsch, S., and Roser, B. T-cells mediated transplantation tolerance. *Nature, 258*:233, 1975.
8. Elfenbein, G. J., Green J., and Paul, W. E. The allogeneic effect: Increased cellular immune and inflammatory responses. *J. Immunol., 112*(6):2166, 1974.
9. Elfenbein, G. J. Anderson, P. N., Humphrey, R. L. et. al. Immune system reconstitution following allogeneic bone marrow transplantation in man: A multiparameter analysis. *Trans. Proc. 8*:641, 1976.
10. Elkins, W. E. The cellular basis of transplantation tolerance. *Trans. Proc. V*(1):685, 1973.
11. Fass, L. Ochs, H. D., Thomas, E. D., Mickelson, E., Storb, R., and Fefer, A. L. Studies of immunological reactivity following syngeneic or allogeneic marrow grafts in man. *Transplantation, 16*:630, 1973.
12. Gale, R. P., and Opelz, G. Immunological and clinical perspectives in human bone marrow transplantation. *Trans. Proc. X*(1):265, 1978.
13. Gengozian, N., and Urso, P., Functional activity of T and B lymphocytes in radiation chimeras. *Transpl. Proc., 8*:631, 1976.
14. Gengozian, N., Congdon, C. C., Allen, E. A., and Toya, R. E. Immune status of allogeneic radiation chimeras. *Trans. Proc., 3*:434, 1971.
15. Gershon, R. K., and Kondo, K. Cell interactions in the induction of tolerance: The role of thymic lymphocytes. *Immunology, 18*:723, 1970.
16. Gorczynski, and R. M., Macrae, S. Differentiation of functionally active mouse T lymphocytes from functionally inactive bone marrow precursors. I. Kinetics of recovery of T cell function in lethally irradiated, bone marrow reconstituted, thymectomized and nonthymectomized mice. *Immunology, 33*(5):697, 1978.
17. Grebe, S. C., and Streilein, J. W., Graft-versus-host reactions: A review. *Adv. Immunol., 22*:137, 1976.
18. Halterman, R. H., Graw, R. G., Jr., Fucillo, D. A., and Leventhanl, B. G. Immunocompetence following allogeneic bone marrow transplantation in man. *Transplantation, 14*:689, 1972.

19. Hellström, J., Hellström, K. E., Storb, R., and Thomas, E. D. Colony inhibition of fibroblasts from chimeric dogs mediated by the dog's own lymphocytes and specifically abrogated by their serum. *Proc. Nat. Acad. Sci. U.S.*, 66:55, 1970.

20. Hellström, J., Hellström, K. E., Trentin, and J. J. Cellular immunity and blocking serum activity in chimeric mice. *Cell. Immunol.* 7:73, 1973.

21. Hirano, T., and Nordin, A. A. Cell-mediated immune responses *in vitro*. I. The development of suppressor cells and cytotoxic lymphocytes in mixed lymphocyte cultures. *J. Immunol.*, 116(4):1115, 1976.

22. Hirano, T., and Nordin, A. A. Cell mediated immune response *in vitro*. II. The mechanism(s) involved in the suppression of the development of cytotoxic lymphocytes. *J. Immunol.*, 117(6):2226, 1976.

23. Humphrey, R. L., and Santos, G. W. Serum protein allotype markers in certain inbred rat strains. *Fed. Proc.*, 30:314, 1971.

24. Komuro, K., and Boyse, E. A., Induction of T lymphocytes from precursor cells *in vitro* by a product of the thymus. *J. Exp. Med.*, 138:479, 1975.

25. Lapp, W. S., Wechsler, A., and Kongshavn, P. A. L. Immune restoration of mice immunosuppressed by a graft-versus-host reaction. *Cell. Immunol.*, 11:419 1974.

26. Nadler, L. M., and Hodes, R. J. Regulatory mechanisms in cell-mediated immune responses. II. Comparison of culture-induced and alloantigen-induced suppressor cells in MLR and CML. *J. Immunol.*, 118:1886, 1977.

27. Neely, J. E., Neely, A. N., and Kersey, J. H. Immunodeficiency following human marrow transplantation: *In vitro* studies. *Trans. Proc.*, X(1):229, 1978.

28. Neiman, P. E., Reeves, W., Roy, G., Flournoy, L., Lerner, K. G., Sale, G. E., and Thomas, E. D., A prospective analysis of interstitial pneumonia and opportunistic viral infection among recipient allogeneic bone marrow grafts. *J. Infect. Dis.*, 136(6):754, 1977.

29. Nordin, A. A., and Farrar, J. J. Studies of the immunological capacity of germ free mouse radiation chimeras. III. *In vitro* reconstitution of the T-helper cell deficiency. *Cell. Immunol.*, 10:218, 1974.

30. Ochs, H. D., Storb, R., Thomas, E. D., Kolb, H. J., Graham, T. C., Mickelson, E., Parr, M., and Rudolph, R. H. Immunologic reactivity in canine marrow graft recipients. *J. Immunol.*, 113:1039, 1974.

31. Ozato, K., Ebert, J. D., and Adler, W. H. The differentiation of suppressor cell populations as revealed by studies of the effects of mitogens on the mixed lymphocyte reaction and on the generation of cytotoxic lymphocytes. *Cell. Immunol.*, 22:323, 1976.

32. Peavy, D. H., and Pierce, C. W. Cell mediated immune responses *in vitro* I. Suppression of the generation of cytotoxic lymphocytes by concanavalin A and concanavalin A-activated spleen cells. *J. Exp. Med.*, 140:356, 1974.

33. Phillips, S. M., Martin, W. J., Shaw, A. R., and Wegmann, T. G. Serum mediated immunological non-reactivity between histoincompatible cells in tetraparental mice. *Nature*, 234:146, 1971.

34. Phillips, S. M., and Wegmann, T. G. Active suppression on a possible mechanism of tolerance in tetraparental mice. *J. Exp. Med.*, 137:291, 1973.

35. Rich, R. R., and Rich, S. S. Biological expressions of lymphocyte activation. IV. Concanavalin A-activated suppressor cells in mouse mixed lymphocyte reactions. *J. Immunol.*, 11:1112, 1975.

36. Rich, R. R., Rich, S. S., and Truitt, G. A. Suppressor T cells in the regulation of immune responses to allogeneic tissues. *Trans. Proc.*, X(1):19, 1978.

37. Rodey, G. E., Bortin, M. M., Bach, F. H., and Rimm, A. A. Mixed leukocyte culture reactivity and chronic graft-versus-host reactions (secondary disease) between allogeneic H-2^K mouse strains. *Transplantation*, 17:84, 1974.

38. Rouse, B. T., and Warner, N. L. Induction of T-cell tolerance in agammaglobulinemic chickens. *Eur. J. Immunol.*, 2:102, 1972.

39. Santos, G. W. Adoptive transfer of immunologically competent cells. III. Comparative ability of allogeneic and syngeneic spleen cells to produce a primary antibody response in the cyclophosphamide pretreated mouse. *J. Immunol.*, 97:587, 1966.

40. Santos, G. W., and Owens, A. H. 19S and 7S antibody production in the cyclophosphamide or methotrexate treated rat. *Nature*, 209:622, 1966.

41. Slavin, R. E., Tutschka, P. J., and Santos, G. W. Contact sensitization to 2-4 Dinitrochlorobenzene (DNCB) in normal, cyclophosphamide (CY) treated and marrow grafted rats. *Fed. Proc.*, 32:879, 1973.

42. Sprent, J., von Boehmer, H., and Nabholz, M. Association of immunity and tolerance to host H-2 determinants in irradiated F$_1$ hybrid mice reconstituted with bone marrow cells from one parental strain. *J. Exp. Med.*, 142:321, 1975.

43. Storb, R., Ochs, H. D., Weichen, P. L., and Thomas, E. D. Immunologic reactivity in marrow graft recipients. *Trans. Proc. VIII*(4):637, 1976.

44. Stutman, O., Yunis, E. J., and Good, R. A. Thymus: An essential factor in lymphoid repopulation. *Trans. Proc.*, 1:614, 1969.

45. Tsoi, M. S., Storb, R., Weiden, P. L. et. al. Canine marrow transplantation: Are serum blocking factors necessary to maintain the stable chimeric state? *J. Immunol.*, 114(2):531, 1975.

46. Tutschka, P. J., and Santos, G. W. The role of suppressor cells in transplantation tolerance. *Trans. Proc.* (in press) 1978.

47. Tutschka, P. J., and Santos, G. W. Marrow transplantation in the Busulfan-treated rat. I. Effect of cyclophosphamide and rabbitt-anti-rat-thymocyte serum as immunosuppression. *Transplantation*, 20:101, 1975.

48. Tutschka, P. J., and Santos, G. W. Bone marrow transplantation in the Busulfan-treated rat. III. Relationship between myelosuppression and immunosuppression for conditioning bone marrow recipients. *Transplantation*, 24(1):52, 1977.

49. Tutschka, P. J., Slavin, R. E., and Santos, G. W. Immunological studies in bone marrow grafted rats. *Exp. Hematol.*, 1:287, 1973.

50. Tutschka, P. J., Schwerdtfeger, R., Slavin, R., and Santos, G. W. Mechanism of donor to host tolerance in rat bone marrow chimeras. In Baum, S. J., and Led-

ney, G. D., eds. *Experimental Hematology Today.* New York: Springer-Verlag, 1977, p. 191.

51. Urso, P. and Gengozian, N. T cell deficiency in mouse allogeneic radiation chimeras. *J. Immunol., 111*(3):712, 1973.

52. Weiden, P. L., Storb, R., Tsoi, M. S. et. al. Infusion of donor lymphocytes into stable canine radiation chimeras. Implications for mechanisms of tranplantation tolerance. *J. Immunol., 116*(5):1212, 1976.

53. Weiss, A., and Fitch, F. W. Macrophages suppress CTL generation in rat mixed leukocyte cultures. *J. Immunol., 119*(2):510, 1977.

54. Willis, J. J. The restorative effect ot thymic epithelial monolayers on lymphoid cells from neonatally thymectomized rats. *Anat. Red., 181*:519, 1975.

55. Witherspoon, R., Noel, D., Storb, R., Ochs, H. D., and Thomas, E. D. The effect of graft-versus-host disease on reconstitution of the immune system following marrow transplantation for aplastic anemia or leukemia. *Trans. Proc., 10*(1):233, 1978.

56. Wonigeit, K., and Pichlmayr, R., Suppression of specific MLC responsiveness during skin allograft rejection in rats. *Trans. Proc., IX*(1):765, 1977.

57. Wright, P. W., Bernstein, I. D., Hamilton, B. et. al. Cell mediated reactivity and serum blocking activity in tolerant rats. *Transplantation, 18*:46, 1974.

17

Re-Analysis of *in Vitro* Lymphocyte Blastogenesis Using the Laser Cytometry Assay

John C. Ruckdeschel,
Lloyd Lininger, Hillary Brzyski,
Mark Miani, and Jerry Becker

The transformation of small peripheral blood lymphocytes into large pyrinophilic lymphoblasts is a biologic event that has been extensively studied over the last two decades (11). The initial discoveries that this transformation could be accomplished both non-specifically by plant lectins (mitogens) and specifically by defined antigens has greatly enhanced our understanding of the immune system (3,13). Initial studies of lymphocyte transformation or blastogenesis were performed by examining stained specimens and manually counting transformed and nontransformed cells. This technique is highly accurate and definitive, but tedious. In addition, per cent calculations of transformed cells were frequently inaccurate due to the failure to recognize increases in small lymphocytes during the period of culture. Figures 17.1 and 17.2 illustrate the changes that occur both in lymphoblast number and small lymphocyte number following either mitogen or antigen stimulation. It can be seen that following mitogen stimulation, lymphoblast numbers rise sharply. This sharp rise in numbers is delayed by a few days for stimulation with streptokinase-streptodornase (SKSD), but the pattern is similar. Shortly after the onset of lymphoblast proliferation, the numbers of small lymphocytes in the cultures begins to increase as well, as shown in Figure 17.2.

Radionucleotide incorporation, particularly [³H]-thymidine incorporation, was shown to be a rapid and reliable means of demonstrating cellular proliferation and soon replaced manual cell counting. The correlation with actual proliferation was well demonstrated, but many of the *caveats* concerning reproducibility of the technique were ignored (4,8,20). In the mid 1960s investigators from numerous laboratories described the inability to reproduce the results of radionucleotide incorporation assays on a serial basis. It is, of course, crucial to our understanding of the function of the immune system to quantify a particular assay over time. For example, antibody titers have long been used serially to indicate both the waxing and waning of immunity to various pathogens. A similar assay of cellular immune function has not been available for serial study.

Beginning in the late 1960s, a number of investigators studied cryopreservation as a way to overcome the difficulties with longitudinal studies of lymphocyte function (6,10,12,19). The inaccuracies related to minor variations in laboratory technique and to variations in cold thymidine levels were essentially eliminated by freezing multiple samples and then studying them all at the same time. Although retrospective, this type of analysis did allow longitudinal study. Cryopreservation has not been widely accepted due to its cost and technical difficulty, and more recently studies have shown

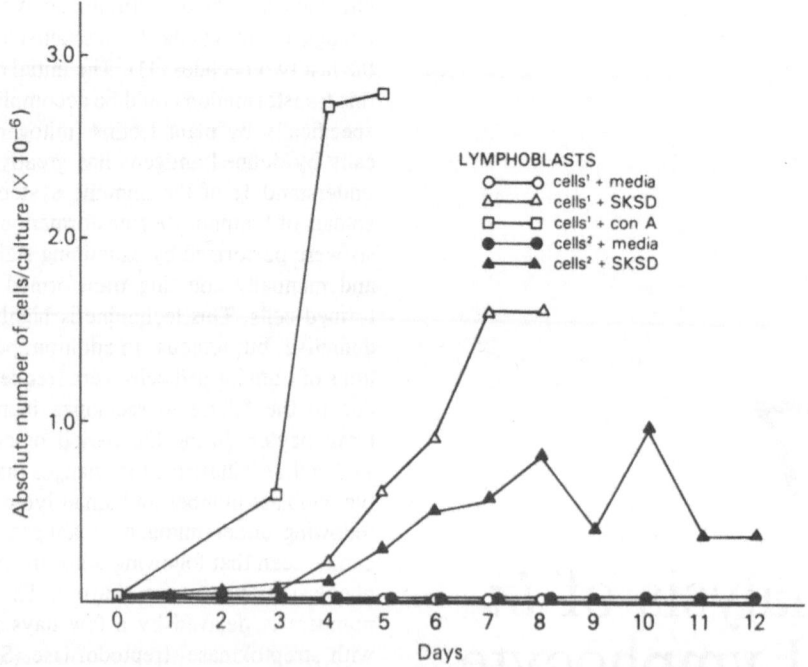

FIGURE 17.1. Lymphoblast numbers following stimulation with con A or SKSD. Percentages of lymphoblasts were determined manually from stained slides and multiplied by the volume specific cell count. Superscripts 1 and 2 refer to separate donors. No proliferation is seen in either donor's control cultures.

selective loss of B-cell populations following cryopreservation (4).

In another approach to overcome the inability to reproduce the results of radionucleotide incorporation assays, various mediator assays have been used (17). Bioassays for various lymphokines have been developed, but have not been demonstrated to be quantifiable serially. In addition, it is thought that some aspects of mediator release may represent partial activation of cell populations without an actual commitment to proliferation (1). Although it is still not clear what the relative roles of partial and com-

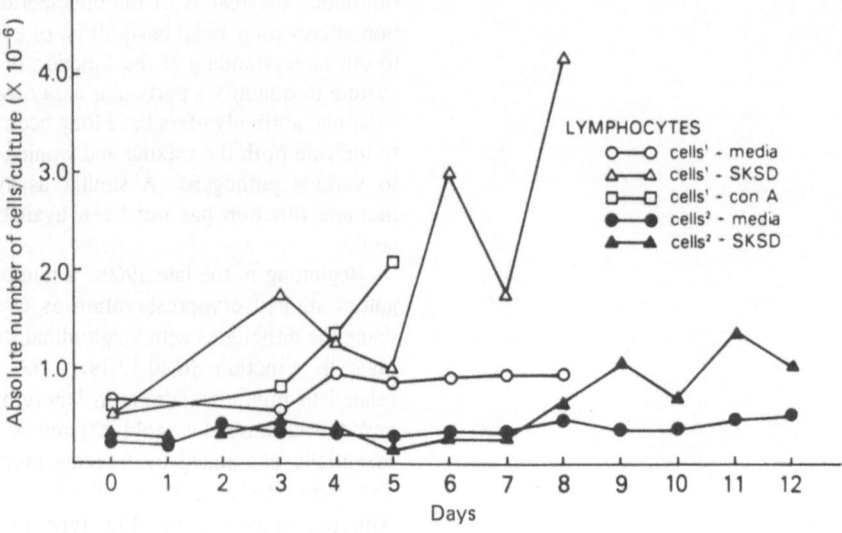

FIGURE 17.2. Small lymphocyte numbers following stimulation with con A or SKSD. The numbers are determined as in Figure 17.1. Comparison with Figure 17.1 shows that small lymphocyte numbers increase sharply within 24 hr of lymphoblast proliferation.

plete activation are in immune system function, it does seem possible that the assays may not be measuring the same phenomena.

In the last few years, we have developed a technique that allows us to directly enumerate numbers of proliferating lymphoblasts rapidly and reliably. As we shall demonstrate, this technique results in much less day to day variation and is significantly superior to radionucleotide incorporation assays for the serial measurement of lymphocyte proliferation.

MATERIALS AND METHODS

Lymphocyte cultures were established routinely by employing Ficoll-Hypaque-separated lymphocytes. The details of our culture technique have been published elsewhere (2,14,15). Radionucleotide incorporation was monitored by the uptake of [³H]-thymidine following a 4-hr pulse. The laser cytometry assay was performed by gently vortexing the cells in culture, diluting an aliquot of the cell suspension in culture media, and analyzing it on the laser system. The major component of the cytometer is a flow system wherein the cell suspension is intro-

duced into a flowing sheath of filtered water such that the column of cells is progressively constricted to a 30- to 50-μ core at the point of interception of the laser beam. Large numbers of cells (10^4 to 10^5) can be counted in under a minute. The cells intersect a laser beam of wavelength 632.8 μ. This wavelength avoids the absorption spectrum of hemoglobin, and consequently, contaminating erythrocytes do not cause undue absorption. The forward angle scatter, or transmittance, is monitored, and a pulse is generated inversely proportional to the transmittance. The height of the pulse is directly proportional to cell size, and we have previously demonstrated that lymphoblasts and lymphocytes can be reliably distinguished using this technique (2,14,15).

RESULTS AND DISCUSSION

Figure 17.3 demonstrates the events occurring in a concanavalin A (con A)-stimulated culture over a period of 5 days. The individual pulses are assigned channel numbers in direct proportion to the pulse height. Small channel numbers are associated with small pulse height (and hence, small cell size) and

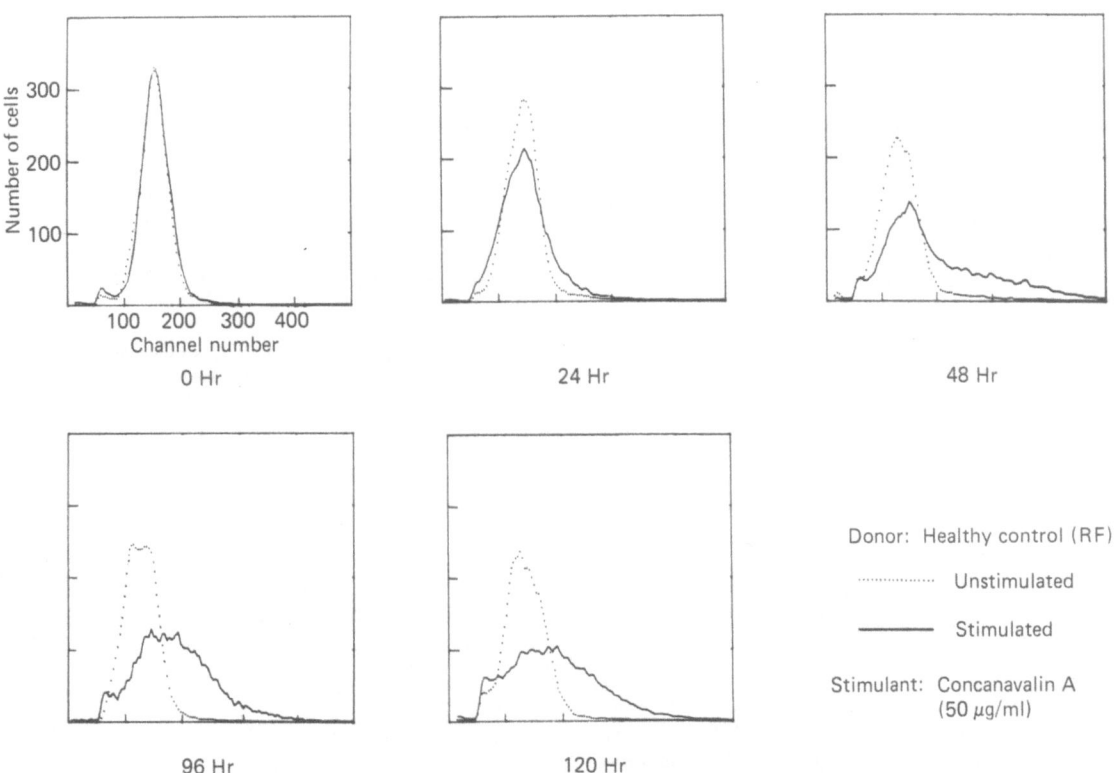

FIGURE 17.3. Lymphocyte transformation following *in vitro* mitogenic stimulation. Ficoll-Hypaque-separated lymphocytes peak in channels 100-190 at time 0. Platelets and debris appear below channel 100 and are windowed out electronically. It is clear that following con A stimulation, a population of large cells (channel number >190) appears in the stimulated cultures.

vice versa. It can be seen that at 0 hr the number of cells peaks in the area between channels 100 and 190. Numerous experiments have demonstrated that this is the small lymphocyte population. Erythrocytes do overlap significantly with this population but are removed by the initial separation process and by acetic acid lysis. Platelets and cell debris appear below channel 100 and can be windowed out electronically. Over the next 48 to 96 hr, it is apparent that a population of larger cells (with channel numbers greater than 190) is appearing in the stimulated cultures. Previous work from our laboratory and from the National Cancer Institute has demonstrated that this is due to lymphoblast proliferation, and a reliable estimate of the numbers of lymphoblasts can be obtained by calculating the percentage of cells beyond channel 190 and multiplying by the volume specific cell count (2,14,15).

In this way, we can obtain numbers of total cells, small lymphocytes (cells in channels 100-190), and lymphoblasts (cells in channels 190-500). Figure 17.4 shows these events. In the first panel, one can see that total cell numbers increase sharply following 48

hr in culture and that this is initially due to lymphoblast proliferation. As was noted in the Introduction, small lymphocytes also increase in absolute number during the course of culture. The time-course of [3H]-thymidine incorporation is also demonstrated in Figure 17.4.

An important and useful aspect of the time-course of lymphocyte proliferation as measured by laser cytometry is the plateau in cell numbers seen after peak response has been reached. Both total cell number and lymphoblast number reach a relatively stable plateau, which does not decay for a period of days. Peak [3H]-thymidine incorporation, however, is known to drop off rapidly as the proliferative stimulus diminshes. This is shown in Figure 17.4 as well. The ability to quantitate the level of peak response by a measurement at a single time point is obviously easier when a particular parameter remains at that peak value for a longer period of time. With this concept in mind, it was postulated that the laser assay would result in a more reliable serial measure of lymphocyte proliferation.

Eight normal donors were studied over a period

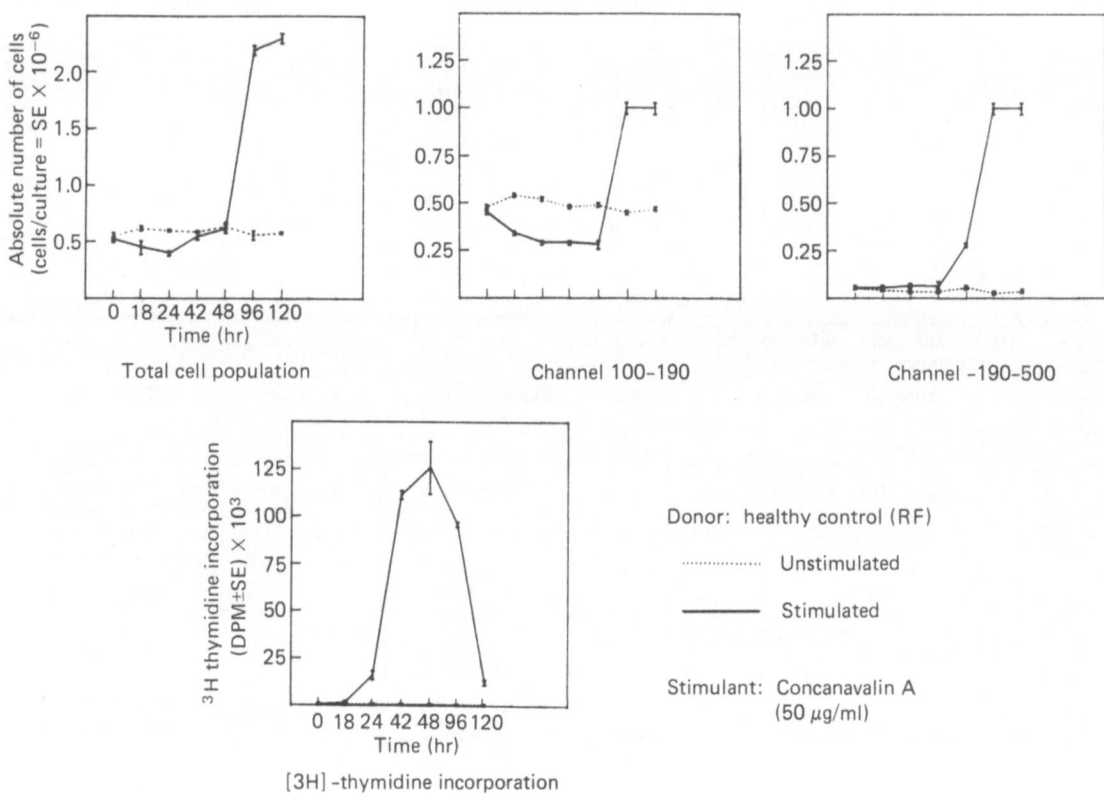

FIGURE 17.4. Time-course of lymphocyte stimulation as measured by laser cytometry. Absolute numbers of lymphoblasts (channels 190-500) and small lymphocytes (channels 100-190) are determined from percentages on the distribution and the volume specific cell counts. All cell populations plateau as the peak response is reached. [3H]-thymidine incorporation precedes cell proliferation by 12 to 24 hr and drops off sharply following peak responsiveness.

of 1 to 5 months using both laser cytometry and radionucleotide incorporation to measure their response to identical doses of con A (50 μg/ml). The supposition was that a normal donor's response to a given stimulus should remain relatively constant when that stimulus is not a pathogenic agent that is likely to be encountered (in this case, con A).

In order to compare assays with different units of measurement, the coefficient of variation was chosen for study. This is defined as $\sigma/\mu \times 100$, where μ is the mean response and σ is the standard deviation (16). The advantage of using the coefficient of variation is that it is a unitless entity. A small coefficient of variation implies greater reproducibility in this set of experiments. The mean deviation and the standard deviation of all responses taken over the study period were obtained for each technique and each donor. Both total cells and total lymphoblasts were studied for the laser assay. Using the Wilcoxan rank sum test (7), it can be seen that the laser cytometry assay is significantly superior to radionucleotide incorporation in terms of reproducibility (Table 17.1). Assays for both total cell number and total lymphoblast number are superior, with total cell numbers being slightly more stable over time.

More recently, we have had an opportunity to analyze the kinetic events occurring during the process of lymphocyte transformation. The availability of direct cell numbers, in particular lymphoblast numbers, allows mathematical modeling of possible kinetic patterns. Table 17.2 illustrates preliminary work in this area. One suggested kinetic pattern of

TABLE 17.2 Observed and Expected Values for Total Cell Number Assuming the Proliferation Curve is Described by a Logistic Function[a]

DAY[b]	EXPECTED[c]	OBSERVED[d]
0	900,000	940,000
1	—	1,080,000
2	—	1,040,000
3	917,700	1,020,000
4	959,700	950,000
5	1,089,900	1,100,000
6	1,415,100	1,450,000
7	1,920,700	2,020,000
8	2,325,500	2,300,000
9	2,510,300	2,500,000
10	—	2,330,000
11	—	2,290,000

[a]Logistic function described as $N = \lambda N_0 (N_{max} - N)$, where N = number of cells at a given time point, N_0 = initial number of responding cells, N_{max} = maximal number of cells present during experiment. A sigmoid curve is obtained.

[b]Standard lymphocyte cultures were established with 9×10^5 lymphocytes and stimulated with SKSD (40 units/ml). Samples were made for each day in culture.

[c]Expected number of cells assuming 900,000 cells in culture at time 0 and growth described by a logistic function and a "final" (day 9) count of 2.5×10^6 cells.

[d]Actual cell numbers determined as described in Materials and Methods.

TABLE 17.1 Comparison of Coefficients of Variation[a] Between Test Dates for the Laser Cytometry and Radionucleotide Incorporation Assays

	COEFFICIENT OF VARIATION		
		Laser Cytometry[d]	
DONOR[b]	Radionucleotide Incorporation[c]	Total Lymphoblasts[e]	Total Cells
1	32.2	10.9	5.0
2	91.3	6.0	7.5
3	45.7	8.0	9.5
4	53.9	10.0	10.6
5	56.7	19.0	8.3
6	46.9	25.0	6.9
7	18.9	16.7	5.9
8	33.8	24.3	17.1

[a]Coefficient of variation defined as mean (μ)/standard deviation (σ) · 100. A smaller coefficient of variation implies greater reproducibility.

[b]Eight healthy donors were tested serially over 1 to 5 months. Concanavalin A at 50 μg/ml from the same lot was used in all experiments.

[c]Original units of measurement disintegrations per minute (DPM) per culture.

[d]Original units of measurement were cells per culture.

[e]Total lymphoblasts and total cell numbers are significantly ($p < .001$ and $p < .0001$, respectively) more reproducible over time.

cell proliferation can be described by a logistic curve. This is usually given in the form $N = \lambda N_0 \cdot (N_{max}-N)$. This describes a rapid exponential growth phase that is slowed as maximal response is reached and is frequently used to describe growth that is constrained by crowding, dilution of stimulus, etc. (9). The end result is a traditional sigmoid-type curve. In Table 17.2, we see the expected cell counts when a culture with 9×10^5 lymphocytes is stimulated with SKSD and presumed to follow a logistic function. The observed column in Table 17.2 are data obtained on total cell number from an actual experiment in which SKSD stimulation was performed. The correlation between observed and expected cell counts is striking and leads to two very important pieces of information. The first is N_0, the initial number of responding cells. Although it is not clear when the actual commitment to proliferation occurs, current evidence suggests the first 24 hr in culture (18). If 4,500 cells at time zero or 9,000 cells at 24 hr began to proliferate exponentially, as described by the logistic function, they could account for the entire proliferation of 1.5×10^6 new cells seen 1 week later. Obviously, cell death and recruitment of other cells into the process may occur and complicate these calculations, but the estimate of 1% of peripheral lymphocytes responding to SKSD is in good agreement with other studies (5,21). Careful analysis of the time at which commitment to proliferation occurs will allow us to eventually determine N_0 from calculations made later in the proliferation curve, when changes are more likely to be greater in magnitude and more reproducible. The mathematical description of the curve can then be calculated for time zero.

The slope of a proliferation curve may also vary and can be described as λ. This may be influenced by a number of enhancing, suppressing, and otherwise modifying influences. As the mathematical description of the response curve is completed, small numbers of time points could be studied (2 to 3 days during a 1-week culture) in order to obtain portions of this curve. The slope (λ) and the initial number of responding cells (N_0) can then be quite easily obtained. This is of particular interest in patients with cancer, since it is not clear whether the initial number of responding cells is diminishing or whether the overall function of a normally responding cell population is diminished by increased suppressor cells or substances.

ACKNOWLEDGMENTS

This work was supported by NIH Research Grants SO7-RR-05394-16, NO1-CM-57032, NO1-CN-45150 and the N.Y.S. Kidney Disease Institue. Albany Medical College Summer Fellowship support is acknowledged for M.M. and J.B.

REFERENCES

1. Cohen, S., David, J., Feldmann, M. et al. Current state of studies of mediators of cellular immunity: A Progress Report. *Cell. Immunol., 33*:233, 1977.
2. Doukas, J. G., Ruckdeschel, J. C., and Mardiney, M. R., Jr. Quantitative and qualitative analysis of human peripheral lymphocyte proliferation to specific antigen *in vitro* by use of the helium neon laser. *J. Immunol. Methods, 15*:229, 1977.
3. Elves, M. W., Roath, S., and Israels, M. C. G. The response of lymphocytes to antigen challenge *in Vitro*. *Lancet, 1*:806, 1963.
4. Jewett, M. A. S., Gupta, S., Hansen, J. A. et al. The use of cryopreserved lymphocytes for longitudinal studies of immune function and enumeration of subpopulations. *Clin. Exp. Immunol., 25*:449, 1976.
5. Jimenez, L., Bloom, B. R., Blume, M. R., and Oettgen, H. F. On the number and nature of antigensensitive lymphocytes in the blood of delayed-hypersensitive human donors. *J. Exp. Med., 133*:740, 1971.
6. Kalmbach, K. W., and Mardiney, M. R., Jr. An improved system for controlled rate cooling of biological material. *Cryobiology, 9*:572, 1972.
7. Lehman, E. H. In *Non-parametrics: Statistical Methods Based on Ranks*. San Francisco: Holden-Day, p. 6, 1975.
8. Mangi, R. J., and Mardiney, M. R., Jr. The *in vitro* transformation of frozen-stored lymphocytes in the mixed lymphocyte reaction and in culture with phytohemagglutinin and specific antigens. *J. Exp. Med., 132*:401, 1970.
9. May, R. M. *Model Ecosystems*. Princeton: Princeton University Press, 1973.
10. Miller, R. A., Bean, M. A., Kodera, Y., and Herr, H. W. Cryopreservation of human effector cells active in antibody-dependent cell-mediated cytotoxicity. *Transplantation, 21*:517, 1976.
11. Nowell, P. C. Phytohemagglutinin: An initiator of mitosis in cultures of normal human leukocytes. *Cancer Res., 20*:462, 1960.
12. Oldham, R. K., and Simmler, M. C. The use of cryopreserved lymphocytes of lymphoblasts in ^{51}Cr lymphocyte cytotoxicity. In Wiener, R. S., Oldham, R. K., and Schwarzenberg, L., eds., *Cryopreservation of Normal and Neoplastic Cells*. Paris: Colloque Villejuif, INSERM, 1973, p. 161.
13. Pearmain, G., Lycett, R. R., and Fitzgerald, P. H. Tuberculin-induced mitosis in peripheral blood leucocytes. *Lancet, 1*:637, 1963.
14. Ruckdeschel, J. C., Doukas, J. G., and Mardiney, M. R., Jr. Parameters of *in vitro* lymphocyte responsiveness to antigen, mitogen and allogeneic cells using a laser based cell sizing apparatus. *Fed. Proc., 34*:995, 1975.
15. Ruckdeschel, J. C., Doukas, J. G., Drake, W. P., and Mardiney, M. R., Jr. Application of laser cytometry to the analysis of *in vitro* lymphocyte blastogenesis. *Transplantation, 23*:396, 1977.
16. Snedecor, G. W., and Gochran, W. G. In *Statistical Methods*, 6th ed., Ames: Iowa State University Press, 1967, p. 62.
17. Waksman, B. H., and Wagshal, A. B. Lymphocytic functions acted on by immunoregulatory cytokines: Significance of the cell cycle. *Cell Immunol., 36*:180, 1978.
18. Weber, T. H., Skoog, V. T., Mattson, A., and Lindahl-Kiessling, K. Kinetics of lymphocyte stimulation *in vitro*. by non-specific mitogens. *Expt. Cell Res., 85*:351, 1974.
19. Wiener, R. S., Breard, J., and O'Brien, C. Cryoperved lymphocytes in sequential studies of immune responsiveness: Problems and prospects. In Wiener, R. S., Oldham, R. K., and Schwarzenberg, L., eds., *Cryopreservation of Normal and Neoplastic Cells*. Paris: Colloque Villejuif, INSERM, 1973, p. 117.
20. Wilson, D. B. Analysis of some of the variables associated with the proliferative response of human lymphoid cells in culture. *J. Exp. Zool., 162*:161, 1966.
21. Wilson, D. B., and Nowell, P. C. Quantitative studies on the mixed lymphocyte interaction in rats. *J. Exp. Med., 131*:391, 1970.

18

Differentiation of Bone Marrow Lymphocytes after Prolonged Administration of Anti-IgM Antibodies to Neonatal and Adult Mice

D. G. Osmond and J. Gordon

As is the case for erythrocytes, granulocytes, monocytes, and platelets, mammalian bone marrow is normally the site of a continuous production and renewal of small lymphocytes (8,13,14,21). In particular, this process concerns the production of B-lymphocytes that exhibit surface IgM molecules and antigen-binding receptors and that form the potential precursors of antibody-producing cells in humoral immune responses (12,15,25,27). Compared with the various, non-lymphoid marrow cell populations, however, relatively little is known of the factors regulating the steady-state production and differentiation of marrow lymphocytes.

Three main cell stages are involved in marrow lymphocytopoiesis, as in other cell lines. Self-renewing stem cells differentiate into committed progenitors, which, after a series of divisions, give rise to non-dividing cells. The latter continue to mature into fully functional forms, and enter the bloodstream. Surface IgM molecules are first readily detectable in the course of the final maturation phase on non-dividing marrow small lymphocytes (15,27). Other B-lymphocyte surface markers (Fc and complement receptors, Ia antigens, IgD) appear either concomitantly with surface IgM or later on the maturing small lymphocytes (5,14,16,27). Surface IgM expression appears to be preceded by the appearance in the cytoplasm of small amounts of IgM that pass rapidly through the cell membrane with little surface accumulation. Thus, proliferating large lymphoid (transitional) cells, which include the immediate precursors of marrow small lymphocytes, do not show a detectable surface IgM by autoradiographic anti-IgM binding, but do show one by sensitive rosetting techniques (26,27). Similarly, rosetting detects surface IgM on certain fetal liver cells several days before it is detected by radiolabeled or fluorescent antiglobulin binding (20). Biosynthetic radiolabeling reveals a rapid turnover of surface IgM, whereas fluorescence antiglobulin binding of fixed cells reveals a weak cytoplasmic IgM in certain large marrow cells regarded as cycling progenitor (pre-B-cells) (7,19) and possibly identical with the proliferating large lymphoid (transitional) cells (8,28).

The administration of anti-IgM antibodies in vivo provides a method of investigating a critical differentiation stage and the control of marrow lymphocyte production (2,4,6,17). Chronic administration of anti-IgM antibodies to neonatal mice eliminates all IgM-bearing small lymphocytes, not only from peripheral lymphoid tissues, but also from the central site of production in the bone marrow (2,17). A similar effect can be achieved in adult mice if anti-IgM treatment is preceded by X-irradiation and bone marrow reconstitution to deplete the lymphoid tissues of both mature and immature lymphocytes (17). The critical target cell for the action of anti-IgM in the marrow and the mechanism by which it is suppressed are not known. These problems have considerable theoretical and practical interest in view of the possibility that the surface binding and elimination of immature B-lymphocytes by anti-IgM may resemble the elimination of clones by specific antigen binding in one form of tolerogenesis, notably to self antigens.

The present experiments were designed to characterize the small lymphocytes remaining in mouse bone marrow after prolonged treatment with anti-IgM antibodies *in vivo* and to determine whether the anti-IgM suppression involves a block in either the production or maturation of these cells. After anti-IgM treatment in neonatal mice and in adult mice depleted either by x-irradiation and marrow reconstitution or by cyclophosphamide, which are reported to eliminate both B-cells and pre-B-cells, marrow small lymphocytes were examined for their incidence, surface markers, and turnover.

MATERIALS AND METHODS

Animals

All mice were (C57BL/6XC3H)F$_1$ from an SPF colony (BioBreeding Laboratories of Canada, Ottawa, Canada).

Anti-IgM Serum

A pool of anti-Ig serum was prepared from 90 rabbits, as described (10,17). The IgM purified from the plasmacytoma 104E (Litton Bionetics, Kensington, Md.) was precipitated in agar with a monospecific rabbit anti-IgM serum, previously prepared. The resulting precipitates were washed, homogenized with Freunds complete adjuvant, and injected into rabbits twice, in doses of 100 μg IgM, 2 weeks apart. Balb/c serum (1 ml) was injected intravenously 1 week after the second injection of the precipitates, and the animals were exsanguinated 5 days later. A gammaglobulin concentrate of pooled serum was prepared by two ammonium sulfate precipitations (50%; 33%), dialyzed against phosphate-buffered saline, and stored frozen. Before use, each aliquot was absorbed out with rat and mouse erythrocytes to remove natural hemagglutinins. The antiserum gave a single band of precipitate with Balb/c serum in immunodiffusion.

Suppression with Anti-IgM

For neonatal suppression, animals were injected intraperitoneally with anti-IgM (5 to 10 mg protein in 0.05 to 0.1 ml) when 1 day old, or later as indicated, and three times weekly thereafter for 6 to 12 weeks. The dose of anti-IgM was sufficient to give a reaction to immunodiffusion with purified IgM at a dilution of 1:64. In control experiments, the mice received equal doses of normal rabbit globulin (NRG).

Two methods of anti-IgM suppression in adult (11-week) mice were used. Groups of females were treated with an intraperitoneal injection of cyclophosphamide (Procytox, 500 mg/kg body weight), whereas males were given wholebody x-irradiation (930 R) and injected with 5 $\times$ 10^6 isologous bone marrow cells. In each case, 5 days later, the mice were given 100 mg anti-IgM globulin intraperitoneally followed by maintenance injections (5 to 10 mg) three times a week for 8 weeks. Control animals received either the appropriate treatment or the anti-IgM injections alone.

Cell Suspensions and Surface Markers

Femoral marrow and spleen cells were suspended in cold Eagle's minimum essential medium (MEM) containing 10% fetal calf serum (FCS) (Gibco, Grand Island, N.Y.), as described (15,27).

To label cell surface IgM, aliquots of 2 $\times$ 10^6 nucleated cells were mixed for 30 min at 0°C with equal volumes of rabbit anti-mouse IgM, purified by chromatography on DEAE-Sephadex and labeled with ^{125}I (specific activity 1.5 $\times$ 10^7 Ci/mg), as described (3,15). After washing through two FCS gradients, the cells were smeared, methanol-fixed, dipped in NTB2 emulsion (Eastman Kodak, Rochester, N.Y.), exposed for 3 days, and stained through the fixed emulsion with MacNeal's tetrachrome. All IgM-bearing small lymphocytes showing 10 autoradiographic grains or more were scored.

The T-lymphocytes were labeled by an indirect method. Cells were incubated for 30 min at 0°C with a rabbit anti-mouse brain serum, specific for T-cells (Cedarlane Laboratories, London, Ontario), washed, then incubated with a ^{125}I-labeled goat anti-rabbit Ig globulin, washed, smeared, and examined after a 3-day autoradiographic exposure. Counts in normal animals at various grain count thresholds established that labeling of 10 grains or more represented T-lymphocyte labeling.

The Fc and complement receptors were detected by cytocentrifuge rosetting techniques, detailed elsewhere (26,27). Cell suspensions were mixed (1:30) with washed sheep red blood cells (SRBC) coated with either a mouse anti-SRBC IgG antibody (for Fc receptors) or rabbit anti-SRBC stroma IgM antibody (Cordis Laboratories, Miami, Fla.) plus normal mouse serum (1:10) (for complement receptor). In control preparations, the cells were mixed with uncoated SRBC. The cell mixtures were incubated in soft centrifuged pellets at 37°C for 30 min, resuspended, cytocentrifuged, and stained. Positive rosettes, indicating the presence of either Fc or complement receptors, respectively, were indicated by the attachment of four SRBC or more per small lymphocyte and corrected for the small number of rosettes in control preparations.

[³H]-Thymidine Labeling of IgM-Negative Small Lymphocytes

Here, DNA-labeling and surface rosetting techniques were combined, as described (26,27). [³H]-Thymidine (specific activity 6.7 Ci/mM; New England Nuclear Corp., Boston, Mass.) was injected subcutaneously (25 μCi/dose) twice daily for five injections. Marrow and spleen cells taken 16 hr after the last injections were mixed (1:30) with washed SRBC coated with goat anti-mouse IgM antibody (Meloy Laboratories, Springfield, Va.) by the chromic chloride method (26), incubated in centrifuged pellets at 0°C for 30 min, cytocentrifuged, and examined autoradiographically after 28 days of exposure. Small lymphocytes were divided into IgM-bearing and IgM-negative cells, according to the formation or lack of rosettes (over four SRBC), and their respective [³H]-thymidine labeling indices (over three grains) were scored.

RESULTS

Elimination of IgM-Bearing Small Lymphocytes from Bone Marrow and Spleen of Anti-IgM Treated Mice

The effects of anti-IgM treatment on the incidence of IgM-bearing small lymphocytes are illustrated in Figures 18.1, 18.2, and 18.3. Small lymphocytes in the marrow and spleen of either normal mice or mice receiving control injections of normal rat globulin (NRG) showed the usual substantial per-

centages of IgM-bearing cells (Figures 18.1 and 18.2). After anti-IgM treatment starting at either day 1 or day 7 of postnatal life, IgM-bearing small lymphocytes were practically absent from both marrow and spleen (Figure 18.1). The rare, IgM-bearing small lymphocytes (0 to 0.6%) were dense and pyknotic in appearance. When anti-IgM treatment was started at 10, 15, or 20 days of age, however, the incidence of IgM-bearing small lymphocytes in both the marrow and spleen remained within normal limits (Figure 18.1).

In adult mice, injections of anti-IgM alone for 8 weeks did not reduce the incidence of IgM-bearing small lymphocytes (Figures 18.2 and 18.3). Similar findings were obtained in mice treated 8 weeks earlier with either x-irradiation and marrow cell transfusion or with cyclophosphamide (Figures 18.2 and 18.3). In contrast, combined treatment and anti-IgM injections eliminated IgM-bearing small lymphocytes from both the marrow and spleen (0 to 2%) (Figures 18.2 and 18.3) as in neonatal anti-IgM suppression. An accumulation of erythroblasts was seen in the spleen of cyclophosphamide-treated mice given either anti-IgM or NRG.

Persistence of Surface IgM-Negative Small Lymphocytes in Bone Marrow of Anti-IgM Treated Mice

The IgM-negative small lymphocytes showed a similar incidence, relative to the total nucleated cell populations, in the marrow of both anti-IgM suppressed and control mice. Examples are given of neonatal suppression in Figure 18.4 and of adult suppression in Figure 18.5. The results of 21 different experiments are combined in Figure 18.6 to show that, despite the elimination of IgM-bearing small lymphocytes in anti-IgM suppressed mice, the mean incidence of IgM-negative small lymphocytes remained essentially unchanged (17%) throughout a series of experiments in which three different batches of anti-IgM antibody were used over a period of 2 years with mice of either sex, ranging in age from 4 to 17 weeks when killed. Low levels of IgM-negative marrow small lymphocytes in anti-IgM treated mice have only been observed when non-SPF animals maintained on antibiotics were used and when continuous anti-IgM treatment was prolonged to 26 to 31 weeks of age (17).

Surface Markers and Turnover of IgM-Negative Small Lymphocytes in Bone Marrow of Anti-IgM Treated Mice

The absence of detectable surface IgM on small lymphocytes in anti-IgM treated mice was not due simply to masking of surface IgM by the injected rabbit anti-mouse IgM antibody. Thus, radiolabeled goat anti-rabbit Ig serum did not detect rabbit globu-

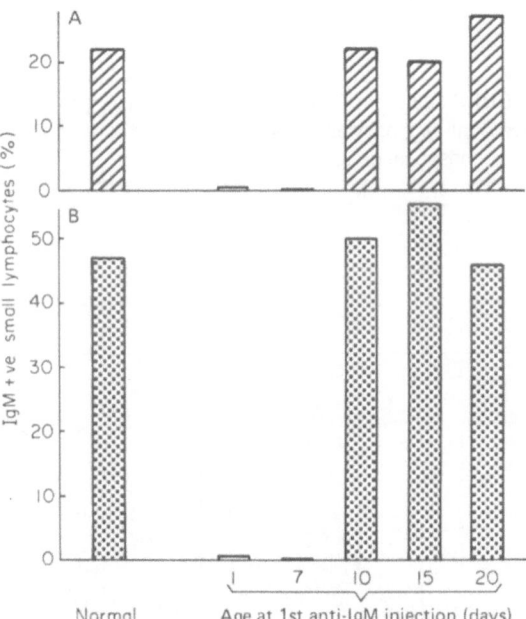

FIGURE 18.1. Percentage of small lymphocytes showing surface IgM in (**A**) bone marrow and (**B**) spleen of mice given anti-IgM starting at various ages.

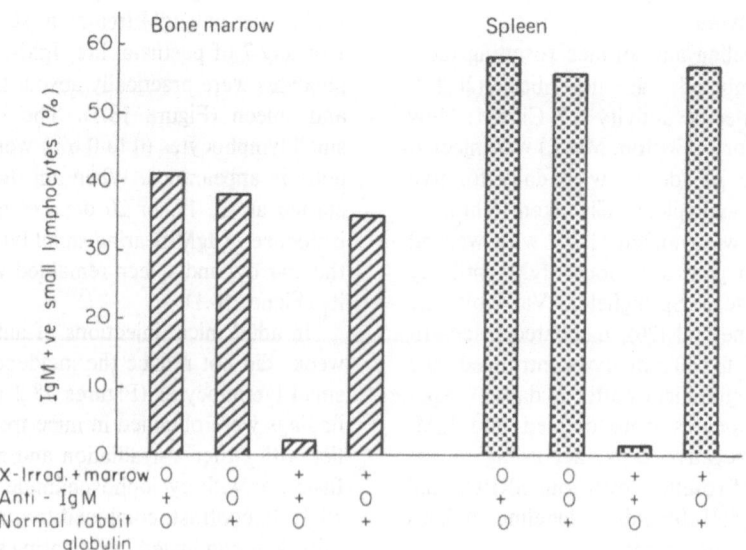

FIGURE 18.2. Percentage of small lymphocytes showing surface IgM in bone marrow and spleen of adult mice given anti-IgM after x-irradiation and bone marrow reconstitution.

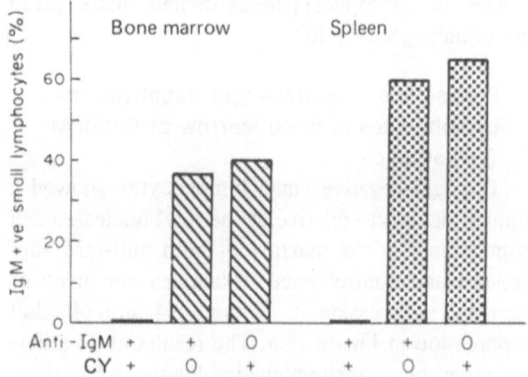

FIGURE 18.3. Percentage of small lymphocytes showing surface IgM in bone marrow and spleen of adult mice given cyclophosphamide (Cy) and anti-IgM.

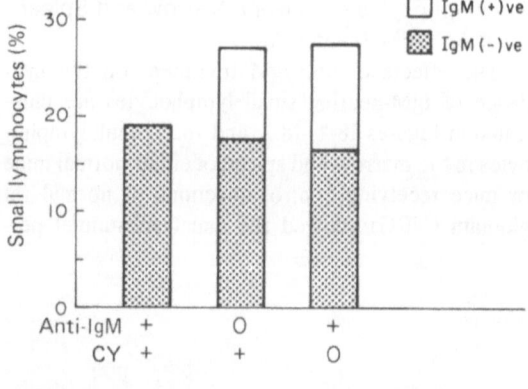

FIGURE 18.5. Incidence of IgM-bearing and IgM-negative small lymphocytes in bone marrow of adult mice given cyclophosphamide (Cy) and anti-IgM.

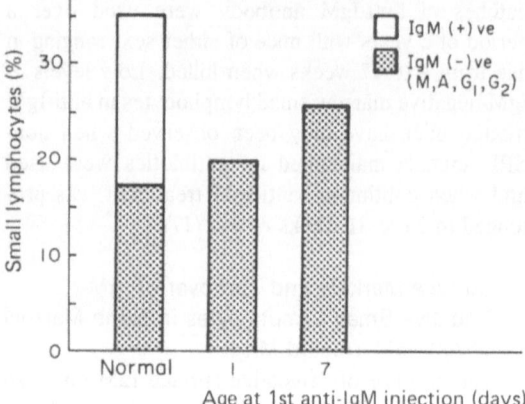

FIGURE 18.4. Incidence of IgM-bearing and IgM-negative small lymphocytes in bone marrow of mice given anti-IgM starting at 1 day and 7 days of age.

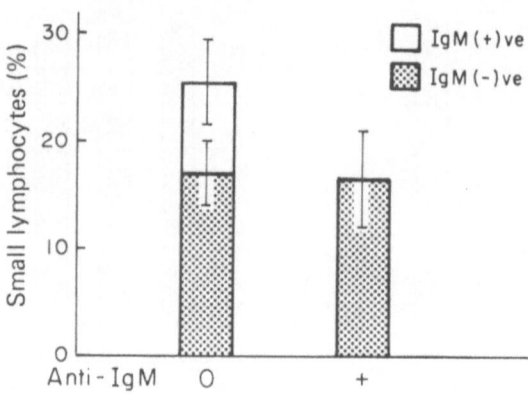

FIGURE 18.6. Incidence of IgM-bearing and IgM-negative small lymphocytes in bone marrow of normal mice and various anti-IgM treated mice. See text for further details.

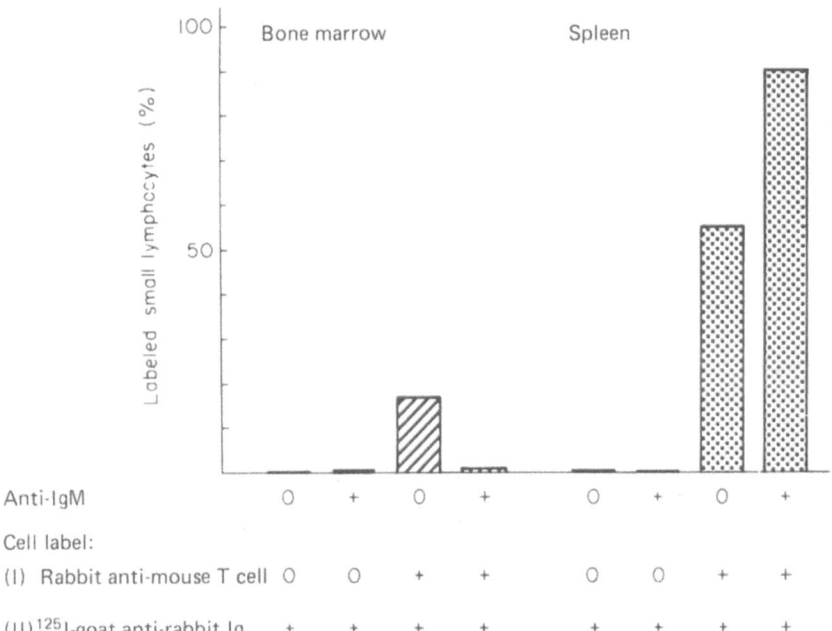

FIGURE 18.7. Lack of IgM masking on small lymphocytes and the incidence of T-lymphocytes in bone marrow and spleen of mice given anti-IgM from 1 day of age. Cells were exposed to ^{125}I-goat anti-rabbit Ig either alone, to detect bound rabbit anti-mouse IgM, or after incubation with a rabbit anti-mouse brain serum, to detect T-lymphocytes.

lin on the surface of these cells (Figure 18.7), though it did label IgM-bearing cells in normal marrow and spleen incubated with the rabbit anti-mouse IgM serum.

Indirect radiolabeling using a rabbit anti-mouse brain serum demonstrated that only a small proportion of the IgM-negative small lymphocytes in the marrow of anti-IgM suppressed mice were T-lymphocytes. In contrast in the spleens of anti-IgM treated mice, the incidence of T-lymphocytes was

markedly increased, and they accounted for the great majority (90%) of residual IgM-negative small lymphocytes (Figure 18.7). Whereas cytocentrifuge rosetting techniques revealed small lymphocytes bearing Fc and complement receptors in normal marrow and spleen, very few such cells remained in anti-IgM suppressed mice (Figure 18.8). Thus, the residual marrow small lymphocytes in anti-IgM suppressed mice appeared to lack the surface markers of both T- and B-lymphocytes.

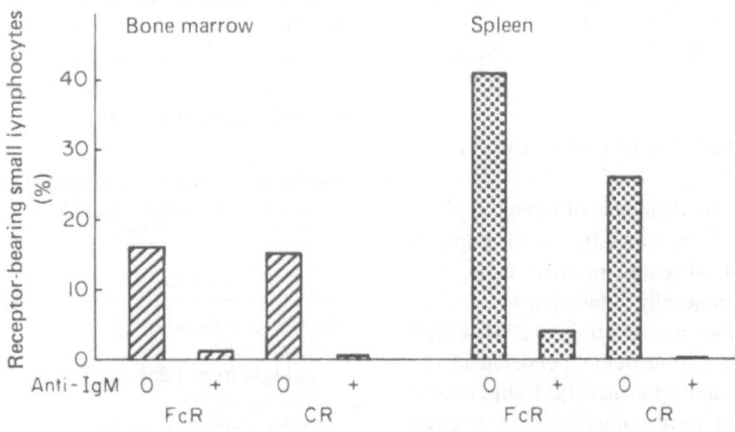

FIGURE 18.8. Percentage of small lymphocytes bearing Fc receptors (FcR) and complement receptors (CR) in bone marrow and spleen of mice given anti-IgM from 1 day of age.

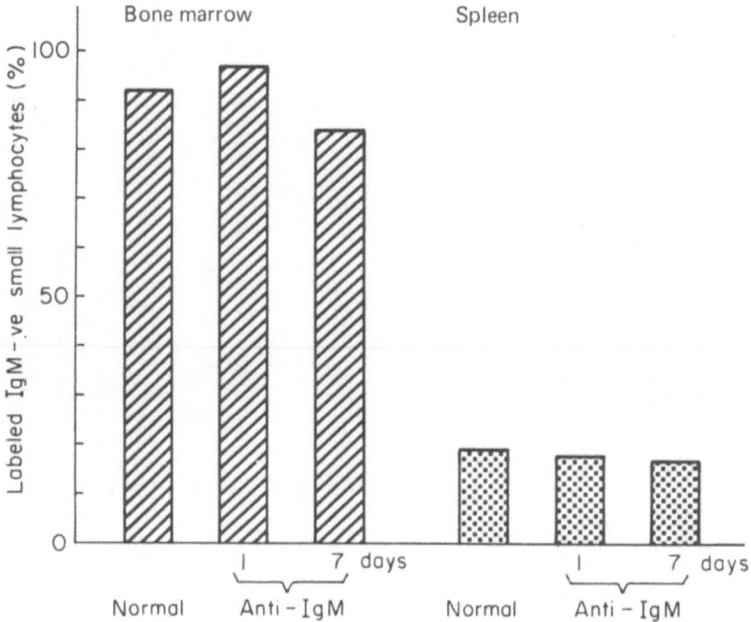

FIGURE 18.9. [³H]-Thymidine labeling of IgM-negative small lymphocytes in bone marrow and spleen of normal mice and of mice given anti-IgM from 1 day and 7 days of age. Following five [³H]-Thymidine injections in 3 days, the labeling indices of IgM-negative small lymphocytes, not forming rosettes with anti-IgM coated SRBC, were determined in cytocentrifuged autoradiographic preparations.

The IgM-negative small lymphocytes in the marrow of anti-IgM suppressed mice showed a rapid turnover. In combined surface- and DNA-labeling experiments, IgM-negative small lymphocytes were recognized by their lack of surface IgM rosetting with anti-IgM coated SRBC. After administration of [³H]-thymidine for 3 days, such cells showed high autoradiographic labeling indices in animals treated with anti-IgM from day 1 and day 7, as in normal mice (Figure 18.9). Because marrow small lymphocytes are themselves non-dividing, these data indicate that nearly all IgM-negative small lymphocytes in the marrow of anti-IgM suppressed mice are young cells, less than 3 days post-mitotic, and are rapidly renewed by the division of proliferating progenitors.

Large Lymphoid Cells in Bone Marrow of Anti-IgM Treated Mice

Table 18.1 shows the incidence of large lymphoid cells (28), based in each case on differential counts of approximately 2,000 nucleated marrow cells. The values tended to be generally somewhat lower than those observed in other mouse strains (C3H, CBA) (9,13,15). Nevertheless, these cells persisted in the marrow of neonatal and adult anti-IgM suppressed mice in numbers that were comparable to or even greater than those in normal and control mice (Table 18.1).

DISCUSSION

The present results demonstrate that in mice treated with anti-IgM antibodies for prolonged periods to eliminate all IgM-bearing small lymphocytes from both the marrow and peripheral lymphoid tissues, the marrow continues to produce large numbers of small lymphocytes lacking the surface markers of both B- and T-lymphocytes. Similar cells are seen whether mice are treated with anti-IgM from birth or as adults following lymphoid cell depletion. These findings are relevant to the control of the production of IgM-bearing B-lymphocytes in the marrow and to the identification of the critical stage in cell differentiation that first becomes sensitive to the suppressive effects of anti-IgM antibodies: this is shown schematically in Figure 18.10.

TABLE 18.1 Incidence of Large Lymphoid Cells in Marrow of Anti-IgM Suppressed Mice

TREATMENT	LARGE LYMPHOID CELLS (%)
Neonatal Suppression	
Normal mice	1.3
Anti-IgM from 1 day	1.0
Adult Suppression	
Cyclophosphamide alone	1.4
Anti-IgM alone	1.1
Cyclophosphamide + Anti-IgM	2.8

To be susceptible to a given sustained concentration of anti-IgM antibody, a differentiating B-lymphocyte may be expected to express an adequate number of IgM molecules to bind the antibody in critical amounts at the cell surface membrane. Normally, surface IgM first becomes detectable by surface anti-IgM binding techniques on non-dividing marrow small lymphocytes within a day or so after their production from proliferating precursors (13,-15,27) (Figure 18.10). When first formed, the marrow small lymphocytes are "null" cells, lacking surface markers of both B- and T-lymphocytes (13,14). The complete elimination of IgM-bearing small lymphocytes together with the persistence of near-normal levels of IgM-negative small lymphocytes in the marrow in the present work suggests strongly that the earliest anti-IgM sensitive stage corresponds with that at which the newly-formed "null" marrow small lymphocytes just begin to exhibit surface IgM detectable by radiolabeled antiglobulin-binding techniques (Figure 18.10). Thus for differentiating marrow B-lymphocytes emerging into an environment containing free anti-IgM antibody, the critical target cell for anti-IgM suppression appears to be the maturing small lymphocyte approximately 1 day post-mitotic. The low density of transient surface IgM molecules on the newly formed marrow small lymphocytes and on cycling progenitors (Figure 18.10) (7,15,27) would therefore appear to be ineffectual in mediating suppression by anti-IgM antibodies. This accords with the findings of Burrows et al. (2) that the marrow of anti-IgM treated mice continues to show normal numbers of large cells with weak cytoplasmic IgM staining by fluorescence, the presumptive B-lymphocyte progenitor cells. The present observations further indicate that, in the presence of anti-IgM, the progenitor cells continue to generate small lymphocytes, which then start their terminal maturation phase. This finding is consistent with the rapidity with which mature B-lymphocytes, characterized by surface IgM and LPS responsiveness, make an appearance when marrow cells from anti-IgM treated mice are incubated *in vitro* free of anti-IgM (2).

The binding of anti-IgM antibodies to the surface of B-lymphocytes *in vitro* results in the aggregation of surface complexes and the modulation of surface IgM, the cell subsequently surviving in an apparently surface IgM-negative state in the continuing presence of the antibodies (1,6,18,22,23). *In vivo*, however, the present studies suggest that anti-IgM exerts a direct cytotoxic effect rather than a modulation of surface IgM alone. Thus, the observed incidences of IgM-bearing and "null" marrow small lymphocytes are in accord with an actual elimination of the former, whereas the latter failed to show the usual incidences of other B-lymphocyte markers, Fc, and complement receptors (26). The few IgM-bearing small lymphocytes detected in anti-IgM suppressed mice showed the morphologic appearances of dying cells. The low incidence of such pyknotic cells in the marrow suggests that small lymphocytes binding anti-IgM antibodies either die and disintegrate rapidly or are released quickly from the marrow to die elsewhere. The killing of emerging virgin B-lymphocytes by anti-IgM antibodies is of considerable theoretical interest, in view of the possible parallel with the killing induced by exposure to antigens. When cultured with haptens coupled to carrier molecules, the marrow or neonatal spleen fail to generate cells specifically capable of responding to the particular hapten, consistent with the "clonal abortion"

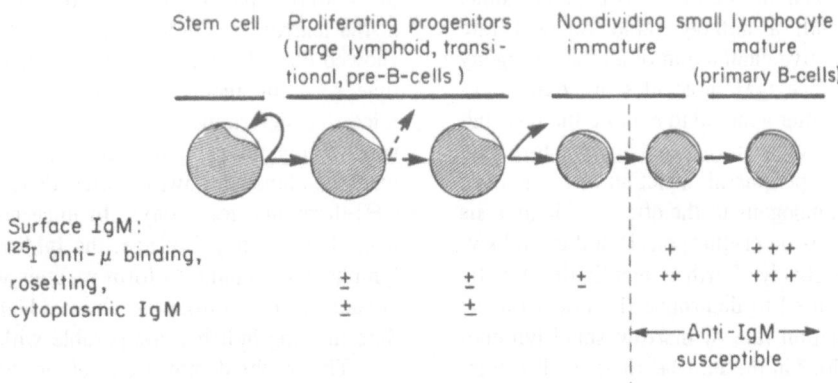

FIGURE 18.10. Scheme of the cell sequence involved in B-lymphocyte production in bone marrow showing the late expression of readily detectable surface IgM, the possible appearance at earlier stages of differentiation of surface IgM molecules with rapid turnover, and of small amounts of cytoplasmic IgM, and the stages susceptible to anti-IgM suppression *in vivo*, as detailed in the text.

hypothesis that maturing B-lymphocytes pass through a critical stage in development when the binding of antigens to surface IgM receptor molecules results in cell death rather than activation (11,-12,24). Thus, potentially self-reactive clones of cells normally may be eliminated before becoming functionally responsive, thus preserving self-tolerance. The kinetics of *in vitro* tolerogenesis are consistent with the results of the current work in suggesting the non-cycling, newly formed small lymphocyte as the target cell in the marrow for antigen-induced clonal abortion (12).

The age-related susceptibility to anti-IgM suppression, limited in normal mice to the first week or so of postnatal life, is confirmed in the present work. This has been attributed to the suppression of ontogenically "immature" B-lymphocytes in view of a susceptibility of neonatal spleen cells to irreversible modulation by anti-IgM *in vitro*, as compared with adult spleen cells (1,18,22,23). Immature B-lymphocytes, however, continue to be produced throughout life in the marrow, and such cells also appear to be sensitive to anti-IgM induced modulation *in vitro* (18). The current findings show that anti-IgM suppression can indeed by induced in suitably treated adults, whereas studies to be detailed elsewhere concerning the clearance of injected anti-IgM antibodies from the circulation are consistent with an alternative explanation, that the apparent resistance to anti-IgM *in vivo* with age may be related mainly to the synthesis of circulating IgM in quantities adequate to prevent effective access of the administered anti-IgM to the vulnerable target cells in the marrow.

Marrow lymphocytopoiesis in conventionally reared laboratory animals is regulated to maintain steady-state conditions. Apart from a possible influence stemming from environmental antigens, the responsible homeostatic mechanisms are largely unknown (14). In anti-IgM treated mice, the marrow lymphocyte progenitors continue to generate small lymphocytes with a notably rapid renewal rate despite the selective elimination of all their progeny as soon as surface IgM molecules are expressed. This appears to offer a model to explore the possibility that feedback or humoral regulatory factors may be stimulated by peripheral depletion of the pool of progeny cells, analogous to the effects of hemolysis and leukophoresis on erythropoiesis and granulocytopoiesis, respectively. Further quantitative kinetic studies are required to determine the effect on the absolute production rate of marrow small lymphocytes of variations in the IgM-bearing, small B-lymphocyte pool and in the products of activated B-lymphocytes, circulating IgM molecules, both of which are absent in anti-IgM treated mice.

Whereas many newly formed "null" small lymphocytes in the marrow are immature B-lymphocytes, others may mature along different lines, which are as yet speculative (27). Thus, maturation studies of marrow small lymphocytes from anti-IgM treated mice also provide an experimental system to explore the potentials of newly generated "null" lymphocytes.

SUMMARY

Small lymphocytes in bone marrow and spleen have been characterized by their surface markers and turnover in mice treated with anti-IgM antibodies. Experimental (C57BL/6XC3H)F_1 SPF mice were injected three times a week with either rabbit anti-mouse IgM gammaglobulin or normal rabbit gammaglobulin. Surface IgM was then detected autoradiographically by the binding of ^{125}I-labeled rabbit anti-mouse IgM. IgM-bearing small lymphocytes were eliminated from both marrow and spleen of mice given anti-IgM for 6 to 12 weeks starting at either day 1 or day 7 of age. When treatment started at 10, 15, and 20 days of age or in adults, the proportions of IgM-bearing small lymphocytes remained normal. But when adult mice were first either x-irradiated and reconstituted with marrow cells or given cyclophosphamide, the anti-IgM antibodies eliminated the IgM-bearing small lymphocytes from the marrow and spleen; normal incidences of IgM-bearing small lymphocytes were found after either x-irradiation and marrow reconstitution or cyclophosphamide alone. The marrow of anti-IgM suppressed mice showed normal incidences of IgM-negative small lymphocytes. These cells showed no masking of surface IgM by rabbit anti-mouse IgM. In the marrow, very few IgM-negative lymphocytes showed T-lymphocyte surface markers by indirect autoradiographic labeling, whereas a great majority did so in the spleen. Virtually no small lymphocytes in the marrow and spleen of anti-IgM treated mice showed Fc and complement receptors by cytocentrifuge rosetting techniques, compared with normal mice. To determine the turnover of IgM-negative small lymphocytes, autoradiography and rosetting were combined following twice daily injections of [^{3}H]-thymidine for 3 days. In mice given anti-IgM from day 1 or day 7 of age, the IgM-negative small lymphocytes, failing to form rosettes with anti-IgM coated sheep erythrocytes, showed high [^{3}H-]thymidine labeling indices, comparable with normal values. The results demonstrate that in the marrow the normal differentiation of small lymphocytes bearing IgM and other surface markers of B-lymphocytes can be prevented by anti-IgM antibodies in both neonatal and adult mice, but the marrow continues to generate many "null" small lymphocytes bearing neither B- nor T-lymphocyte surface markers. This suggests that in the marrow the critical differentiation stage at which cells become susceptible to anti-IgM suppression *in vivo* is that of the expression of readily detectable surface IgM on newly formed small lymphocytes in the first day or so after their production. The findings are relevant to the development of primary B-lymphocytes, their proposed sensitivity to antigen-induced "clonal abortion," the regulation of marrow lymphocytopoiesis, and the potentials of marrow "null" lymphocytes.

ACKNOWLEDGMENTS

The technical assistance of Mrs. Pat Young and Miss Els Schotman is gratefully acknowledged.

The work was supported by the Medical Research Council of Canada and the National Cancer Institute of Canada.

REFERENCES

1. Bruyne, C., Urbain-Vansanten, G., Planard, C., DeVos-Cloetens, C., and Urbain, J. Ontogeny of mouse B lymphocytes and inactivation by antigen of early B lymphocytes. *Proc. Nat. Acad. Sci. U.S.*, 73:2463, 1976.

2. Burrows, P. D., Kearney, J. F., Lawton, A. R., and Cooper, M. D. Pre-B cells: Bone marrow persistence in anti-μ-suppressed mice, conversion to B lymphocytes, and recovery after destruction by cyclophosphamide. *J. Immunol.*, 120:526, 1978.

3. Hunter, W. M., and Greenwood, F. C. Preparation of iodine-131-labeled human growth hormone of high specific activity. *Nature*, 194:495, 1962.

4. Kearney, J. F., Cooper, M. D., and Lawton, A. R. B lymphocyte differentiation induced by lipopolysaccharide. III. Suppression of B cell maturation by anti-mouse immunoglobulin antibodies. *J. Immunol.*, 116:1664, 1976.

5. Lala, P. K., Layton, J. E., and Nossal, G. J. V. Maturation of B lymphocytes. II. Sequential appearance of increasing IgM and IgD in the adult bone marrow. *Eur. J. Immunol.* (in press).

6. Manning, D. D. Heavy chain isotype suppression. A review of immunosuppressive effects of heterologous anti-Ig heavy chain antisera. *J. Reticuloendoth. Soc.*, 18:63, 1975.

7. Melchers, F., von Boehmer, H., and Phillips, R. A. B Lymphocyte subpopulations in the mouse. Organ distribution and ontogeny of immunoglobulin-synthesizing and mitogen-sensitive cells. *Transplant. Rev.*, 25:26, 1975.

8. Miller, S. C., Kaiserman, M., and Osmond, D. G. Small lymphocyte production and lymphoid cell prolif-

eration in mouse bone marrow. *Experientia, 34*:129, 1978.

9. Miller, S. C., and Osmond, D. G. Quantitative changes with age in bone marrow cell populations of C3H mice. *Exp. Hematol.* 2:227, 1974.

10. Murgita, R. A., Mattioli, C. A., and Tomasi, T. B., Jr. Production of a runting syndrome and selective γA deficiency in mice by the administration of anti-heavy chain antisera. *J. Exp. Med., 138*:209, 1973.

11. Nossal, G. J. V., and Pike, B. L. Evidence for the clonal abortion theory of B lymphocyte tolerance. *J. Exp. Med., 141*:904, 1975.

12. Nossal, G. J. V., Shortman, K., Howard, M., and Pike, B. L. Current problems areas in the study of B lymphocyte differentiation. *Immunological Rev., 37*:187, 1977.

13. Osmond, D. G. Formation and maturation of bone marrow lymphocytes. *J. Reticuloendoth. Soc., 17*:99, 1975.

14. Osmond, D. G. Potentials of bone marrow lymphocytes. In Cairnie, A. B., Lala, P. K., and Osmond, D. G., eds., *Stem Cells of Renewing Cell Populations*, New York: Academic Press, 1976.

15. Osmond, D. G., and Nossal, G. J. V. Differentiation of lymphocytes in mouse bone marrow. II. Kinetics of maturation and renewal of antiglobulin-binding cells studied by double labelling. *Cell. Immunol., 13*:132, 1974b.

16. Osmond, D. G., and Rahal, M. D. Differentiation of bone marrow lymphocytes: expression of surface Ia and H-2K antigens. *Anat. Rec., 190*:497, 1978.

17. Osmond, D. G., Wherry, P. E., Daeron, M., and Gordon, J. Immunoglobulin-bearing cells in bone marrow of mice after prolonged treatment with anti-IgM antibodies. *Nature, 260*:328, 1976.

18. Raff, M. C., Owen, J. J. T., Cooper, M. D., Lawton, III, A. T., Megson, M., and Gathings, W. E. Differences in susceptibility of mature and immature mouse B lymphocytes to anti-immunoglobulin-induced immunoglobulin suppression *in vitro*. *J. Exp. Med., 142*:1052, 1975.

19. Raff, M. C., Megson, M., Owen, J. J. T., and Cooper, M. D. Early production of intracellular IgM by B lymphocyte precursors in mouse. *Nature, 259*:224, 1976.

20. Rosenberg, Y. O., and Parish, C. R. Ontogeny of the antibody forming cell line in mice. IV. Appearance of cells bearing Fc receptors, complement receptors, and surface immunoglobulin. *J. Immunol., 118*:612, 1977.

21. Rosse, C. Small lymphocyte and transitional cell populations of bone marrow: their role in the mediation of immune and hemopoietic progenitor cell functions. *Int. Rev. Cytol., 45*:155, 1976.

22. Sidman, C. L., and Unanue, E. R. Receptor-mediated inactivation of early B lymphocytes. *Nature (London), 257*:149, 1975.

23. Sidman, C. L., and Unanue, E. R. Development of B lymphocytes. I. Cell populations and a critical event during ontogeny. *J. Immunol., 114*:1730, 1975.

24. Stocker, J. W. Tolerance induction in maturing B cells. *Immunology, 32*:283, 1977.

25. Stocker, J. W., Osmond, D. G., and Nossal, G. J. V.

Differentiation of lymphocytes in the mouse bone marrow. III. The adoptive response of bone marrow cells to a thymus cell-independent antigen. *Immunology*, 27:795, 1974.

26. Yang, W. C., and Osmond, D. G. Maturation of bone marrow lymphocytes. I. Quantitative rosetting methods of detecting Fc and complement receptors and surface immunoglobulin. *J. Immunol. Meth.*, (in press).

27. Yang, W. C., Miller, S. C., and Osmond, D. G. Maturation of bone marrow lymphocytes. II. Development of Fc and complement receptors and surface immunoglobulin studied by rosetting and radioautography. *J. Exp. Med.* (in press).

28. Yoshida, Y., and Osmond, D. G. Identity and proliferation of small lymphocyte precursors in cultures of lymphocyte rich fractions of guinea pig bone marrow. *Blood*, 37:73, 1971.

19

The Thymic Microenvironment

R. K. Jordan, D. A. Crouse,
C. M. Harper, E. B. Watkins,
and J. G. Sharp

All the mature cellular elements of the blood (erythrocytes, granulocytes, platelets, monocytes, and lymphocytes) require the continuous replacement of their respective populations throughout the life-span of the animal. Contemporary concepts point to the pluripotential hemopoietic stem cell as the ancestral cell for all the various differentiating populations. Recently, a model of the process of hemopoiesis from the stem cell to mature progeny has been described as consisting of three overlapping pyramidal-shaped hierarchies of cell compartments (24). Using this model of hemopoietic differentiation, which has been extensively investigated and described for myeloid elements, we wish to propose a parallel pyramidal system of differentiation for the T-lymphocyte series (Figure 19.1).

Since the work of Miller (26) and Martinez et al (23), there have been numerous demonstrations that the presence of the thymus is essential for the proper development and maintenance of cell-mediated, as well as certain aspects of humoral immunity (9,19). Indeed, the cells participating in cell-mediated immunity are lymphocytes derived from the thymus (T-lymphocytes). The development of such T-lymphocytes involves the migration of stem cells into the thymus, the differentiation from stem cells into thymocytes and further into non-immunocompetent small lymphocytes that leave the thymus, and finally the maturation of these precursors to immunocompetent T-lymphocytes in the periphery (28,38).

In the hemopoietic system, the term stem cell is generally reserved for cells capable not only of extensive self-replication but also the generation of various progenitor cells for all the blood cells types, i.e., they are multipotential. Although lymphocytes arise from hemopoietic stem cells as do all other hemopoietic cells (25), stem cells of restricted potentiality have also been demonstrated and recently stem cells restricted to T-lymphocyte differentiation have been detected in mouse bone marrow and spleen (1,2). In the three-tiered hierarchy of hemopoiesis, progenitor cells are defined as being capable of a limited degree of self-replication, but they differ from stem cells in that they and their progeny are restricted to a specific pathway of hemopoiesis. The T-restricted stem cells therefore should be more properly termed T-lymphocyte progenitors (see Figure 19.1). Thus, the lymphoid component of the mouse thymus appears to be derived from such progenitors that migrate into the thymus from the bone marrow, or in the embryo, from the developing liver rather than from pluripotent migrant stem cells per se (24,25).

The period of intrathymic differentiation is characterized at the morphological level by a graded reduction in cell size from large through medium to small lymphocytes (34). Furthermore, cell prolifera-

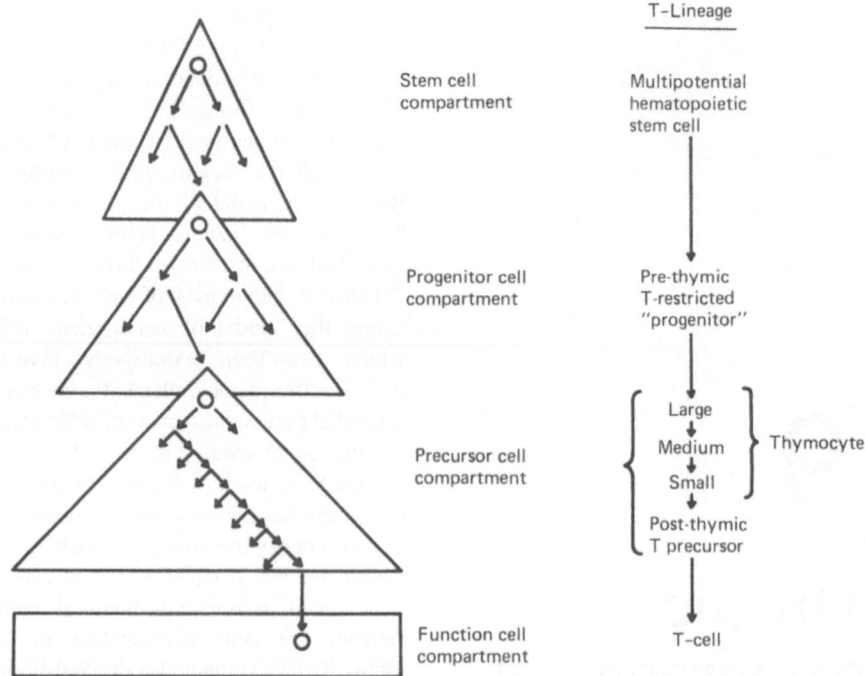

FIGURE 19.1. A schematic representation of the three-tiered hierarchy of hemopoiesis applied to the T-cell.

tion is a major feature, although there is a lack of precise kinetic information at this time. Basically, two major populations of small thymic lymphocytes may be distinguished: cortical small lymphocytes and medullary small lymphocytes (30). These populations differ from one another with respect to surface alloantigens, steroid and radiation sensitivity, and functional competence. It had been thought for some time that the steroid-resistant, immunocompetent small medullary lymphocytes were derived from "immature" cortical small lymphocytes and represented the precursors of peripheral T-cells. But more recently, independent pathways of maturation for cortical and medullary thymocytes have been proposed, with both contributing directly to the peripheral T-lymphocyte pool (8). In keeping with the idea that cortical small lymphocytes are the precursors of large numbers of peripheral T-cells is the demonstration by Stutman and his colleagues of a post-thymic T-precursor population, consisting of non-immunocompetent cells that are functionally mature in the periphery under the humoral influence of the thymus (38). Thus, we consider that this post-thymic precursor cell should be included with the thymocytes (large, medium, and small) in comprising the T-lymphocyte precursor compartment (see Figure 19.1).

In the myeloid series, complex regulatory mechanisms are known to operate at all levels of the hierarchy of differentiation. Thus, (a) multipotential stem cells are committed to a specific pathway of differentiation (i.e., stem cell to progenitor cell transition) by the influence of the hemopoietic-inducing microenvironment (HIM) (41) and/or by entirely random hemopoiesis (HER) (39), (b) recruitment and proliferation in the progenitor compartment is controlled by specific humoral factors [e.g., colony stimulating factor (CSF), erythropoietin (EPO)] (40), (c) the continuing proliferation and maturation of cells within the precursor compartment also appears to be dependent upon the presence of the same or similar humoral factors (47), and (d) feedback controls are known to operate at least at some levels (18).

In the T-lymphoid series, it is probable that the initial step (i.e., stem cell to T-restricted progenitor) is similarly controlled by HIM or HER; in subsequent stages, however, there appear to be significant differences. This is not surprising if one considers that, in myeloid differentiation, populations of mature cells of functional homogeneity are produced (e.g., erythrocytes for oxygen transport), whereas during lymphoid differentiation there is an additional post-progenitor generation of diversity in function with respect to antigen recognition and response.

In the myeloid progenitor compartments, specific humoral factors play a major regulatory role, but the regulation of prethymic T-progenitor cells encompasses more than humoral mechanisms. *In vivo* studies of the restoration of neonatally thymectomized mice obtained using thymus grafts have revealed that hemopoietic cell populations that contain prethymic progenitor cells (but no post-thymic T-lymphocyte precursors) will cooperate only with free-thymic grafts in restoring cell-mediated function (38), that is, prethymic T-progenitors must interact directly with the non-lymphoid cells of the thymus for their proliferation and differentiation. If the thymus graft is enclosed in a cell-tight diffusion chamber, reconstitution is not achieved (38). Thus it appears that, within the T-lineage, the transition from progenitor to precursor is a discrete step,

dependent upon the thymic microenvironmental influence. It would thus appear that the thymic non-lymphoid cells provide not only a simple framework for the organ but also the cellular basis of a specific microenvironment capable of sustaining and directing the proliferation and differentiation of the T-lymphocyte precursor compartment. These regulatory concepts as they apply to the T-lineage are embodied in Figure 19.2.

The concept of a specific thymic microenviron-

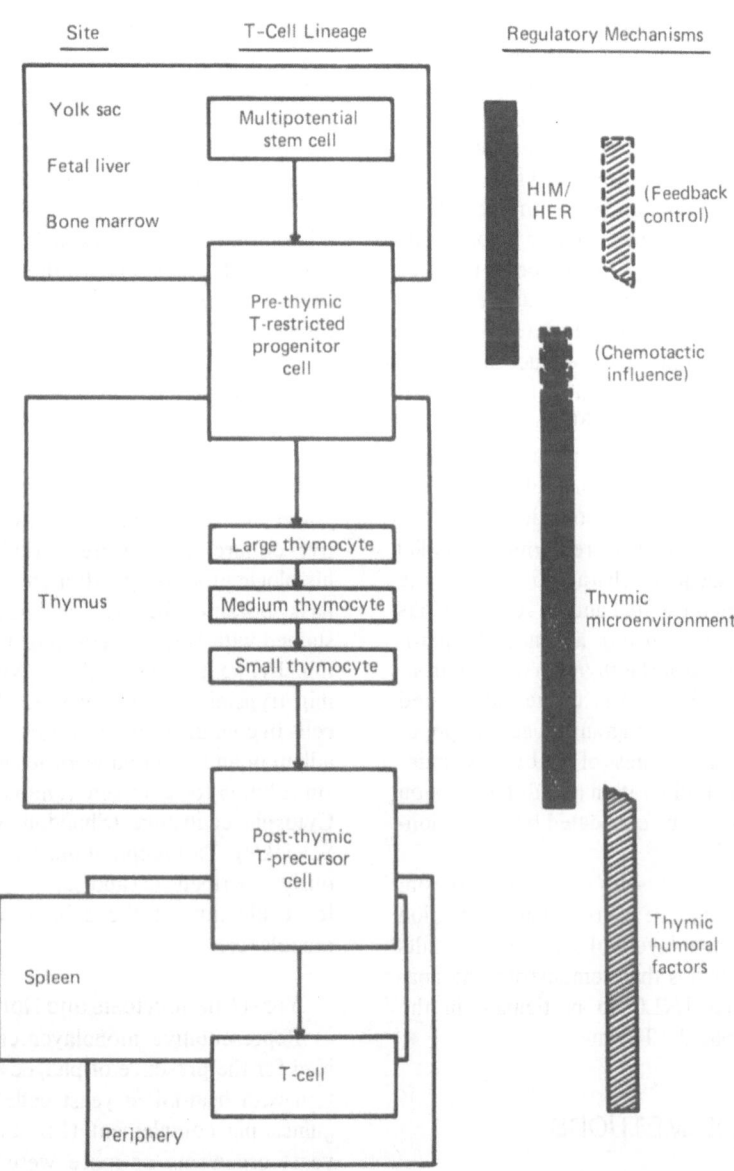

FIGURE 19.2. A diagrammatic presentation of the regulatory influences upon the T-cell lineage. The transition of the pluripotential stem cell to the progenitor series is probably controlled by HIM/HER. The proliferation and maturation of precursor populations is regulated by the thymic microenvironment, whereas the final step of maturation occurs in the periphery under the influence of thymic hormonal factors.

ment does not necessarily require that the inductive events regulating thymocyte maturation involve cell contact, since focal production and limited diffusion of specific factors are not precluded. But the type of critical analysis required for the resolution of this and other questions regarding the nature of the cell/ cell interactions within the thymic milieu depends upon the isolation, characterization, and functional testing of the cellular elements that comprise the non-lymphoid population. Eventually this means that individual cell types must be cloned and examined for functional activity as pure populations, but at present, most studies are based on the simple "dissection" of the two components of the interaction (i.e., the microenvironment and the immigrant progenitors), and their subsequent recombination under controlled conditions. The most direct approach to the initial dissection is to use simple *in vitro* monolayer cultures. Most previous studies have also used recombination procedures but have employed only *in vitro* protocols. These involved the co-culture of thymic non-lymphoid cells (TNLC) with putative T-precursor cells (27,44) as well as the evaluation of thymic humoral factors (17,32). Such techniques have had only limited success in the induction of T-cell function. In all of these *in vitro* studies, it has been assumed, without independent assessment, that TNLC derived and maintained in culture express the microenvironmental milieu. Furthermore, it has been generally considered that the cultured TNLC in these studies are thymic epithelial cells, although no adequate characterization of the monolayers has been presented and no reference has been made as to the presence or absence of macrophages. We believe that if the thymic microenvironment (i.e., TNLC) is isolated in tissue culture and then "rejoined" with the immigrant progenitor populations *in vivo,* the effectiveness of TNLC in sustaining and directing the proliferation and differentiation of such progenitors may be elucidated both functionally and histologically.

In this chapter we present a summary of our studies to date. We include a brief characterization of TNLC grown in monolayer culture and the results of transplantation studies that demonstrate the ability of such cultured TNLC to participate in the formation of a lymphoid "Thymus."

MATERIALS AND METHODS
Animals

Inbred CBA, BABL/c, and C57BL/6 mice were used throughout these studies unless otherwise stated. Newborn mice obtained from inbred matings were used as thymus donors within 24 hr of birth. All grafting studies were performed in fully syngeneic systems with adult mice used as recipients at 6 to 8 weeks of age. Mice were maintained under conventional conditions with food and water supplied *ad libitum.*

Monolayer Cultures

Thymic non-lymphoid cells were established in monolayer cultures by explanting approximately 25 fragments of newborn mouse thymus into each 25 cm² tissue culture flask. Initial explants were set up in a small amount of RPMI 1640 supplemented with 10% fetal calf serum, 200 mM 1-glutamine, and antibiotics (100 μg/ml penicillin and 100 mcg/ml streptomycin). After 48 to 72 hr, the original media was carefully aspirated from the flask and fresh complete media was added without disturbing the thymic explants. Subsequent media changes were made as required and increased in relative frequency as the monolayers approached confluency. All cultures were incubated at 37°C in a humidified atmosphere of 5% CO_2 in air. Under these conditions, confluent monolayers were observed by 21 to 28 days post-initiation. In the course of the experiments reported here, over 300 such cultures have been initiated and observed.

Monolayer Histology and Cytocentrifuge Procedure

During the period of monolayer development, cultures were regularly observed with an inverted phase microscope (Leitz, Diavert) and representative culture flasks were periodically removed for histologic processing. After carefully cutting free the flask bottom, the growing surface was routinely stained with Wright-Giemsa or toluidine blue. Additionally, some flasks were harvested by a 5- to 10-min trypsinization (Trypsin-EDTA) at 37°C. Many cells in confluent monolayers were found to be quite adherent and required vigorous agitation in the trypsin solution for complete removal of the monolayer. Cytospin centrifuge (Shandon Southern Inst. Co., Sewickley, Pa.) preparations were obtained as previously described (13) and stained with Wright-Giemsa for evaluation of the cellular components of the monolayer.

Yeast Phagocytosis and Nonspecific Esterases

Representative monolayer cultures were examined for the presence of phagocytic cells by incubation with heat-killed yeast cells in the presence of guinea pig complement (12). Cells ingesting three yeast organisms or more were considered phagocytic. Similar representative cultures, as well as trypsinized cytospin preparations, were stained with α-napthylacetate (33) to evaluate the relative frequency of nonspecific esterase-positive cells. Some cultures were stained for nonspecific esterases following the yeast-phagocytosis assay.

Thymic Non-Lymphoid Cell Function

The protocols employed in these studies are outlined in Figure 19.3. The experiments were of three basic designs: (a) transplantation, (b) organ culture, and (c) co-culture. For the transplantation and organ culture studies, TNLC were trypsinized from confluent 28-day-old monolayers, as previously described. The cellular content of a single culture vessel was then spun down (400 g for 10 min) in a small hand-forged conical bottom tube. The media was decanted and the tube was then scored and broken near the pellet surface. In grafting studies, intact 6- to 8-week-old mice, syngeneic to the monolayer, were used as recipients. Mice were anesthetized with ether and kidney grafted with the TNLC pellet. Details of this procedure have been presented elsewhere (13). Thirty days after grafting, mice were sacrificed for histologic assessment of the graft site. The graft and surrounding kidney were dissected in chilled phosphate buffer and fixed in 2% glutaraldehyde (0.2 M phosphate buffer, pH 7.3) for 40 min at 4°C, dehydrated, and embedded in Araldite. Sections (1 μm) were stained with toluidine blue and examined with the light microscope. Additionally, ultrathin sections were stained with lead citrate and uranyl acetate for examination with an *AEI-EM6B* electron microscope.

In organ culture studies, the pellets of TNLC were cultured for 7 days according to the method described by Owen et al. (31). Organ cultures were examined histologically for signs of organotypic development.

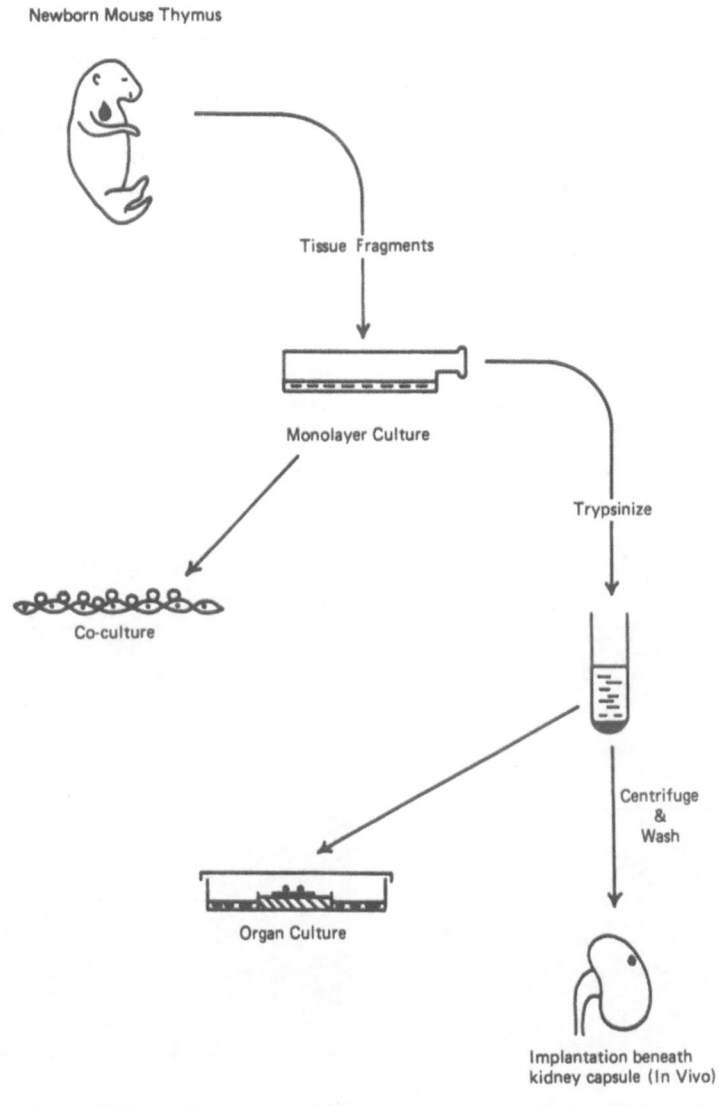

Newborn Mouse Thymus

Tissue Fragments

Monolayer Culture

Co-culture

Trypsinize

Centrifuge & Wash

Organ Culture

Implantation beneath kidney capsule (In Vivo)

FIGURE 19.3. Protocol employed in the assessment of TNLC function *in vivo* and *in vitro*.

In co-culture studies, target cell preparations (*nu/ nu* spleen or neonatally thymectomized mouse marrow) were cultured on confluent syngeneic TNLC monolayers for 24 hr. Cells were then recovered by aspiration and washing of the monolayer culture. Recovered populations were counted and adjusted in concentration for microwell cultures of mitogen stimulation [Phytohemaglutinin (PHA), concanavalin-A(Con-A)]. Lipopolysaccharide (LPS), routine 3-day microcultures were processed and harvested by standard techniques (10).

RESULTS

Characteristics of Thymic Non-Lymphoid Cell Monolayers

Neonatal mouse thymus explants established in tissue culture produce cellular outgrowths that give rise to a confluent monolayer in about 21 days. The morphology of such confluent monolayers is heterogeneous. Although for the most part they are composed of cells that resemble macrophages in culture (37), there are areas in which the cells are epithelial

in appearance, being arranged in small groups or islands in which there is contiguity of adjacent cell margins (Figures 19.4 and 19.5).

In order to confirm the macrophage nature of the majority of cultured TNLC, we have tested their phagocytic capactiy *in vitro*. Upward of 80% of monolayer cells are actively phagocytic toward yeast in the presence of guinea pig complement. Furthermore, these same cells contain large amounts of nonspecific esterase as demonstrated histochemically by the α-naphthylacetate method (Figure 19.6). Strong positive-staining reactions are considered to be a good indication of cells from the macrophage/ monocyte cell series (14). Some distortion of the phagocytes results from the phagocytosis of yeast, but it is not sufficient to preclude their identification with the phase-contrast microscope or in stained preparations. They compromise a population of cells that may modulate their form from a rounded profile, or spread cytoplasm with a central nucleus, to a shape we describe as dendritic because of the extended cytoplasmic processes they exhibit (see Figure 19.4). The arrangement of closely associated "spread" and "dendritic" cells in culture is a feature characteristic of epitheloid macrophages, and as this

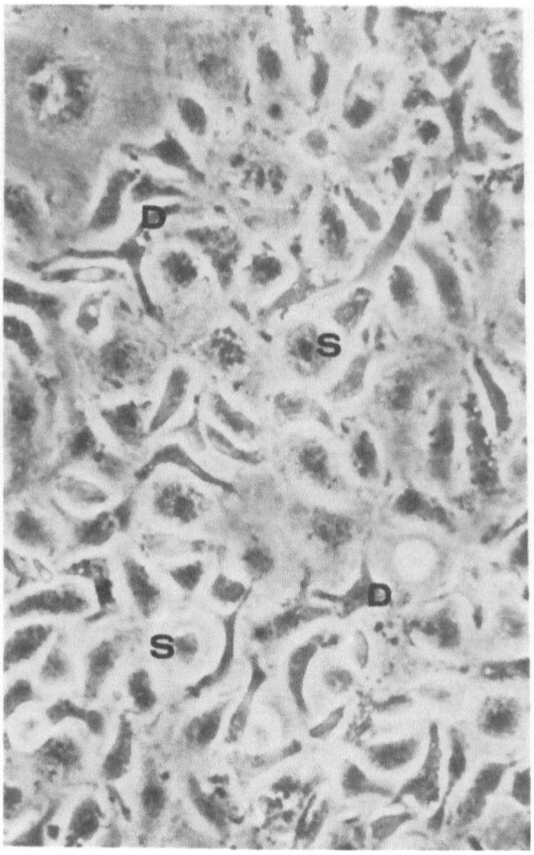

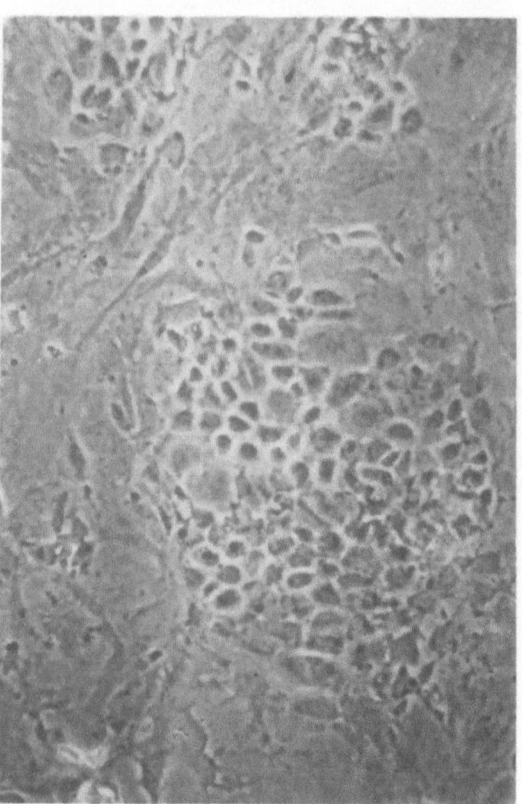

FIGURE 19.4. Phase contrast. A 28-day-old TNLC monolayer composed primarily of spread (S) and dendritic (D) cell types.

FIGURE 19.5. Phase contrast. An area of a confluent TNLC monolayer exhibiting focal epithelial growth and a few associated fibroblasts.

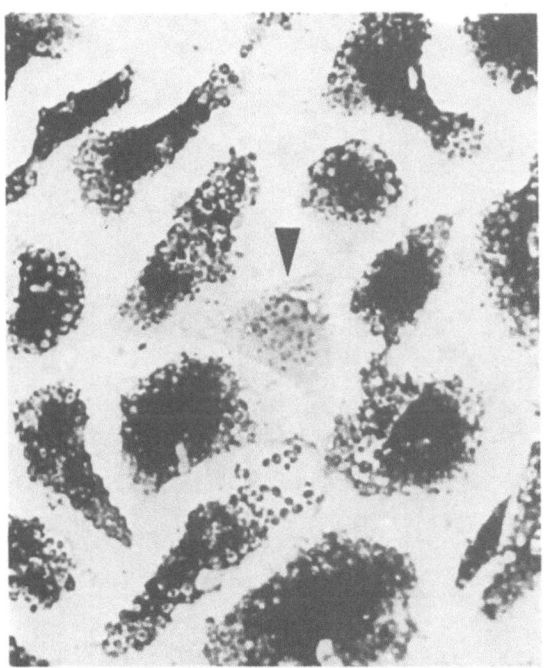

FIGURE 19.6. α-Napthylacetate stained preparation of a confluent TNLC monolayer after the yeast phagocytosis assay. Note the predominance of cells that have ingested numerous yeast organisms and stained intensely (brown/black) for nonspecific esterase. A single cell (arrow), which is negative for both end points, is also shown.

name suggests, these may be readily confused with true epithelial cells by the unwary.

Scattered among the phagocytically active cells are other cells that, together with those cells forming the epithelial islands, are neither phagocytic nor positive for nonspecific esterase even after stimulation with endotoxin (5) (see Figure 19.6).

Although fibroblasts per se are never anything but a minor contaminent in our confluent TNLC cultures, after about 28 days of culture, structures resembling the "fibroblastic" plaques (FCFC) described by Friedenstein (7) appear. These were first seen as small aggregates of cells arranged concentrically around an amorphous eosinophilic center; they increase in size and come to closely resemble the plaques obtained in monolayer cultures derived from other lymphohemopoietic tissues (7,46). Although these TNLC plaques consist of nonphagocytic cells, we are not convinced of their fibroblastic nature; they may, at present, also be related to the epithelial component of TNLC. In this context, it is interesting to note that such plaques have been considered as promising candidates for colonies of hemopoietic stromal cells (24). Furthermore, the radiation survival curves of TNLC demonstrate a biphasic shape characteristic of a heterogeneous cell population. One, a radiosensitive population

with the survival parameters of macrophages ($D_0 = 125$ rad, $N = 1.35$ and $D_q = 40$ rad) and the other, a very radioresistant population with characteristics similar to those reported by Wilson et al. (46) for PFU-c and which, morphologically, resemble the FCFC of Friedenstein (7).

Depletion of Thymocytes

Although the initial cultures contain many thymocytes, these largely disappear in the first few days of culture. After varying periods of incubation, we have checked for the presence or absence of lymphocytes and for the capacity of monolayer cells to reform a lymphoepithelial structure in organ culture. Examination of cytocentrifuge preparations shows that a small number of lymphocytes (1 to 2%) persist after 15 days of culture; but we have found no morphologic evidence for the presence of lymphocytes beyond 20 days (Figure 19.7).

Newborn mouse thymus can be disaggregated into single cells and following reaggregation will form a lymphoepithelial structure in organ culture (31). Using this as a base line, we have examined the ability of monolayer cells to form lymphoepithelial organ cultures after reaggregation. Reaggregates of 15-day monolayer cells contain lymphoepithelial areas, whereas reaggregates of 20-day monolayer cells do not. We therefore conclude, that after 3 weeks in monolayer, TNLC are free of lymphocytes and thus such cultures have lost their intrinsic capacity of organotypic growth when reaggregated as organ cultures.

Thymic Non-Lymphoid Cell Function

In co-culture experiments with TNLC monolayers, we have confirmed a number of previous studies that purport to demonstrate the functional capacity of thymic monolayers in the induction of mitogen responsiveness in putative precursor cells (27,44) (see Table 19.1). The minimal responses obtained from such studies, similar in magnitude to those obtained with purified thymic extracts, are unconvincing. We believe that a more definitive method of testing for thymic function is to transplant the cultured TNLC back to an *in vivo* environment and to study (a) their ability to reform a lymphoid "thymus" and (b) their ability to participate in the restoration of immunocompetence in immunodeficient hosts.

In a preliminary series of experiments, 29 6-week-old intact mice (CBA or BALB/c) were grafted with spin-mediated aggregates of syngeneic cultured TNLC. The TNLC were obtained by trypsination of 28-day monolayer cultures. These preparations were screened for the presence of lymphoid cells by morphologic appraisal of cytocentrifuge preparations. All 29 TNLC preparations were free of lymphoid

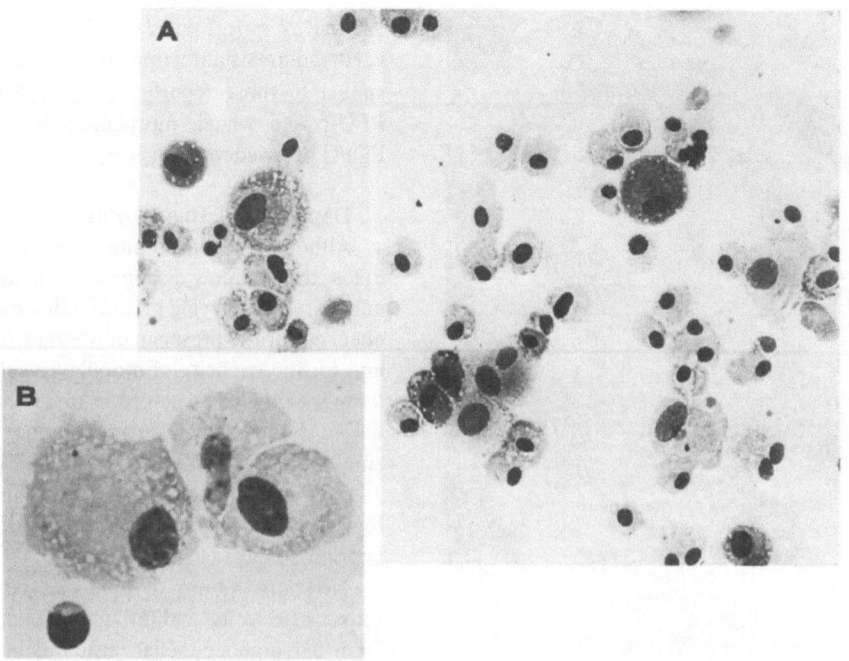

FIGURE 19.7. Wright-Giemsa stained cytocentrifuge preparations of trypsinized TNLC cultures. (**A**) Low-power view of 28-day preparation with no lymphoid cells. (**B**) A 15-day-old preparation with lymphoid contamination.

cells and were subsequently grafted under the renal capsule for approximately 30 days. In 16 recovered grafts, lymphoepithelial structures were found (Figure 19.8). In addition, all 29 grafts contained cystic structures (Figure 19.9). The walls of such cysts consist of multinucleated giant cells that bear a strong morphologic resemblance to fused macrophages, fusion being a feature demonstrated when macrophages encounter insoluble material.

The lymphoepithelial structures that developed in the grafts (Figure 19.8) resembled developing thymus with a central area of epithelium containing a few lymphocytes ("medulla") and a peripheral zone consisting predominantly of lymphocytes but inter-

spersed with occasional "epithelial" cells ("cortex"). The epithelial nature of the cells in the central area was confirmed by the demonstration of desmosomal junctions in electronmicroscopic studies (13). Therefore, we propose that the epithelial population, which comprises a small minority of cultured TNLC, is responsible for the microenvironmental effects of thymic traffic on precursor cells.

In an initial attempt to reconstitute deficient animals with TNLC grafts, 24 6-week-old CBA mice were used. These animals were divided into three groups: (a) Control—no treatment; (b) thymectomized, irradiated (800 rad, 60 Co-γ), and reconstituted with 1×10^7 syngeneic bone marrow cells

TABLE 19.1 Mitogenic Responsiveness of Target Populations Following 24 hr of Co-Culture with TNLC or Control Fibroblast Monolayers

INDUCING MONOLAYER	TARGET POPULATION							
	Bone Marrow from Normal Thymectomized Mice				Spleen from nu/nu Mice			
	No Mitogen	PHA	CON A	LPS	No Mitogen	PHA	CON A	LPS
TNLC	8,507 ±[a] 1,245	15,128 ± 2,238	8,725 ± 1,193	18,758 ± 1,687	2,443 ± 385	4,593 ± 273	2,119 ± 435	12,207 ± 3,810
Fibroblast	18,999 ± 1,814	25,692 ± 7,852	19,914 ± 2,414	20,378 ± 3,232	5,826 ± 438	5,290 ± 735	3,134 ± 447	13,069 ± 1,419

[a]Blastogenic response of target cells expressed as a mean absolute DPM ± S.D (DPM, disintegration per minute; S.D. = standard deviation). (N = 9 for each point) with background (i.e., response of cells *not* concultured with monolayers) subtracted.

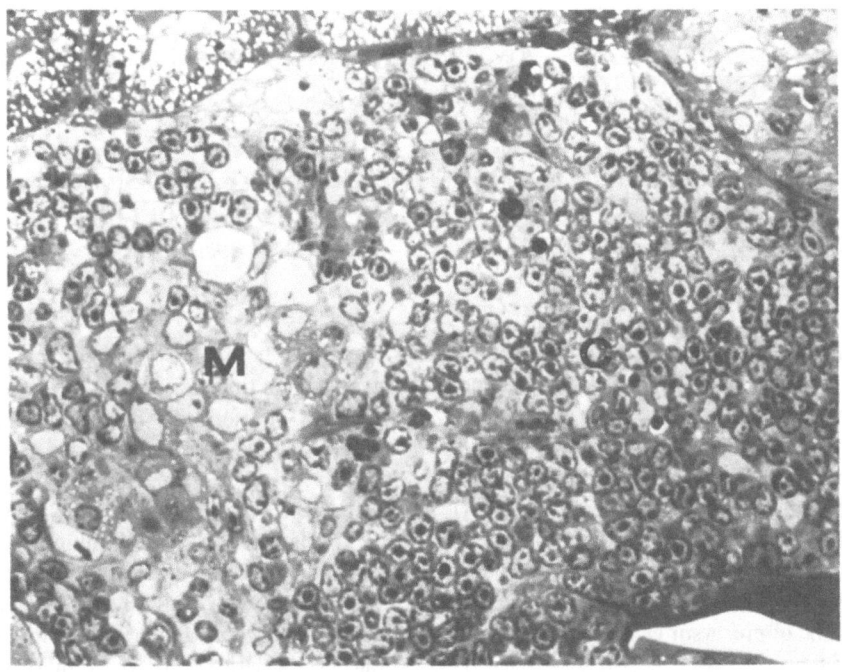

FIGURE 19.8. Araldite-embedded, toluidine blue-stained section of a graft of TNLC recovered at 30 days post-transplantation. The lymphoepithelial nature of the graft is evident and can be divided into two morphologically different areas; (1) an area rich in lymphocytes with a few interspersed epithelial cells, "cortex" (C), and (2) an area with fewer lymphocytes and an abundance of epithelial cells, "medulla" (M).

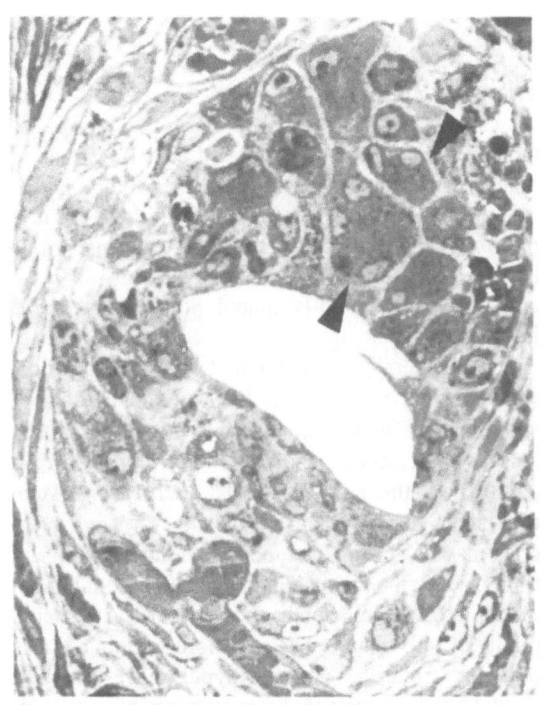

FIGURE 19.9. Araldite-embedded, toluidine blue-stained section of a typical cystic structure formed at the TNLC graft site. Several multi-nucleated giant cells are present (arrows).

(ATxBM); and (c) thymectomized, irradiated (800 rad, ^{60}Co-γ), reconstituted with 1×10^7 syngeneic bone marrow cells, and grafted with cultured TNLC (ATxBM/TNLC). All animals were maintained in filter-top cages under conventional conditions. At 34 days post-grafting, all animals were orbitally bled and complete blood counts were obtained (see Table 19.2).

In view of the significant difference in the absolute lymphocyte counts of the ATxBM and the ATxBM/TNLC mice, the animals were sacrificed. At the time of sacrifice (i.e., 35 days), the ATxBM/TNLC mice appear to have been afforded a measure of "protection," which was reflected in the general health of the animals and their body weights (see Table 19.2). At sacrifice, all grafts were recovered for histologic examination, and cell preparations were made from the lymph nodes and spleens for mitogen stimulation studies. In addition, all experimental mice were autopsied to confirm complete thymectomy.

Our attempts to demonstrate restoration of T-cell responsiveness in ATxBM/TNLC mice, as indicated by responses to specific T cell mitogens (PHA, Con A), were inconclusive. Histologic examination of the recovered grafts, however, demonstrated the same two major features seen in the preliminary experiments. Cystic structures with multinucleated giant

TABLE 19.2 Absolute Differential Counts and Body Weights in TNLC Reconstituted Mice.

	LYMPHOCYTES	MONOCYTES	NEUTROPHILS	EOSINOPHILS	BODY WEIGHTS
Control	4,085 ± 100	308 ± 35	897 ± 67	71 ± 18	18.8 ± 0.3
AT X BM	1,183 ± 431[a,b]	327 ± 145	5,766 ± 826[a]	75 ± 13	14.5 ± 0.3[a,c]
AT X BM – TNLC	3,005 ± 564	387 ± 67	4,792 ± 435[a]	132 ± 49	18.0 ± 0.5
	Results are expressed as means ± SEM; significant differences are indicated in footnotes				

[a]Significantly different from control $p < .01$.
[b]Significantly different from ATxBM/TNLC $p < .05$.
[c]Significantly different from ATxBM/TNLC $p < .01$.

cells were found in great abundance. Infrequent epithelial areas were also found (Figure 19.10); but such epithelial areas contained only small numbers of large basophilic cells (arrow). Such large basophilic cells are seen as the first signs of thymic lymphopoiesis in the fetal mouse (29), the fetal sheep (11), and the chick embryo (21). Similar large basophilic cells are also seen in regenerating thymus (6). Thus, the presence of the large basophilic cells indicate to us that the traffic of precursor cells is only just commencing at 35 days.

Further reconstitution studies are in progress in

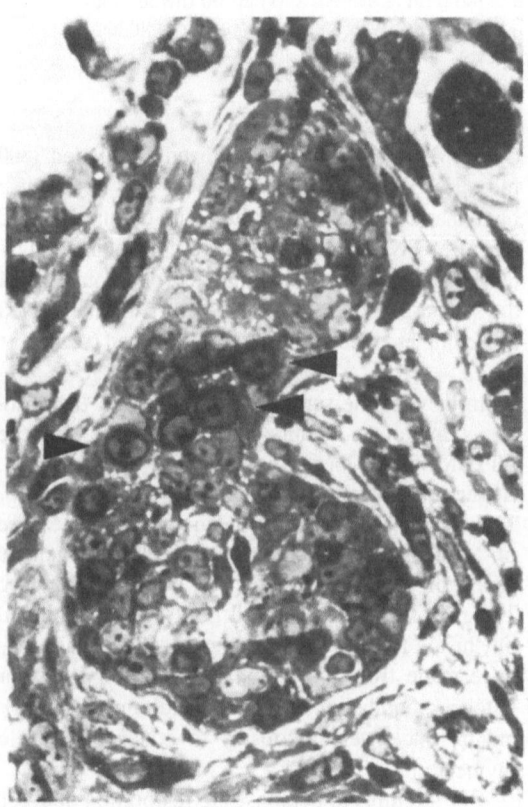

FIGURE 19.10. Araldite-embedded toluidine blue-stained section of a "lymphoepithelial" structure found 35 days after TNLC grafting of an immunodeficient host. Several large basophilic cells are present (arrows).

which assessment of functional reconstitution is evaluated at times beyond 35 days. In addition to the ATxBM mouse, other immunodeficient models are incorporated in these studies.

Macrophage/Lymphocyte Clusters

In the first 48 hours of culture, large numbers of lymphocytes are exuded from the explanted thymic fragments. Some of the lymphocytes ahhere to the earliest non-lymphoid migrants to form distinctive cluster formations. Many of these clusters, which are stable in nature, may be harvested from the forming monolayer by aspiration. In cytocentrifuge preparations, the clusters are seen to consist of a central mononuclear cell surrounded by adherent lymphocytes (Figure 19.11). Staining of such clusters with α-naphthylacetate demonstrates that the central cell contains large amounts of the enzyme, nonspecific esterase (inset, Figure 19.11).

DISCUSSION

Quantitatively, the proportions of cells (i.e., macrophages versus epithelial cells) in the 28-day TNLC monolayer is probably not representative of the normal *in vivo* thymus. But in qualitative terms, we do believe that such TNLC monolayers can, on occasion, reflect the non-lymphoid populations of the intact organ.

It is generally assumed that the principal non-lymphoid cell of the thymus is the epithelial reticular cell. This assumption, although possibly correct, ignores the complexity of the intact thymus for, in addition to the epithelial cells, there are a large number of resident macrophages (3,16). Their presence may be strikingly demonstrated by the application of specific histochemical or functional techniques (16,35), and further supported by the large number of macrophage colony-forming cells obtained from the normal thymus (22). Thus, it should be no surprise that macrophages always comprise the predominant component of the TNLC monolayer, especially since tissue culture often favors macrophage proliferation.

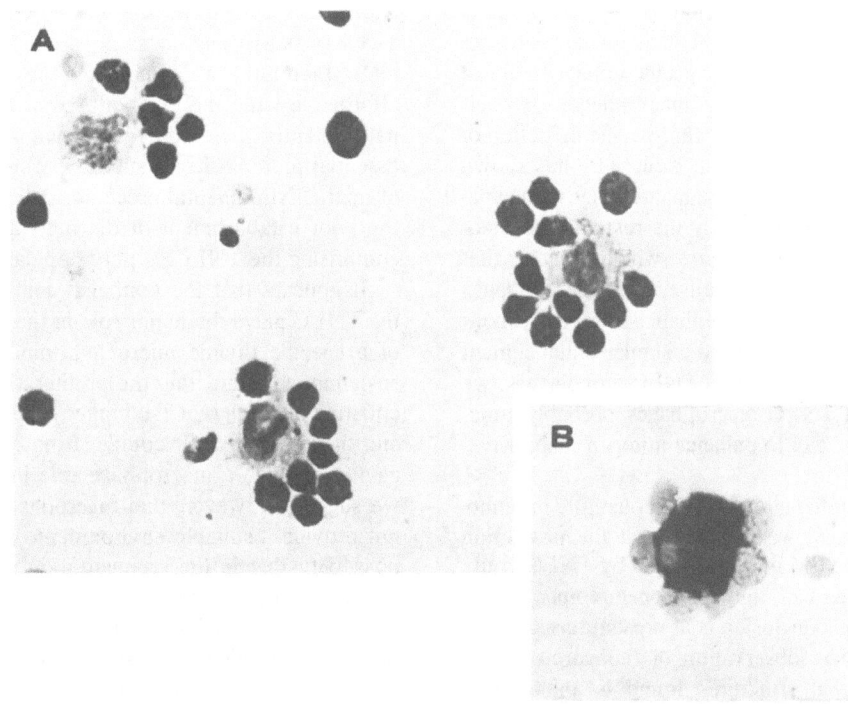

FIGURE 19.11. Cytocentrifuge preparations of cell clusters formed in fresh thymus explant cultures. (A) Wright-Giemsa stained clusters with a central large mononuclear cell surrounded by numerous lymphocytes. (B) α-Napthylacetate-stained clusters demonstrating the nonspecific esterase-positive nature of the central cell while the lymphocytes remain negative.

The development of lymphoid structures within the grafts of TNLC suggest that, when grown under our conditions, thymic monolayers may retain the cells responsible for the thymic microenvironment. Since only approximately 50% of the grafts contain such lymphoid structures, however, we suggest that some TNLC monolayers are lacking an essential component. The consistent finding of multinucleated giant cell cysts in all TNLC grafts together with the demonstration that the predominant cell type in all our monolayer cultures is the macrophage lead us to suggest that the epithelial cells provide the essential cellular basis for the thymic microenvironment. At best, the epithelial cells comprise a small minority of TNLC in primary cultures. When a TNLC graft fails to produce lymphoid structures, it may be that the initial TNLC preparation contained too few epithelial cells, these having been eliminated from the monolayers while growing in competition with the macrophages (12).

The suggestion that it is primarily the epithelial cells that provide the cellular basis for the microenvironment is supported by our electronmicroscopic studies of the recovered grafts (13). These demonstrate the epithelial nature of the cells closely associated with the lymphocytes in the lymphoid struc-

tures as evidenced by the presence of desmosomes. We have yet to assess the functional capabilities and surface alloantigenic phenotypes of the lymphoid cells. Similarly, we have not proven, as yet, that these lymphocytes are of host origin. But because we believe that the 28-day cultured TNLC preparations are free of lymphocytes prior to grafting, we suggest that the lymphocyte component of the lymphoepithelial structures is derived from immigrant host prethymic progenitor cells. Chromosomal marker studies are presently being performed in order to determine the origin of the lymphocytes found within the grafts.

The results of our own co-culture experiments are very similar to those obtained by other investigators (27,44,45). In most previous studies, it was concluded that the observed enhancement of mitogen responsiveness, although limited, is significant and indicates the induction of T-cell differentiation. In view of our demonstration that the majority of the cells in TNLC monolayers are macrophages, we propose that the limited enhancement of mitogen responsiveness seen in our experiments is most likely due to the inductive capabilities of macrophages or macrophage factors. Recent evidence obtained from a number of studies has indicated that macrophages may enhance mitogen responsiveness *in vitro* (reviewed in ref. 43). It is unclear whether this effect is due to the nonspecific enhancement of mitogen responsiveness by macrophages or their factors or whether it reflects the specific induction of T-lymphocyte differentiation. Keller (15) has shown that monolayers of peritoneal macrophages are capable of enhancing significantly the responsiveness of rat spleen cells to LPS *in vitro,* which suggests that the former may be the cause, i.e., macrophages function nonspecifically to enhance mitogen responsiveness. Since we observed a similar enhancement of LPS responsiveness with TNLC monolayers, we conclude that TNLC macrophages probably function nonspecifically to enhance mitogen responsiveness *in vitro.*

From our initial attempts to reconstitute immunodeficient animals, we conclude that the protection afforded to ATxBM mice at 35 days by TNLC grafts is not associated with fully developed lymphoepithelial grafts. This conclusion is at present based solely on the histologic observation of recovered grafts. Lymphoepithelial structures found in the grafted material resembled early embryonic thymus at the stage of initial colonization of the thymic rudiment by immigrant progenitor cells. Thus, we cannot envisage the TNLC grafts as having made any real contribution by processing of immigrant progenitors. This, however, does not preclude the possibility of a humoral influence provided by the grafted TNLC. Such a humoral influence may be related to the epithelial cells (i.e., a putative thymic hormone) or to a factor liberated from the macrophage component of the graft. The latter is more likely because protection was observed in animals from which no lymphoepithelial grafts were recovered.

Willis and St. Pierre (45) have pursued a similar rationale in a rat system. In their studies using neonatally thymectomized rats as recipients, they claimed restoration of T-cell function. But this restoration was assessed only by the *in vitro* measurement of lymphocyte responses to mitogens or in the mixed lymphocyte reaction, and no direct *in vivo* correlation was made. Moreover, histologic examination of the recovered implants failed to show thymic morphology; apparently lymphocytes or their progenitors had not entered, remained, or proliferated within the implant. Furthermore, from their published photomicrographs, it appears that cystic structures similar to those observed in our studies were formed. From this we conclude that macrophages were a significant component of their TNLC graft, and thus their minimal enhancement of responsiveness could be an accessory cell effect. The mitogen data obtained in our reconstitution studies were highly variable and inconclusive for all mitogens

tested, including LPS. No specific enhancement of T-cell responsiveness was apparent in grafted animals; therefore, we conclude that the protection afforded by the TNLC grafts was nonspecific in nature. Thus, it becomes increasingly clear that an essential prerequisite of such experimental analysis of microenvironmental-precursor cell interactions is the prior establishment of the individual cell types comprising the TNLC as pure populations.

It appears that the epithelial cell component of the TNLC plays the major role in the establishment of a specific thymic microenvironment capable of sustaining and directing the proliferation and differentiation of immigrant T-lymphocyte progenitor cells and their progeny. Of course, from our studies we cannot exclude a macrophage role in this process; we suggest, however, that macrophages alone cannot provide a suitable environment. This raises the possibility that in the reconstitution and co-culture studies the nonspecific role played by these accessory cells is an artifact of the system with no real physiologic significance. Against this idea, that the thymic macrophages play no significant role in the thymic microenvironment, is our observation of the early formation of lymphocyte-macrophage clusters similar to those observed by other workers (21,35,-36). Although such clusters may reflect a specific induction of T-lymphocyte differentiation by macrophages (4,42), it has also been postulated that adherence of autologous macrophages and thymocytes may participate in mechanisms that limit the number and circulation of thymocytes (36). In this context, the recent demonstration by Zinkernagel et al. (48) that the non-lymphoid cells of the thymus play an important role in the differentiation of anti-self-H_2 specificities T-cells is of interest. A speculative role for the TNLC macrophage may be to act as a "guardian" cell, preventing self-reactive T-precursors from escaping from the thymus to the periphery. The macrophage cell type would appear ideally suited for such a role, and *in vivo* thymic macrophages are ideally situated, being found in their heaviest concentration at the cortico-medullary junction. The demonstration of T-precursor cell and autologous non-thymic macrophage clusters *in vitro* (21,36,42) together with our finding of spontaneous lymphocyte-macrophage clusters in spleen cell cultures (unpublished observation) lead us to propose a second role for macrophages in T-cell differentiation—a role as a "nurse" cell. This "nurse" cell role would involve the differentiation of post-thymic T-precursor cells to functional T-lymphocytes and would depend upon the presence of thymic humoral factors. These speculative roles for macrophage action in T-cell differentiation, together with the widely accepted role of the macrophage in antigen

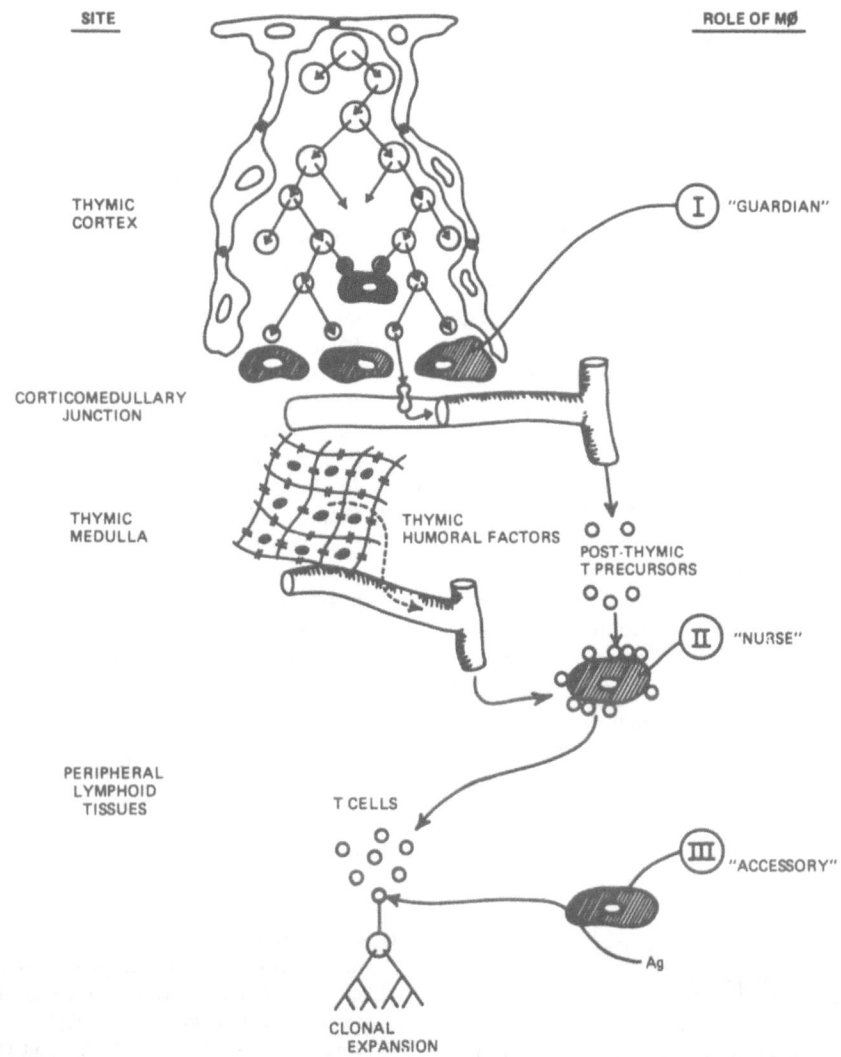

SITE   ROLE OF MØ

THYMIC
CORTEX

I   "GUARDIAN"

CORTICOMEDULLARY
JUNCTION

THYMIC
MEDULLA

THYMIC
HUMORAL FACTORS

POST-THYMIC
T PRECURSORS

II   "NURSE"

PERIPHERAL
LYMPHOID
TISSUES

T CELLS

III   "ACCESSORY"

Ag

CLONAL
EXPANSION

FIGURE 19.12. A schematic representation of the known and proposed roles of macrophages (MØ) in the maturation, differentiation, and function of T-lymphocytes.

presentation and clonal expansion, are illustrated diagrammatically in Figure 19.12.

SUMMARY

Prethymic T-restricted stem cells require contact with the thymic stroma through traffic for their further differentiation. We used a monolayer culture technique to obtain thymic non-lymphoid cells (TNLC) free of lymphoid elements as a first step in investigating the nature of this interaction. In co-culture experiments using TNLC monolayers, we confirmed a number of previous studies that purport to demonstrate the functional capacity of thymic monolayers in the induction of mitogen responsive-

ness in putative precursor cells. But the minimal responses obtained from such studies, similar in magnitude to those obtained with purified thymic extracts, are unconvincing. We believe that a more definitive way of testing for thymic function is to transplant the culture TNLC back to an *in vivo* environment. In initial attempts to reconstitute immunodeficient adult thymectomized, irradiated, and bone marrow grafted (ATxBM) mice, a measure of "protection" was afforded by a syngeneic TNLC graft. This protection was evident at 35 days, post-grafting, and was reflected in the general health, body weight, and differential white cell count of the recipients. Histologically recovered grafts contained (a) cystic structures consisting of multinucleated giant cells that resembled fused macrophages and (b) lymphoepithelial structures that resembled developing thymus. The degree of development of such lymphoepithelial structures varied between intact and deficient recipients; in the latter, the epithelial areas were infrequent and at 35 days contained only small numbers of large basophilic cells. The morphology and cytochemistry of TNLC monolayers, together with the consistent findings of giant cell cysts in grafted material, led us to surmise that the majority of the TNLC derived from monolayers are macrophages. Other cells, including epithelial cells, comprise a small minority of culture TNLC and are responsible for the microenvironmental effects of traffic on precursors. In this context it is of interest that "fibroblast" plaques of colony-forming cells (FCF-c) are present in TNLC monolayers. Such plaques have been described as promising candidates for colonies of hemopoietic stromal cells. We conclude that the early protection afforded to deficient mice by TNLC grafts is not associated with fully developed lymphoepithelial structures, but with some form of humoral activity. This may be related to the epithelial cells, i.e., a putative thymic hormone, to the macrophage component of TNLC, or to FCF-c. In view of recent data suggesting macrophage involvement in T-cell maturation, we feel that TNLC macrophages may play a role not only in the early protection provided by transplantation but also in the induction in co-culture of mitogen responses.

ACKNOWLEDGMENTS

We wish to thank Sandra Grazulewicz, Shona Bohbrink, and Jill Hagadorn for excellent technical assistance and Chris Bergum who typed the manuscript. We would like to thank Dr. Carl T. Hansen of the National Institutes of Health for the supply of nude mice.

R.K.J. thanks the University of Newcastle-upon-Tyne, England, for a sabbatical fellowship and the University of Nebraska for a visiting professorship. This research was supported by a grant from the National Cancer Institute #CA 18548, NIH grant HD 07097, NSF RIAS SER 77-06922 and UNMC general research funds. This support is gratefully acknowledged.

REFERENCES

1. Abramson, S., Miller, R. G., and Phillips, R. A. The identification in adult bone marrow of pluripotent and restricted stem cells of the myeloid and lymphoid systems, *J. Exp. Med., 145*:1567, 1977.

2. Basch, R. S., and Kadish, J. L. Hematopoietic thymocyte precursors and properties of precursors. *J. Exp. Med., 145*:405, 1977.

3. Bearman, R. M., Levine, G. D., and Bensch, K. G. The ultrastructure of the normal human thymus: A study of 36 cases. *Anat. Rec., 190*:755, 1978.

4. Beller, D. I., Farr, A. G., and Unanue, E. R. Regulation of lymphocyte proliferation and differentiation by macrophages. *Fed. Proc., 37*:91, 1978.

5. Cline, M. J., Rothman, B., and Golde, D. W. Effect of endotoxin on the production of colony-stimulating factor by human monocytes and macrophages. *J. Cell Physiol., 84*:193, 1974.

6. Everett, N. B., and Tyler, R. W. Radioautographic studies of the stem cell in the thymus of the irradiated rat. *Cell Tissue Kinet., 2*:347, 1969.

7. Friedenstein, A. F. Precursor cells of mechanocytes. *Int. Rev. Cytol., 327*, 1976.

8. Goldschneider, I. Antigenic relationship between bone marrow lymphocytes, cortical thymoctyes and a subpopulation of peripheral T-cells in rat; description of a bone marrow lymphocyte antigen. *Cell Immunol., 24*:289, 1976.

9. Good, R. A. Structure-function relations in the lymphoid system. *Clin. Immunobiol. 1*:1, 1972.

10. Hirschhorn, R., Hirschhorln, K., and Waithe, W. I. Methods in lymphocyte transformation studies. In McCluskey, R. T. and Cohen, S., eds., *Mechanisms of Cell-mediated Immunity*. New York: Wiley, 1974, p. 115.

11. Jordan, R. K. Development of sheep thymus in relation to *in utero* thymectomy experiments. *Eur. J. Immunol., 6*:693, 1976.

12. Jordan, R. K., and Crouse, D. A. Studies on the thymic microenvironment: I. Morphological and functional characterization of thymic non-lymphoid cells grown in tissue culture. *J. Reticuloendothel. Soc.* In press.

13. Jordan, R. K., Crouse, D. A., and Owen, J. J. T. Studies on the thymic microenvironment: II. Non-lymphoid cells responsible for transferring the microenvironment. *J. Reticuloendothel. Soc.* (submitted).

14. Kass, L. Nonspecific esterase activity in "Hairy Cells." *Acta. Haemat., 58*:103, 1977.

15. Keller, R. Major changes in lymphocyte proliferation evoked by activated macrophages. *Cell Immunol. 17*:542, 1975.

16. Kostowiecki, M. The thymic macrophages *Z. mikr.-anat. Forsch. 69*:585, 1963.

17. Kruisbeek, A. M., Astaldi, G. C. B., Blankwater, M. J., Zijlstra, J. J., Levert, L. A., and Astaldi, A. The *in vitro* effect of a thymic epithelial culture supernatant on mixed lymphocyte reactivity and intracellular cAMP levels of thymocytes and on antibody production to SRBC by Nu/Nu spleen cells. *Cel. Immunol., 35*:134, 1978.

18. Kurland, J. L., and Moore, M. A. S. The regulatory role of the macrophage in normal and neoplastic hemopoiesis. In Baum, S. J. and Ledney, G. D., eds., *Experimental Hematology*. New York: Springer-Verlag, 1977, p. 51.

19. Law, L. W., Dunn, T. B., Trainin, N., and Levey, R. H. Studies of thymic function. In Defendi, V., and

Metcalf, D., eds., *The Thymus*. Philadelphia: Wistar Press, 1964, p. 105.

20. LeDouarin, N. M., and Jotereau, F. V. Tracing of cells of avian thymus through enbryonic life in interspecific chimeras. *J. Exp. Med., 142*:17, 1975.

21. Lipsky, P. E., and Rosenthal, A. S. Macrophage-lymphocyte interaction. I. Characteristics of the antigen-independent-binding of guinea pig thymocytes and lymphocytes to syngeneic macrophages. *J. Exp. Med., 138*:900, 1973.

22. MacVittie, T. J., and Weatherly, T. L. Characteristics of the *in vitro* monocytemacrophage colony-forming cells detected with the mouse thymus and lymph nodes. In Baum, S. J., and Ledney, G. D., eds., *Experimental Hematology Today* New York: Springer-Verlay, 1977, p. 147.

23. Martinez, C., Kersery, J., Papermaster, B. W., and Good, R. A. Skin homograft survival in thymectomized mice. *Proc. Soc. Exp. Biol. Med. 109*:193, 1962.

24. Metcalf, D. *Hemopoietic Colonies, Recent Results in Cancer Research.* Vol. 61, New York: Springer-Verlay, 1977, p. 1.

25. Metcalf, D., and Moore, M. A. S. *Haemopoietic Cells.* Amsterdam: North-Holland, 1971.

26. Miller, J. F. A. P. Effect of neonatal thymectomy on the immunological responsiveness of the mouse. *Proc. Roy. Soc. B., 156*:415, 1962.

27. Mosier, D. E., and Pierce, C. W. Functional maturation of thymic lymphocyte populations *in vitro. J. Exp. Med., 136*:1484, 1972.

28. Owen, J. J. T. The origins and development of lymphocyte populations, In Porter, R., and Knight, J., eds., *Ontogeny of Acquired Immunity*, Ciba Foundation Symposium. Amsterdam: Associated Scientific Publishers, 1972, p. 35.

29. Owen, J. J. T., and Ritter, M. A. Tissue interaction in development of thymus lymphocytes. *J. Exp. Med., 129*:431, 1969.

30. Owen, J. J. T., and Raff, M. C. Studies on differentiation of thymus-derived lymphocytes. *J. Exp. Med., 132*:1216, 1970.

31. Owen, J. J. T., Jordan, R. K. J., Robinson, J. H., Singh, U., and Wilcox, H. N. A. *In vitro* studies on the generation of lymphocyte diversity. *Cold Spring Harbor Symp., 41*:129, 1977.

32. Pazmino, N. H., Ihle, J. N., and Goldstein, A. L. Induction *in vivo* and *in vitro* of terminal deoxynucleotidyl transferase by thymosin in bone marrow cells from athymic mice. *J. Exp. Med., 147*:708, 1978.

33. Pearse, A. G. E. *Histochemistry, Theoretical and Applied,* 3rd ed. Boston: Little Brown, 1972.

34. Sainte-Marie, G., and Leblond, C. P. Thymus-cell populations dynamics. In Good, R. A., and Gabrielsen, A. E., eds., *The Thymus in Immunobiology.* New York and London: Harper & Row, 1964, p. 207.

35. Sharp, J. A. The association of lymphocytes with larger motile cells in cultures of mammalian thymus. *J. Pathol. 103*:87, 1971.

36. Siegel, I. Natural and antibody-induced adherance of guinea-pig phagocytic cells to autologous and heterologous thymocytes. *J. Immunol., 105*:879, 1970.

37. Stewart, C. C., Lin, H. S., and Adles, C. Proliferation and colony-forming ability of peritoneal exudate cells in liquid culture. *J. Exp. Med., 141*:1114, 1975.

38. Stutman, O. The posthymic precursor cell In van Bekkum, D. W., ed., *Biological Activity of Thymic Hormones.* Rotterdam: Kooyker Scientific Publications, 1976, p. 87.

39. Till, J. E., McCulloch, E. A., and Siminovitch, L. A stochastic model of stem cell proliferation, based on the growth of spleen colony-forming cells. *Proc. Nat. Acad. Sci. U.S.A., 51*:29 1964.

40. Till, J. E., Price, G. B., Mak, T. W., and McCulloch, E. A. Regulation of blood cell differentiation. *Fed. Proc., 34*:2279, 1975.

41. Trentin, J. J. Hemopoietic microenvironments. *Transplant. Proc., 10*:77, 1978.

42. Van den Tweel, J. G., and Walker, W. S. Macrophage-induced thymic lymphocyte maturation. *Immunology, 33*:817, 1977.

43. Unanue, E. R. The regulation of lymphocyte functions by the macrophage. *Immunol. Rev., 40*:227, 1978.

44. Waksal, S. D., Cohen, I. R., Waksal, H. W., Wekerle, H., St. Pierre, R. L., and Feldman, M. Induction of T-cell differentiation *in vitro* by thymus epithelial cells. *Ann. N.Y. Acad. Sci. 249*:492, 1975.

45. Willis, J. I., and St. Pierre, R. L. Immunological reconstitution of neonatally thymectomized rats following implantation of thymic epithelial cells. *Adv. Exp. Biol. Med., 73A*:111, 1976.

46. Wilson, F. D., Stitzel, K. A., Klein, A. K., Shifine, M., Graham, R., Jones, M., Bradley, E., and Rosenblat, L. S. Quantitative response of bone marrow colony forming units (CFU-C and PFU-C) in weanling beagles exposed to acute whole body gamma irradiation. *Radiat. Res., 74*:289, 1978.

47. VanZant, G., Goldwasser, E., and Pech, N. Studies of the erythroid inductive microenvironment *in vitro.* In Baum, S. J. and Ledney, G. D., eds., *Experimental Hematology Today.* New York: Springer-Verlag, 1977, p. 71.

48. Zinkernagel, R. M., Callahan, G. N., Althage, A., Cooper, S., Klein, P. A., and Klein, J. On the thymus in the differentiation of "H-2 Self Recognition" by T cells: Evidence for dual recognition? *J. Exp. Med., 147*:882, 1978.

20

Characterization of the Myeloma Colony-Forming Cell

Anne W. Hamburger,
Mary B. Kim,
and Sydney E. Salmon

Multiple myeloma has served as a model neoplasm in both mouse and man (21). Studies of myeloma immunoglobulin synthesis and metabolism have been applied to quantitate the total-body number of myeloma cells and to follow changes in tumor mass with treatment. Such serial tumor kinetic studies, and those of the tritiated thymidine labeling index of the tumor, have provided insights on the kinetics of growth and regression of myeloma and on approaches to treatment (21). Such studies, however, did not directly assess the clonigenic tumor cells, the cells that provide the basic renewal system of the tumor.

Although primary explants of plasma cells in transplantable mouse myeloma have been successfully cloned in soft agar (12,17), the *in vitro* cultivation of human plasma cells has met with little success. Chan Park, using methods similar to those he devised for *in vitro* cultivation of mouse myeloma cells, was able to culture human myeloma cells in soft agar. But he was unable to establish linearity between numbers of cells plated and the number of colonies formed and thus could not use this technique as an assay system. The lack of an *in vitro* bioassay for myeloma colony-forming cells (CFU-c myeloma) has hampered the study of the natural history of myeloma and the quantitation of myeloma cell mass.

Our recent studies (4) have demonstrated that tumor cells from bone marrows of patients with multiple myeloma could proliferate in semi-solid agar to generate colonies of monoclonal plasma cells. Such cells required the presence of a medium conditioned by the adherent spleen cells of mineral oil-primed BALB/c mice and by 2-mercaptoethanol (2-ME). Between 0.001 to 0.1% of the nucleated bone marrow cells were able to form colonies.

Cells picked from the colonies had the morphology of immature plasmablasts and mature plasma cells. Colony cells were capable of immunoglobulin synthesis *in vitro* and contained intracytoplasmic monoclonal immunoglobulin.

The present studies were undertaken to obtain more information on the nature of the cells generating these plasma cell colonies *in vitro*. Additonal studies characterizing factors necessary for cell growth are reported.

MATERIALS AND METHODS
Patient Studies
Patients with well-documented multiple myeloma were selected for study. Detailed clinical and immunologic criteria for diagnosis and clinical staging of myeloma have been described previously (2).

Collection of Cells

Bone marrow cells were obtained from patients after informed consent was obtained. Cells were aspirated into a heparinized syringe, mixed with an equal volume of 3% dextran-saline, and sedimented at room temperature for 45 min. Cells in the supernatant were collected and washed three times in Hank's balanced salt solution (HBSS) with 10% fetal calf serum (FCS) (Flow Lab, Anaheim, Ca.). The viable nucleated cell counts, as determined in a hemocytometer using trypan blue, were routinely more than 95%. Bone marrow differential counts were performed on slides prepared with a cytocentrifuge and routinely stained with WRight-Giemsa.

Culture Assay for Myeloma Colony-Forming Cells

Cells were cultured as described (4). Briefly, 1-ml underlayers containing 0.25 ml of medium conditioned by the adherent spleen cells of mineral oil-primed BALB/c mice (BALB/c cm) in 0.5% Bacto Agar (Difco, Detroit, Mich.) were prepared in 35-mm plastic Petri dishes. Bone marrow cells to be tested were suspended in 0.3% agar in enriched CRML 1066 medium. 2-Mercaptoethanol was added to give a final concentration of 50 μM immediately before culture. Each culture received 5×10^5 cells in a 1-ml agar-media mixture.

Cultures were incubated at 37°C in 5% CO_2 in a humidified incubator.

Scoring of Cultures

Cultures were examined using a Nikon MS inverted phase microscope at 100 and 200X. Final colony counts were made 14 to 21 days after plating. Aggregates of 40 cells or more were considered colonies. Individual colonies were removed from the dishes using a fine capillary pipette and were suspended in a drop of heat-inactivated FCS. Colonies were air dried 3 to 4 hr. Cells were stained routinely with Wright-Giemsa, 0.6% orcein in 60% acetic acid, and methyl green pyronin for morphology, and for peroxidase and plasma cell acid phosphatase activity (4). In additon, entire culture plates were routinely monitored for contaminating granulocyte-macrophage colonies by perioxidase staining (28) prior to colony counting.

Effect of Substitution of Sulfhydryl Compounds

Reagent grade dithiothreitol (DTT), monothioglycerol (MTG), cystein, and 2-ME, were used (Sigma, St. Louis, Mo.). Solutions of sulfhydryl compounds were prepared in double distilled water (Gibco) shortly before use and diluted in tissue culture media. Cysteine-methyl disulfide was provided by Dr. John Toohey (UCLA) (25).

Effect of Hydrocortisone on Myeloma Cell Growth

Preservative-free hydrocortisone sodium succinate (Solu-Cortef, Upjohn, Kalamazoo, Mich.) was reconstituted with RPMI 1640 medium (Gibco) and added to cultures in various concentrations as indicated.

Separation of Myeloma Colony-Forming Cells by Velocity Sedimentation

Cell separation by sedimentation rate at unit gravity was performed on single cell suspensions of bone marrow and peripheral blood leukocytes as described by Miller and Phillips (14) against a 15 to 30% FCS gradient in phosphate buffered saline (PBS) (ph 7.2).

Adherence Separation Studies

Cells were separated by their ability to adhere to a variety of substrates. Freshly washed bone marrow cells were separated by their ability to adhere to plastic by the method of Messner et al. (11). Other samples were separated by their ability to adhere to nylon wool columns as described by Julius et al. (7). Finally, cells were separated by their adherence to glass beads at 37°C according to the active adherence technique of Shortman et al. (22).

Depletion of E-rosetting cells

Rosettes were formed at 4°C using sheep erythrocytes and rosetted cells were separated from non-rosette-forming cells by centrifugation through gradients of Ficoll-Hypaque (Pharmacia, Piscataway, N.J.) as described by Wahl et al. (26). Rosetted cells were freed of adherent erythrocytes by incubation of the cell suspension at 37°C for 7 min in 0.17 M NH_4Cl. Cells were immediately centrifuged through 100% FCS to remove red cell ghosts.

Effect of RPMI 8226 Antiserum on Colony Formation by Myeloma Colony-Forming Cells

An antiserum generated in rabbits against an established myeloma cell line from a patient with IgG myeloma (RPMI 8226) was prepared and provided by Dr. Robert Krueger (Christ's Hospital, Cincinnati, Oh.) (10). The lot of antiserum reacted with RPMI 8226 cells to an end-point titer of 1:128 as determined by positive immunofluorescence of at least 5% of the cells. The antiserum had an immunofluorescent titer of 1:16 against normal bone marrow plasma cells. In the experiments presented, 0.1 ml of appropriate dilutions of antiserum was added to the cultures.

Assay for Granulopoietic Activity

The assay used to measure granulocyte colony formation by human bone marrow cells was as described by Pike and Robinson (19).

TABLE 20.1 Effect of Substitution of Alternative Feeder Layers on Myleoma Colony Formation

SOURCE OF CONDITIONED MEDIUM[a]	NUMBER OF COLONIES/ 5 × 10⁵ CELLS[b]
BALB/c (oil primed)	94 ±5
BALB/c (untreated)	15 ±6
CD-1 (oil primed)	7 ±4
CD-1 (untreated)	10 ±6
DBA/2 (oil primed)	9.8 ±4
DBA/2 (untreated)	20 ±7
Media alone	8 ±3

[a]See Methods.
[b]Mean ± SE (five plates).

TABLE 20.2 Effect of Thiols on Myeloma Colony-Forming Cell Growth

SUBSTANCE	CONCENTRATION [μM]	M-CFU-c/5 × 10⁵ CELLS
2 Mercaptoethanol	50	28 ± 5[a]
	10	5 ± 3
Cysteine	50	2 ± 2
	10	8 ± 3
Dithiothreitol	50	3 ± 2
	10	3 ± 3
Monothioglycerol	50	35 ± 6
	10	0 ± 0
Cysteine-methyl disulfide	50	4 ± 4
	10	2 ± 2

[a]Mean ± SE (eight plates).

RESULTS

Factors Affecting Cell Growth

We previously demonstrated that cells from over 90% of myeloma patients tested required media conditioned by the adherent spleen cells of mineral oil-primed BALB/c mice for growth (16). Media conditioned by adherent spleen cells of other strains of mice, or untreated BALB/c, did not support growth (Table 20.1).

Figure 20.1 depicts the effect of different concentrations of BALB/c cm on colony formation by myeloma cells from the bone marrow of a multiple myeloma patient in relapse. Myeloma colony growth was maximally stimulated when conditioned medium was present at a dilution of ¼ in the underlayer. Dilutions below ¼ or above ½ did not support myeloma colony formation. Three additional experiments with cells from other myeloma patients have yielded similar results.

The sulfhydryl compounds cysteine, dithiothreitol, and cysteine-methyldisulfide did not promote proliferation of CFU-c myeloma at the concentrations tested (Table 20.2). But monothioglycerol, at concentrations of 50 μ/M, proved as effective as 2-ME in supporting myeloma colony formation. Neither 10 μM 2-ME nor 10 μM MTG supported CFU-c myeloma growth.

Hydrocortisone produced a dose-dependent inhibiton of myeloma colony formation by marrow cells of three myeloma patients over the dose range tested. The number of colonies was reduced 80% at a dosage of 5 μg/ml (about 10 μM) (Figure 20.2).

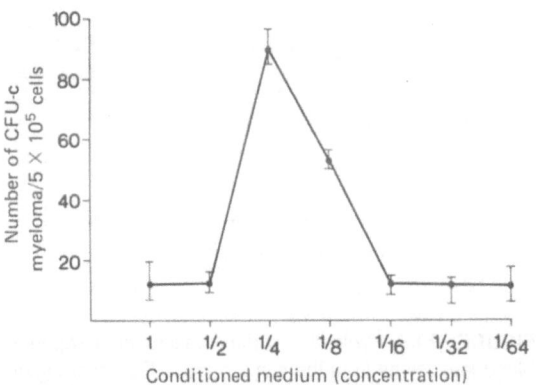

FIGURE 20.1. Effect of different concentrations of conditioned medium on CFU-c myeloma growth. Bone marrow cells from a myeloma patient in relapse were plated on top of underlayers containing conditioned medium in the proportions indicated. Each point represents the mean of four plates ± SE. Three separate trials on different patients have yielded similar results.

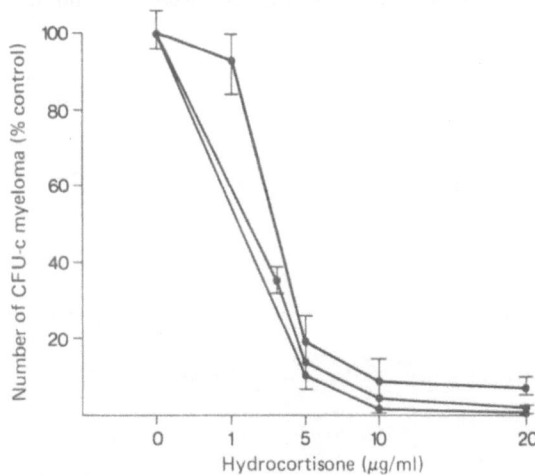

FIGURE 20.2. Effect of hydrocortisone on myeloma colony formation. Cells were plated in agar medium containing the indicated concentrations of the drug. Each line represents the percent survival of colonies grown from bone marrow cells of three different myeloma patients in relapse. Each point represents the mean of four plates ± SE.

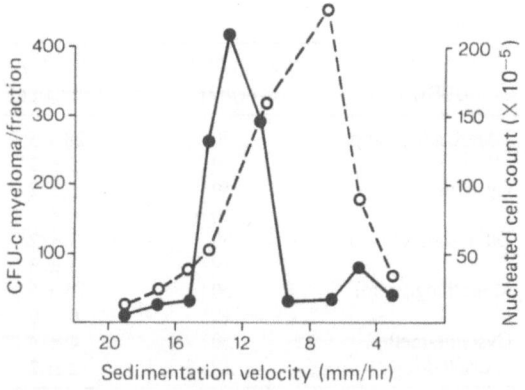

FIGURE 20.3. Results of a typical sedimentation velocity separation of cells from a bone marrow of a myeloma patient in relapse. Here, 10_8 cells were sedimented through a 15 to 30% fetal calf serum (FCS) gradient at 4°C for 150 min. Mean colony counts from four replicate cultures. Nucleated cell profile (o––––o); CFU-c myeloma profile (•———•).

Sedimentation Velocity Analysis of Myeloma Colony-Forming Cells

The sedimentation velocity of bone marrow CFU-c myeloma was determined using the Staput apparatus. A representative experiment using bone marrow cells from an untreated myeloma patient is depicted in Figure 20.3. Three additional experiments have yielded similar results. Myeloma colony-forming cells sedimented as a single broad band with a peak sedimentation velocity of 13 mm/hr.

In an additional experiment, marrow cells from a second myeloma patient were separated on a Staput gradient, after which the fractions were split and the

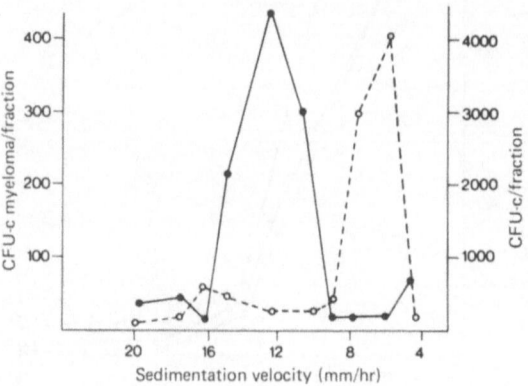

FIGURE 20.4. Results of a sedimentation velocity separation of cells from a bone marrow of a second myeloma patient in relapse. Cells were sedimented as described, fractions were split, and cells were grown in either a granulocyte or a myeloma culture assay. Mean colony counts four replicate cultures. Number of CFU-c myeloma per fraction (•———•); number of CFU-c per fraction (o———o).

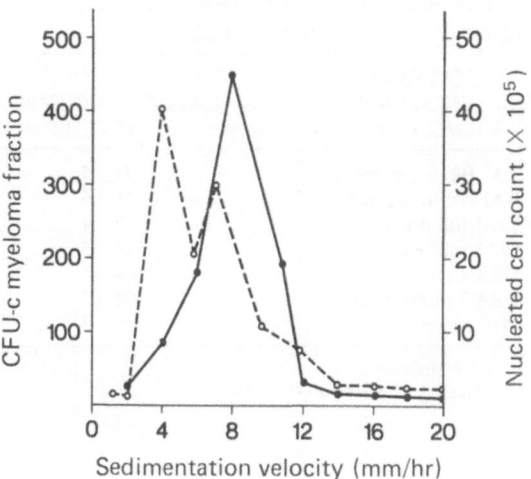

FIGURE 20.5. Velocity sedimentation separation of peripheral blood leukocytes from a myeloma patient in relapse. Here, 5×10^7 cells were sedimented through a 15 to 30% FCS gradient at 4°C for 150 min. The dotted line represents the nucleated cell profile; the unbroken line the CFU-c myeloma profile. Mean colony counts from four replicate cultures. Nucleated cell profile (o––––o); CFU-c myeloma profile (•———•).

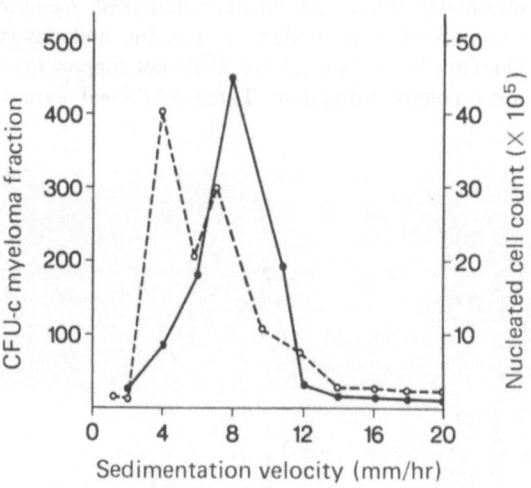

FIGURE 20.6. Velocity sedimentation of peripheral blood leukocytes from the same patient. Cells were sedimented as described in Figure 20.3. The fractions were then split and the cells were grown in either a granulocyte or a myeloma culture assay. The unbroken line represents the CFU-c myeloma per fraction; the broken line represents the number of CFU-c per fraction. Mean colony count from four replicate cultures. Number of CFU-c myeloma per fraction (•———•); number of CFU-c per fraction (o––––o).

TABLE 20.3 Myeloma Colony Formation by Adherent and Non-Adherent Cell Populations

SEPARATION TECHNIQUE	TRIAL NUMBER	FRACTION	TOTAL CELLS/ FRACTION	TOTAL CFU-c MYELOMA/ FRACTION
Glass Bead Column	1	Filtrate	22×10^6	3,960
		Adherent	14×10^6	196
	2	Filtrate	14×10^6	1,110
		Adherent	8×10^6	128
	3	Filtrate	6×10^6	2,840
		Adherent	3×10^6	120
Nylon wool	1	Filtrate	5×10^6	40
		Adherent	6×10^6	545
	2	Filtrate	25×10^6	300
		Adherent	15×10^6	3,500
Plastic dish	1	Non-adherent	10×10^6	192
		Adherent	4×10^6	8
	2	Non-adherent	4×10^6	415
		Adherent	6×10^6	80

cells were plated in either a granulocyte culture assay or the myeloma culture assay. The number of granulocyte (G) or CFU-c myeloma were evaluated in each fraction. The results are shown in Figure 20.4. The CFU-c sedimented much more slowly than the CFU-c myeloma. There were approximately 10 times as many CFU-c in this marrow as CFU-c myeloma.

In addition, 5×10^7 peripheral blood leukocytes obtained by leukopheresis of a patient with IgA myeloma were separated at 1 g. This patient had 9% plasma cells in his peripheral blood. Peripheral blood CFU-c myeloma from this patient sedimented at about 8 mm/hr (Figure 20.5).

A sample of peripheral leukocytes from each fraction was also assayed for granulocyte colony formation. The results are shown in Figure 20.6. The CFU-c, with a peak sedimentation velocity of 5 mm/hr, sedimented more slowly that the CFU-c myeloma.

Myeloma colony formation from samples of peripheral blood was only observed when plasma cells were present in the peripheral blood. Peripheral blood leukocytes from 20 other myeloma patients with normal peripheral blood smears failed to form colonies.

Adherence Separation Studies

Between 50 to 70% of the bone marrow cells adhered to plastic dishes, 30 to 40% to glass bead columns, and 35 to 40% to nylon wool. Viability was over 80% in all cases.

Myeloma colony-forming cells were enriched in populations that failed to adhere to plastic tissue culture dishes or glass beads (Table 20.3). The morphologic composition of these cell populations is summarized in Table 20.4. The cell population that failed to adhere to plastic dishes was enriched for plasma cells. In contrast, plasma cells were depleted in the cell population that failed to adhere to glass bead columns.

The CFU-c myeloma adhered to nylon wool (Table 20.4). Morphologically recognizable plasma cells were enriched in this adherent fraction (Table 20.4).

TABLE 20.4 Composition of Cells Separated by Adherence Procedures

SEPARATION TECHNIQUE	FRACTION	MORPHOLOGIC TYPE (%)[a]					
		PG	NPG	PC	Lymphocytes	Monocytes	Erythroid
Glass beads	Control	13	44	29	2	4	8
	Non-adherent	59	7	12	3	0	19
	Adherent	2	31	50	2	4	10
Nylon wool	Control	5	21	55	4	2	13
	Non-adherent	42	6	2	10	0	40
	Adherent	22	6	90	0	3	15
Plastic dishes	Control	8	38	26	8	5	18
	Non-adherent	2	13	41	10	3	31
	Adherent	13	72	5	2	1	9
Results are the mean of three experiments; 500 cells counted/slide							

[a]PG, proliferating granulocyte (myeloblasts, promyelocytes, myelocytes); NPG, non-proliferating granulocyte (metamyelocytes, "band" neutrophils, segmented neutrophils); PC, plasma cell.

TABLE 20.5 Effect of Depletion of E-Rosetting Cells on Myeloma Colony-Forming Cell Growth

TRIAL #	PLASMA CELLS %	CELLS FORMING ROSETTES (%)	CFU-c MYELOMA/ 10^6 UNFRACTIONATED CELLS	CFU-c MYELOMA/ 10^6 ROSETTED CELLS	CFU-c MYELOMA/ 10^6 ROSETTE DEPLETED CELLS
1	15	34	16 ± 4	8 ± 6	60 ± 4
2	85	10	8 ± 4	6 ± 3	20 ± 4
3	35	26	50 ± 9	16 ± 6	126 ± 18
		Overall recovery of cells varies from 65 to 82%			

Effect of Depletion of E-rosetting Cells

As described in Table 20.5, E-rosette-depleted cells were enriched for CFU-c myeloma and rosetted populations contained reduced numbers of colony-forming cells. About 80% of CFU-c myeloma were found in rosette-depleted populations.

Effects of Anti-RPMI 8226 Serum

A rabbit antiserum with specificity for membrane and cytoplasmic antigens of RPMI 8226 cells was added to cultures to determine if it could reduce myeloma colony formation as compared to controls. Control experiments of myeloma colony formation by the 8226 myeloma cell line indicated that 50% inhibition of colony formation could be observed at an antiserum dilution of 1/32. As determined by trypan blue exclusion, the antiserum was not cytotoxic to the majority of bone marrow cells from myeloma patients. Incubation of myeloma bone marrow cells with antiserum at dilutions of 1/2 to 1/128 with or without guinea pig complement for 1 hr at 37°C did not result in more than 10% cytotoxicity as measured by exclusion of trypan blue.

Figure 20.7 shows the results of addition of anti-RPMI 8226 serum to a culture of bone marrow cells from a myeloma patient in relapse. The bone marrow sample contained 30% plasma cells. At a dilution of 1/8, the antiserum reacted with about 15% of cells as determined by indirect immunofluoresence on fixed smears. The results of the colony assay showed a 63% inhibition of myeloma colony formation at an antiserum dilution of 1/8. The addition of 0.1 ml of a 1:20 dilution of guinea pig complement to the antibody-containing culture did not change these results significantly. In contrast, the antiserum reduced granulocyte formation from the same marrow only 10% at this dilution. Normal rabbit serum resulted in a slight depression (5) of both CFU-c myeloma and CFU-c at dilutions of 1/4 and 1/8.

DISCUSSION

Colonies grown *in vitro* in soft agar from bone marrow cells of patients with multiple myeloma have been shown to be composed of monoclonal plasma cells. Under optimal conditions, 0.1% of myeloma cells form colonies.

Only conditioned medium prepared from the adherent spleen cells of mineral oil-primed BALB/c mice supported myeloma colony formation. Conditioned medium prepared from other strains of mice, or untreated BALB/c, could not support growth. This might be expected, since mineral oil induces plasmacytomas in BALB/c mice only (20). We have previously observed the development of plasmacytomas in mice injected according to our protocol after 8 months to 1 year. The growth-promoting effects of the conditioned media appeared to be dose dependent, with peak activity at concentrations of 1/4 and 1/8. The "H-factor," a heat-labile protein isolated by Namba and Honoka (15) that supported growth of cultured murine myeloma cells, displayed a peak activity at a concentration of 1/8. That factor was also produced by murine adherent spleen cells. Preliminary experiments in our laboratory indicate that the active fraction(s) in our conditioned medium

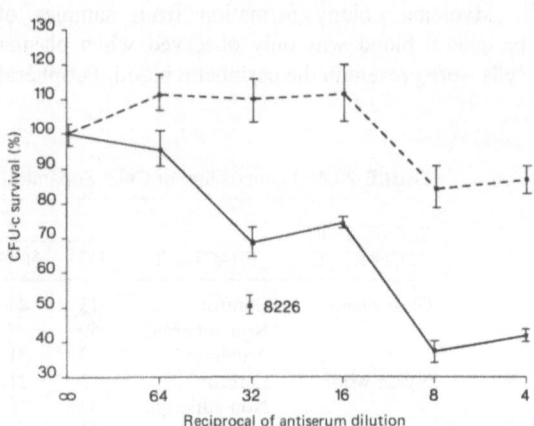

FIGURE 20.7. Effect of RPMI 8226 cell antiserum on myeloma and granulocyte colony growth. Each point represents the mean of four plates ± SE. Number of myeloma colonies (●——●); number of granulocyte colonies "8226" (●---●) represents the antibody dilution at which 8226 colony formation was inhibited 50%.

is also a heat-labile protein(s). The exact relationship of our factor(s) to the "H-factor" remains to be determined. Myeloma colony growth required the presence of sulfhydryl compounds. Only monothioglycerol or 2-ME promoted myeloma cell proliferation at the concentrations tested. The role of these sulfhydryl compounds remains unclear, although the 2-ME helper effect in a variety of immunologic reactions (5) and a clonal assay for mouse B-lymphocytes (13) has been well documented. Monothioglycerol has also proven to be a potent stimulator of hemopoietic cell growth (6,9). It is possible that such compounds substitute for specific metabolites or induce the formation of a growth regulator by the same mechanism that 2-ME promotes the release of Colony Stimulating Activities (CSA) from lymphoid cells.

Myeloma colony formation was inhibited by hydrocortisone at concentrations of around 10 μM. Similarly, hydrocortisone suppressed the *in vitro* synthesis of immunoglobulins by bone marrow cells of some patients with Bence-Jones myeloma (23). In contrast, hydrocortisone at these pharmacologically achievable concentrations markedly enhanced the pokeweed mitogen-induced plaque-forming response of normal human peripheral blood B-lymphocytes (3) and was not cytolethal to human lymphoblastoid cell lines (1). Our results may have been different from theirs, since CFU-c myeloma comprised only a small fraction of total plasma cells (i.e., 0.001 to 0.1%). Their death would not have resulted in a detectable decrease in cell counts or overall immunoglobulin synthesis. Only a colony-forming assay such as this would have reflected changes in CFU-c myeloma numbers. High doses of corticosteroids have previously been observed to have some utility in myeloma therapy (21) and the assay may be reflecting this sensitivity *in vitro*.

Myeloma colony-forming cells in the peripheral blood had a slower sedimentation velocity than those in the marrow, indicating that the CFU-c myeloma in the blood is smaller than that in the marrow. Similarly, peripheral blood CFU-c were smaller than bone marrow CFU-c (24). The fractions containing the greatest number of CFU-c myeloma were enriched for plasma cells in both peripheral blood and bone marrow separations. This, coupled with the fact that CFU-c myeloma colony formation only occurred when plasma cells were present in peripheral blood, would suggest that the CFU-c myeloma was a recognizable plasma cell.

The result of experiments in which E-rosetting cells were depleted from bone marrow cells prior to plating indicated that E-rosetting cells were not necessary for myeloma colony formation. Similarly, colony formation by murine B-lymphocytes was not obviously enhanced by T-cells (8). In contrast,

Nathan et al. (16) found that proliferation of human erythroid precursors (BFU-e) in semi-solid media required a soluble product of T-cells. Similarly, Wiktor-Jehrecjczak et al. (27) noted that the presence of θ bearing cells were required for reconstitution of erythropoiesis by bone marrow cells in W/W^v anemic mice.

The antiserum generated against RPMI 8226 cells reduced meyloma colony formation. This would indicate that the antiserum detected an antigenic site common to both a long-term myeloma tissue culture line and freshly explanted human CFU-c myeloma. The fact that the antibody inhibited growth of CFU-c myeloma *in vitro* may have diagnostic or therapeutic significance.

Future studies of the CFU-c myeloma may identify regulatory factors controlling the proliferation of neoplastic plasma cells or their unique susceptibilities to nutritional deprivation.

SUMMARY

Freshly explanted human myeloma cells formed colonies of monoclonal plasma cells in soft agar in the presence of medium conditioned by the adherent spleen cells of mineral oil-primed BALB/c mice. Conditioned media derived from adherent spleen cells of other strain of mice, or untreated BALB/c, did not support growth. The medium showed peak activity at a dilution of 1/4. 2-Mercaptoethanol, or monothioglycerol, was necessary for colony formation. Other thiols tested were ineffective in promoting colony growth. Hydrocortisone at concentrations of 10 μM to 100 μM significantly inhibited myeloma colony formation. Peripheral blood leukocytes from patients with myeloma formed colonies only when plasma cells were present in the peripheral blood. Velocity sedimentation at 1 g showed that the peripheral blood myeloma colony-forming cells sedimented at 8 mm/hr. In contrast, myeloma colony-forming cells derived from the bone marrow sedimented at 13 mm/hr. Colony-forming cells adhered to nylon wool, but not glass beads or plastic dishes. The presence of E-rosetting cells was not required for myeloma colony formation. Antibody prepared against a human myeloma cell line, RPMI 8226, reduced colony formation.

These studies demonstrate the usefulness of this bioassay for determining functional properties of the myeloma colony forming cell.

ACKNOWLEDGMENTS

This work was supported by grants CA21839 and CA 17094 from the National Cancer Institute.

REFERENCES

1. Bird, C., Robertson, A., Read, J., and Currie, A. Cytolethal effects of glucocorticoids in human lymphoblastoid cell lines. *J. Pathol., 123*:145, 1977.

2. Durie, B., and Salmon, S. A clinical staging system for multiple myeloma. *Cancer, 36*:842, 1970.

3. Fauci, A., Pratt, K., and Whalen, G. C. Activation of human B lymphocytes, IV. Regulatory effects of corticosteroids on the triggering signal in the plaque-forming cell responses of human peripheral blood B lymphocytes to polyclonal activation. *J. Immunol., 119*:598, 1977.

4. Hamburger, A., and Salmon, S. Primary bioassay of human myeloma stem cells. *J. Clin. Invest., 60*:846, 1977.

5. Heber-Katz, E., and Click, R. Immune response *in vitro*. V. Role of mercaptoethanol in the mixed leukocyte reaction. *Cell. Immunol., 112*:502, 1972.

6. Iscove, N., and Sieber, F. Erythroid progenitors in mouse bone marrow detected by macroscopic colony formation in culture. *Exp. Hematol., 3*:32, 1975.

7. Julius, M., Simpson, E., and Herzenberg, L. A. A rapid method for the isolation of functional thymus-derived murine lymphocytes. *Eur. J. Immunol., 3*:645, 1973.

8. Kincade, P. W., Paige, C. S., Parkhouse, M. E., and Lee, G. Characterization of murine colony forming B cells. I. Distribution, resistance to anti-immunoglobulin antibodies, and expression of Ia antigens. *J. Immunol., 120*:1289, 1978.

9. Koeffler, H. P., and Golde, D. W. Acute myelogenous leukemia: A human cell line response to colony stimulating activity. *Science, 200*:199, 1978.

10. Krueger, R., Staneck, L., and Boehlecke, J. Tumor associated antigens in human myeloma. *N. Nat. Cancer Instit., 56*:711, 1976.

11. Messner, H., Till, J., and McCulloch, E. A. Interacting cell populations affecting granulopoietic colony formation by normal and leukemic human marrow cells. *Blood, 42*:701, 1973.

12. Metcalf, D. Colony formation in agar by mouse plasmacytoma cells: potentiation by hemopoietic cells and serum. *J. Cell. Physiol., 81*:397, 1973.

13. Metcalf, D., Nossal, G., Warner, N., Miller, J., Mandell, J., Layton, J., and Gutman, G. Growth of B-lymphocyte colonies *in vitro*. *J. Exp. Med., 142*:1539, 1975.

14. Miller, R., and Phillips, R. Separation of cells by velocity sedimentation. *J. Cell Physiol., 73*:191, 1969.

15. Namba, Y., and Hanoka, M. Immunocytology of cultured IgM forming cells of mice. *J. Immunol., 109*:1193, 1972.

16. Nathan, D., Chess, L., Hillman, D., Clarke, B., Bread, J., Merlez, E., and Housman D. E. Human erythroid burst forming unit: T-cell requirement for proliferation *in vitro*. *J. Exp. Med., 147*:324, 1978.

17. Park, C. H., Bergsagel, D., and McCulloch, E. Mouse myeloma tumor stem cells: A primary cell culture assay. *J. Nat Cancer Inst., 46*:411: 1971.

18. Parker, J., and Metcalf, D. Production of colony stimulating factors in mitogen stimulated lymphocyte cultures. *J. Immunol., 112*:502, 1974.

19. Pike, B., and Robinson, W. Human bone marrow colony growth in vitro. *J. Cell Physiol., 76*:77–81.

20. Potter, M. Immunoglobulin producing tumors and myeloma proteins of mice. *Physiol. Rev., 52*:631, 1972.

21. Salmon, S. E. Immunoglobulin synthesis and tumor kinetics of multiple myeloma. *Sem. Hematol., 10*:135, 1973.

22. Shortman, K., Williams, W., Jackson, H., Russel, P., Bynt, P., and Diener, E. The separation of different cell classes from lymphoid organs. IV. The separation of lymphocytes from phagocytes on glass bead columns, and its effect on subpopulations of lymphocytes and antibody forming cells. *J. Cell Biol., 48*:566, 1971.

23. Solomon, A. Bence-Jones proteins and light chain of immunoglobulins. IV. Effect of corticosteroids on synthesis and excretion of Bence-Jones protein. *J. Clin. Invest., 61*:97, 1978.

24. Tebbi, K., Rubin, S., Cowa, D., and McCulloch, E. A. A comparison of granulopoiesis in cultures from blood and marrow cells of nonleukemic individuals and patients with acute leukemia. *Blood, 48*:235, 1976.

25. Toohey, J. Sulfhydryl dependence in primary explant hematopoietic cells. Inhibition of growth *in vitro* with Vitamin B_{12} compounds. *Proc. Nat. Acad. Sci. U.S., 72*:73, 1975.

26. Wahl, D., Rosenstreich, D., and Oppenheim, J. Separation of human lymphocytes by E-rosette sedimentation. In Bloom, B., and David, J., eds., *In Vitro Methods in Cell Mediated and Tumor Immunity*. New York: Academic Press, 1976, p. 231.

27. Wiktor-Jedrecjczak, W., Shakis, S., Ahmed, A., and Santos, G. The θ-sensitive cells erythropoiesis: Identification of a defect in W/W^v anemic mice. *Science, 91*:313, 1977.

28. Zucker-Franklin, D., and Grusky, G. The identification of eosinophil colonies in soft agar cultures by differential staining for peroxidases. *J. Histochem. Cytochem., 24*:1270, 1976.

21

Anti-Host Immune Reactivity after Allogeneic Bone Marrow Transplantation

E. A. J. Wolters, N. H. C. Brons,
R. Benner, and O. Vos

Transplantation of allogeneic bone marrow cells into lethally irradiated mice may produce radiation chimeras who suffer from secondary disease. The pathogenesis of this disease is primarily a graft versus host (GvH) reaction. Such a GvH reaction may be caused by immunocompetent T-lymphocytes that reside in the bone marrow inoculum. This hypothesis is supported by experiments involving velocity sedimentation and density separation of these bone marrow cells (1,5,23). In these studies, a maximal GvH activity in the lymphocyte-rich fractions (23) and in the fractions with the greatest *in vitro* proliferative responsiveness to irradiated allogeneic cells was observed (1). Treatment of donor mice with anti-lymphocyte serum (ALS) completely abolished the GvH-inducing capacity of the spleen and lymph node cells (2) and greatly reduced the GvH-eliciting capacity of the bone marrow cells (9); this suggests that the bone marrow T-cells have the potential to enter the circulation (8).

On the other hand, there is the possibility that T-cells, newly formed under influence of the recipient's thymus, are also involved. Goedbloed and Vos (6) studied the influence of thymectomy of the host before irradiation and allogeneic bone marrow transplantation on the incidence of secondary disease, and Van Putten (21) studied this effect in a xenogeneic combination. In both these studies, it was found that thymectomy of the recipients prior to the bone marrow transplantation can delay mortality. The authors, however, could not distinguish between death caused by secondary disease and death caused by a wasting syndrome due to thymectomy. Chen et al. (4) also found that the thymus influences the rate of development of secondary disease, but not the mortality. Simmons (15), however, found no difference in the survival patterns of sham-operated mice and thymectomized, allogeneically reconstituted mice. An important drawback in all these studies is the fact that there is no quantitative assay for the GvH reaction in irradiated recipients.

In previous studies (22), we have found that after transplantation of C57BL/Rij spleen cells into lethally irradiated (C57BL/Rij X CBA/Rij) F_1 mice, a specific delayed type hypersensitivity (DTH) to the host histocompatibility antigens develops in the spleen and lymph nodes of the recipient mice. This DTH reactivity can be demonstrated by injecting recipient spleen and lymph node cells into normal C57BL/Rij mice. Challenge of these secondary recipients with CBA/Rij spleen cells then evokes an easily measurable DTH response. Using this new assay, we then studied the development of anti-host immune reactivity after semi-allogeneic bone marrow transplantation in lethally irradiated thymectomized mice and sham-operated control mice.

MATERIALS AND METHODS

Animals

These included C57BL/Rij (H-2^b) male mice, 10 to 15 weeks old, CBA/Rij (H-2^q) female mice, 30 to 40 weeks old, and (C57BL/Rij × CBA/Rij) F$_1$ (H-2$^{b/q}$) male and female mice, 10 to 20 weeks old. They were purchased from the Medical Biological Laboratory TNO, Rijswijk (ZH), and the Laboratory Animals Center of the Erasmus University, Rotterdam, The Netherlands.

Preparation of Cell Suspensions

Cell suspensions were prepared in a balanced salt solution (BSS) as described previously (22). Bone marrow cells were collected by flushing femurs and tibiae with BSS. For reconstitution, 1×10^7 bone marrow cells, suspended in a volume of 0.5 ml BSS, were injected intravenously into mice. The bone marrow cells were always obtained from 10-week-old C57BL/Rij donor mice. Nucleated cells were counted with a Coulter Counter, Model B.

Irradiation

The recipient (C57BL/Rij x CBA/Rij) F$_1$ mice received 925 rad wholebody irradiation, generated in a Philips Müller MG 300 x-ray machine as described in detail previously (22). Radiation control mice died in 10 to 16 days.

Adult Thymectomy

Adult thymectomy and sham thymectomy were always performed at 6 weeks of age. The surgery was performed as described by Miller (12). The mice were anaesthetized with Nembutal (Abbott S. A., Saint-Rémy-sur-Avre, France) (70 mg/kg body weight). The adult thymectomy mice were allowed to recover for a minimum of 10 weeks before experimental use. All adult thymectomy mice were examined for thymic remnants at the end of each experiment.

Selective Elimination of Thy-1.2 Positive Cells

The production of anti-Thy-1.2 sera and their use for the selective elimination of T-cells have been described previously (22). Bone marrow cells were treated for 30 min at 4°C with anti-Thy-1.2 serum. The amount of anti-Thy-1.2 serum used was two or three times more than was needed to kill at least 95% of corticosteroid-resistant thymocytes (CRT). These CRT were obtained by intraperitoneal injection of dexamethasone sodium phosphate (Merck & Co., Rahway, N.J.) (30 mg/kg body weight) 2 days before harvest. After incubation, the cells were centrifuged, resuspended in BSS, and incubated with guinea pig complement for 15 min at 37°C. Thereafter, the cells were washed three times and resuspended in BSS.

Anti-Thymocyte Serum

Anti-thymocyte serum (ATS) was prepared in New Zealand white rabbits with two intravenous injections of 5×10^8 Balb/c thymocytes, according to the method of Jooste et al. (7).

Before use in the *in vivo* experiments, ATS and normal rabbit serum (NRS) were absorbed once with an equal volume of mouse erythrocytes. The C57BL/Rij bone marrow donors were injected subcutaneously with 0.2 ml ATS or NRS; injection sites were equally distributed over the inguinal and axillary areas. These injections were given 5 and 2 days before the bone marrow was obtained.

The immunoglobulin fraction of the ATS, which was to be used in the immunofluorescence staining technique, was precipitated with saturated (NH$_4$)$_2$SO$_4$. This fraction was absorbed three times with Balb/c IgG2b plasma cell tumor cells and once with spleen cells from (C57BL/Rij × CBA/Rij) F$_1$ mice, which were thymectomized, lethally irradiated, and reconstituted with 2×10^6 syngeneic fetal liver cells. The specificity of the absorbed antiserum for T-lymphocytes was determined as described in a previous paper from our laboratory (20).

Immunofluorescence Staining of T- and B-Cells

The cells were washed with 5% bovine albumin (Poviet, Amsterdam, The Netherlands) in 0.01 M phosphate buffered saline (5% BA-PBS; pH 7.8). Thereafter, 1×10^6 cells in 25 μl were incubated for 30 min at 4°C with 25 μl of an appropriate dilution of ATS, washed twice with 1% BA-PBS, resuspended in a final volume of 25 μl, and mixed with 25 μl FITC-goat anti rabbit-immunoglobulin conjugate and 25 μl TRITC-goat anti mouse-immunoglobulin conjugate (both from Nordic, Tilburg, The Netherlands). After a 30-min incubation, the cells were washed three times with 1% BA-PBS and mounted on a glass slide in an equal volume of buffered glycerol (nine parts glycerol and one part PBS). The slides were examined with a Zeiss standard microscope equipped with a vertical illuminator IV/F and an Osram HBO 50 mercury lamp. On each slide, 300 cells were scored.

Assay for Delayed Type Hypersensitivity

The delayed type hypersensitivity (DTH) assay has been described in detail in a previous paper (22). In the assay, a number of cells equivalent to one whole spleen from one irradiated and reconstituted (C57BL/Rij × CBA/Rij) F$_1$ recipient mouse are injected intravenously into male C57BL/Rij recipient mice. The DTH reactivity of these recipient mice is determined by measuring the difference in thickness of the hind feet 24 hr after a subcutaneous injection of 2×10^7 CBA/Rij spleen cells into the instep of the right hind foot. The specific increase in foot thickness was calculated as the relative increase in foot thickness of the immune mice minus the relative increase in foot thickness of the control mice. The swelling in challenged normal C57BL/Rij control mice varied between 15 and 22%.

RESULTS

Cellular Changes in the Thymus, Spleen, and Lymph Nodes After Irradiation and Bone Marrow Transplantation

The cellular changes in the thymus, spleen, and peripheral lymph nodes (inguinal and axillary) of lethally irradiated (C57BL/Rij × CBA/Rij) F_1 mice inoculated with 1×10^7 C57BL/Rij bone marrow cells were studied.

Following a sharp decrease in the total number of nucleated cells in the thymus due to irradiation damage, recovery started around day 7. This was found after both semi-allogeneic and syngeneic reconstitution (Figure 21.1). Syngeneically reconstituted mice showed a faster increase, with greater numbers of thymocytes, than mice reconstituted with C57BL/Rij bone marrow cells.

The numbers of T- and B-cells in the spleen and peripheral lymph nodes were determined by the membrane immunofluorescence technique during the first 40 days after irradiation and semi-allogeneic

bone marrow transplantation in sham thymectomized and thymectomized recipients. These figures were compared with the recovery of T- and B-cells in the spleen and lymph nodes of non-thymectomized recipients who had been irradiated and syngeneically reconstituted. In the sham thymectomized mouse spleen, a temporary increase and subsequent decrease of T-cells were found between day 10 and day 15 after semi-allogeneic bone marrow transplantation (Figure 21.2). The increase in T-cells in these sham thymectomized recipients was higher than in similarly reconstituted adult thymectomized recipients by day 12; after day 23, the number of T-cells in sham thymectomized recipient mice again increased. In syngeneically reconstituted mice, the increase in T-cell numbers in the spleen occurred as late as 2 weeks after transplantation. But B-cell recovery was much faster in these syngeneically reconstituted mice (Figure 21.2). No difference in B-cell recovery could be detected between semi-allogeneically reconstituted sham thymectomized mice and adult thymectomized recipient mice.

In the peripheral lymph nodes of both sham and adult thymectomized semi-allogeneically reconstituted mice, an increase and a subsequent decrease in T-cell number was found between day 8 and day 15 (Figure 21.3). Again, in syngeneically reconstituted mice this early increase of T-cell numbers was not observed. In these mice, there was a gradual T-cell recovery after about 15 days. No significant differ-

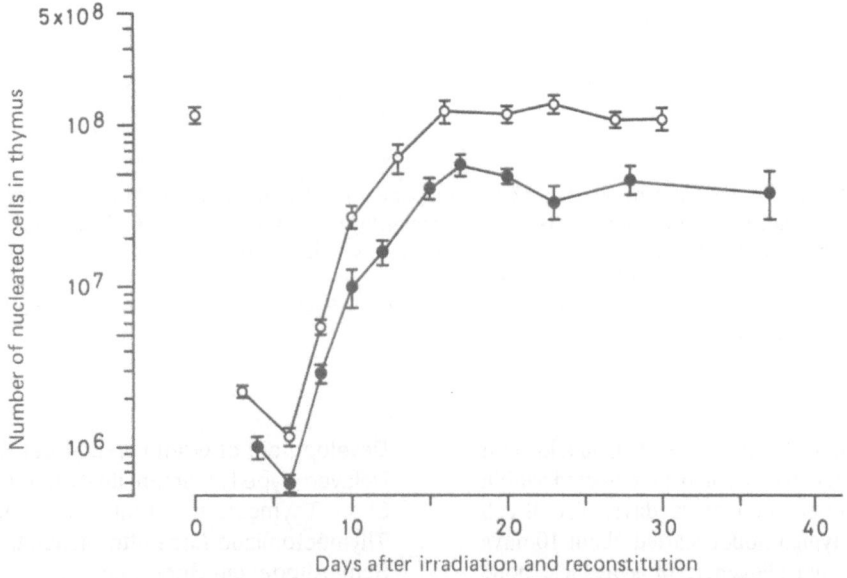

FIGURE 21.1. Recovery of the number of nucleated cells in the thymus of lethally irradiated (C57BL/Rij X CBA/Rij) F¹ mice, reconstituted with either 1×10^7 C57BL/Rij bone marrow cells (●——●) or 1×10^7 syngeneic bone marrow cells (○——○). The figures representing the values immediately after irradiation were obtained from non-irradiated control mice. Each experimental point represents the mean ± 1 SEM of five mice.

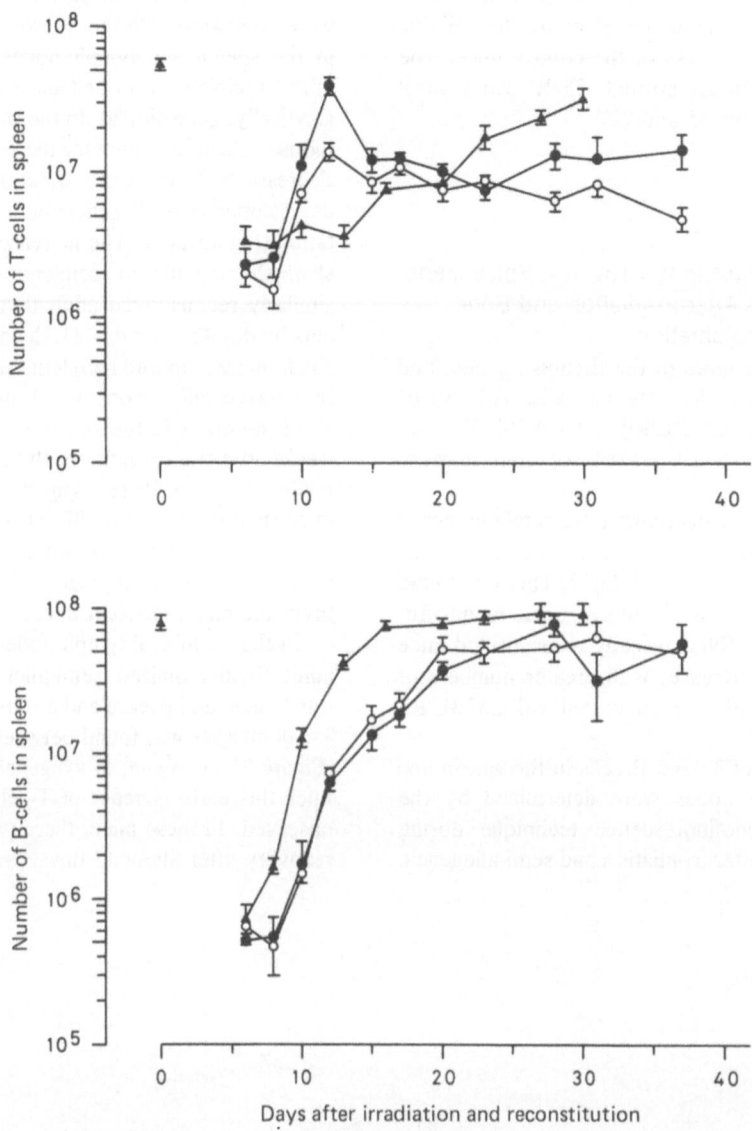

FIGURE 21.2. Recovery of the number of T-cells (upper part) and B-cells (lower part) in the spleen of lethally irradiated sham thymectomized (●——●) and adult thymectomized (○——○) (C57BL/Rij X CBA/Rij) F₁ mice reconstituted with 1 × 10⁷ C57BL/Rij bone marrow cells, and of similarly, but syngeneically reconstituted (C57BL/Rij X CBA/Rij) F₁ mice (▲——▲). The figures representing the values immediately after irradiation were obtained from non-irradiated control mice. Each experimental point represents the mean ± 1 SEM of five mice.

ence in T-cells in the lymph nodes of sham and adult thymectomized recipients could be detected within the period of observation of 40 days. The B-cell recovery in the lymph nodes started about 10 days after irradiation and allogeneic or syngeneic bone marrow transplantation (Figure 21.3). At day 16–17 after allogeneic reconstitution, however, the B-cells in the lymph nodes decreased sharply, probably because of a developing GvH reaction.

Development of Graft-versus-Host Related Delayed-Type Hypersensitivity in the Spleen of Sham Thymectomized Mice and Adult Thymectomized Mice after Irradiation and Semi-Allogeneic Bone Marrow Transplantation

At various intervals after irradiation and reconstitution of (C57BL/Rij x CBA/Rij) F₁ mice with 1 × 10⁷ C57BL/Rij bone marrow cells, the recipient

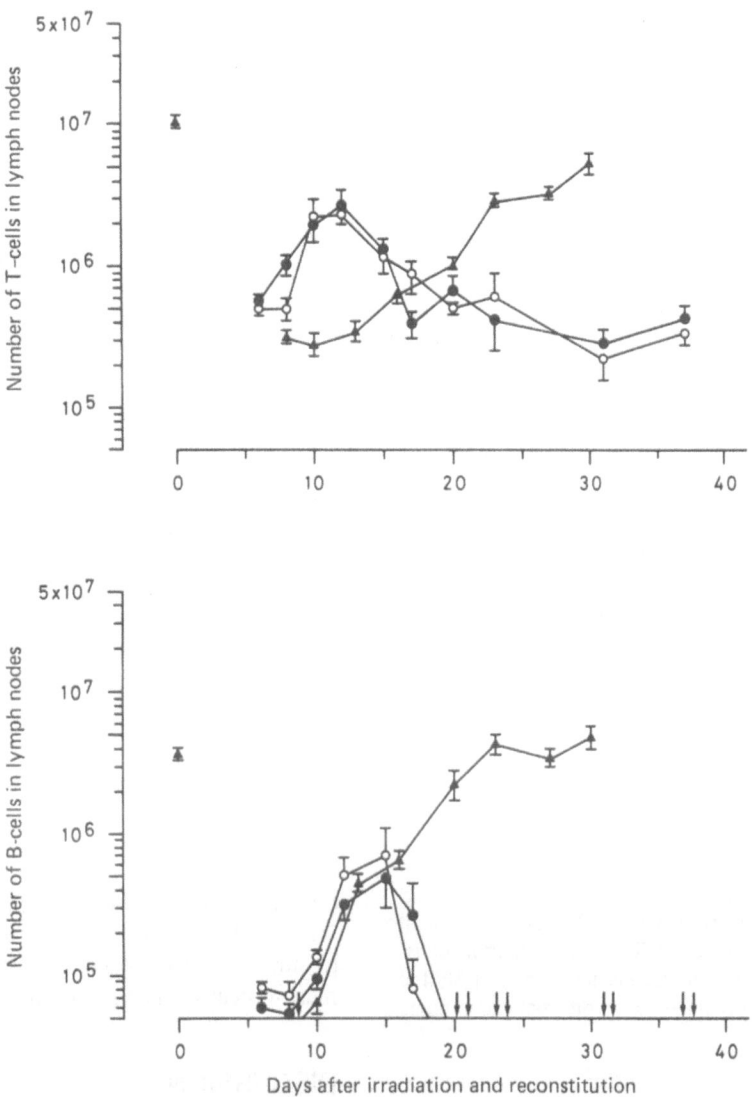

FIGURE 21.3. Recovery of the number of T-cells (upper part) and B-cells (lower part) in the peripheral lymph nodes of lethally irradiated sham thymectomized (●——●) and adult thymectomized (○——○) (C57BL/Rij X CBA/Rij) F₁ mice reconstituted with 1 × 10⁷ C57BL/Rij bone marrow cells, and of similarly, but syngeneically reconstituted (C57BL/Rij × CBA/Rij) F₁ mice (▲——▲). (Arrows indicate cell numbers below the abscissa.) The figures representing the values immediately after irradiation were obtained from non-irradiated control mice. Each experimental point represents the mean ± 1 SEM of five mice.

spleens were injected into normal C57BL/Rij mice. Immediately thereafter, these C57BL/Rij mice were challenged with 2 × 10⁷ CBA/Rij spleen cells. At 24 hr after challenge, the DTH reaction was measured. Figure 21.4 (A) shows the development of DTH reactivity in sham thymectomized recipient mice. As soon as 8 days after transplantation, a DTH reactivity could be demonstrated. Maximum DTH was found to occur at about day 12. Thereafter, it declined until no significant DTH could be detected on day 20. In adult thymectomized recipient mice, on the other hand [Figure 21.4 (B)], DTH started to increase around day 15, reached maximum values around day 20, and steadily decreased thereafter. The course of the DTH in adult thymectomized recipient mice in different experiments was more variable than the course in sham thymectomized recipients.

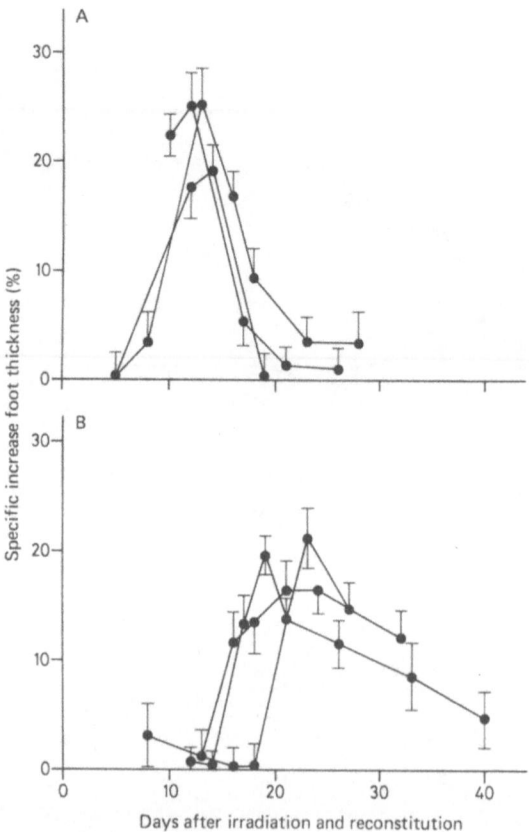

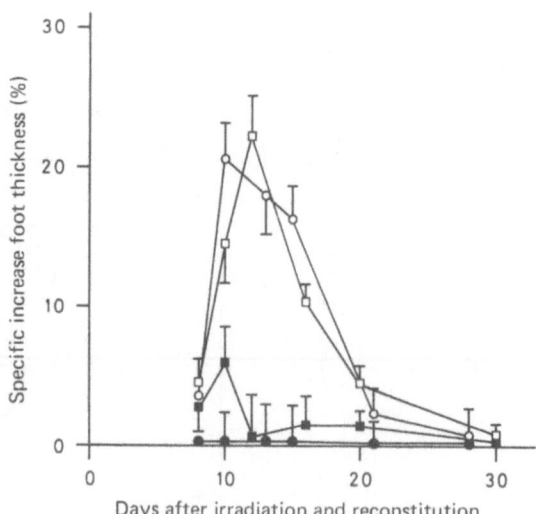

FIGURE 21.5. Development of GvH-related DTH in the spleen of lethally irradiated (C57BL/Rij X CBA/Rij) F_1 mice, reconstituted with 1×10^7 C57BL/Rij bone marrow cells, which were treated *in vitro* with either anti-Thy-1.2 serum (●——●) and C or NMS (○——○) and C^1 before transplantation or reconstituted with 1×10^7 bone marrow cells from C57BL/Rij donors treated *in vivo* with ATS (■——■) or NRS (□——□) before harvesting bone marrow cells. Each experimental point represents the mean ± 1 SEM of a group of at least five mice.

FIGURE 21.4. Development of GvH-related DTH in the spleen of lethally irradiated sham thymectomized (**A**) and adult thymectomized (**B**) (C57BL/Rij X CBA/Rij) F_1 mice, transplanted with 1×10^7 C57BL/Rij bone marrow cells. Each experimental point represents the mean ± 1 SEM of a group of at least five mice. Three different experiments are shown.

Dependence of Graft-versus-Host Related Delayed-Type Hypersensitivity on T-Cells in the Bone Marrow

In order to test whether the anti-host DTH depends on the Thy-1.2 positive cells in the bone marrow inoculum, the C57BL/Rij bone marrow cells were treated *in vitro* with anti-Thy-1.2 serum and complement (C') or normal mouse serum (NMS) and C' before transplantation into irradiated (C57BL/Rij × CBA/Rij) F_1 recipient mice. As can be seen in Figure 21.5, treatment of the bone marrow cells with anti-Thy-1.2 serum and C' before transplantation completely prevented the development of anti-host DTH during the observation period of 30 days. The question of whether these bone marrow T-cells belong to the pool of mature recirculating T-cells was studied by treatment of the C57BL/Rij donor mice with anti-thymocyte serum (ATS) *in vivo* 5 and 2 days before the bone marrow cells were harvested.

Control donor mice were treated with normal rabbit serum (NRS). As can be seen in Figure 21.5, ATS treatment greatly reduced the capacity of the bone marrow cells to elicit an anti-host DTH response.

DISCUSSION

The present study shows that after semi-allogeneic bone marrow transplantation into lethally irradiated mice, an anti-host DTH responsiveness is built up in the spleen of the recipient mice. This DTH starts to increase 8 days after reconstitution, reaches maximum values on day 12–13, and decreases sharply thereafter. By day 20 this significant response could no longer be detected. (Figure 21.4). This correlates well with the increase and subsequent decrease of number of T-cells in the spleen (Figure 21.2) and the lymph nodes (Figure 21.3) of the sham thymectomized recipient mice and the time-course of the anti-host response in irradiated CBA mice inoculated with C57BL bone marrow cells, as measured with the Simonsen splenomegaly assay by Van Bekkum et al. (19). These authors found an anti-host response by day 10 and 15 after reconstitution. Thereafter, they could no longer detect anti-host reactivity.

In contrast to this early anti-host immune reactivity, the symptoms of secondary disease in this C57BL/Rij → (C57BL/Rij × CBA/Rij) F$_1$ combination appear about 30 days after irradiation and reconstitution. The severity of this disease varies considerable. Some of the chimeras die, others recover after about 3 months, and some of the animals have no signs of the disease. This variation does not occur in the anti-host immune response measured in our DTH system. This response develops in all animals, at the same time, and to about the same extent.

An old, but still inconclusively answered question is whether delayed graft-versus-host disease (GvHD) in mice is caused by mature T-cells in the transplanted inoculum or whether it is due to newly formed T-cells arising from primitive precursor cells under the influence of the recipient thymus (18). Treatment of the bone marrow inoculum with anti-Thy-1.2 serum and complement in vitro before transplantation eliminated its ability to induce GvHD mortality in lethally irradiated (semi-) allogeneic recipients (13,16). As can be seen in Figure 21.5, after treatment of the bone marrow cells with anti-Thy-1.2 serum and complement in vitro no anti-host DTH reactivity could be detected. Similarly, elimination of the mature recirculating T2-cells from the bone marrow by ATS in vivo (8) almost completely prevented the development of anti-host DTH (Figure 21.5). Ledney and Van Bekkum (9) could enhance the survival rate of the recipient mice from 10 to 50% by treatment of the allogeneic bone marrow donors with purified ALS. But as stated by Thierfielder and Rodt (17), the parent → F$_1$ combination is an "easy system" with regard to GvH suppression in the mouse. The above-mentioned results suggest that mature T-cells in the bone marrow inoculum have a predominant role in the development of secondary disease in the parent → F$_1$ combination studied.

Löwenberg (10) described a delayed form of GvHD after transplantation of allogeneic fetal liver cells in the C57BL/Rij → CBA/Rij combination. Since the fetal liver is considered to contain virtually no immunocompetent T-cells at the chosen moment of harvest, this study of Lydyard and Ivanyi (11) suggests that newly formed T-cells contribute to the development of GvHD. In the earlier studies of Goedbloed and Vos (6) and Van Putten (21), the contribution of T-cells newly formed under influence of the recipient's thymus also became apparent, since thymectomy of the recipients before allogeneic and xenogeneic bone marrow transplanation could delay mortality. Osmond and Nakatsui (14) showed that thymectomy diminishes the recovery of GvH reactivity in the spleen of syngeneically bone marrow reconstituted rats, when assayed 3 weeks after reconstitution. In our experiments, thymectomy was found to delay the onset of GvH-related DTH reactivity. Furthermore, after thymectomy the maximum DTH response was somewhat diminished. These results suggest that mature T-cells elicit the anti-host DTH, whereas T-cells newly formed under the influence of the thymus, which starts to recover before the onset of GvH-related DTH (Figure 21.1), accelerate and enhance this response. In contrast to the sham thymectomized recipients, the occurrence of anti-host DTH in the spleen of adult thymectomized recipients did not coincide with a temporary increase of the number of splenic T-cells, which suggests that lymphocyte numbers are not a reliable measure for the anti-host immune reactivity in this semi-allogeneic combination.

In the splenomegaly GvH assay, synergy occurs between newly formed T1-cells and long-lived T2-cells (3). In that system, the eliciting cell is the T1-cell, whereas the T2-cell amplifies the response of these short-lived cells (3). Obviously, this is the reverse of our anti-host DTH assay. Although this difference might be attributed to the use of irradiated, adult recipients versus non-irradiated, newborn recipients, the possibility has to be considered that it is due to the different parameter studied.

SUMMARY

During initiation of a delayed graft versus host (GvH) reaction by injection of C57BL/Rij bone marrow cells into lethally irradiated (C57BL/Rij × CBA/Rij) F$_1$ hybrid mice, a state of delayed type hypersensitivity (DTH) against host histocompatibility antigens occurs. This DTH was maximal 12 days after bone marrow transplantation and could no longer be demonstrated on day 20. In adult thymectomized recipients, DTH appeared about 16 days after transplantation and reached a maximum at about 20 days. The maximum DTH responses of adult thymectomized mice were lower than the maximum responses of non-thymectomized mice. These findings were studied in relation to cellular changes in the thymus, spleen, and lymph nodes. Treatment of the C57BL/Rij bone marrow cells with anti-Thy-1.2 serum and complement in vitro, and treatment of the C57BL/Rij donor mice in vivo with anti-thymocyte serum before harvesting the bone marrow cells, could completely or almost completely prevent GvH–DTH.

These results suggest that mature T-cells in the bone marrow inoculum play a major role in the development of anti-host DTH and that T-cells newly formed under influence of the recipient's thymus can accelerate and enhance this GvH–DTH responsiveness.

ACKNOWLEDGMENTS

We are indebted to Mr. F. Luiten for excellent technical assistance. We thank Mrs. Cary Meijerink-Clerkx for typing the manuscript.

The investigation is part of a project program on the regulation of hemopoiesis, subsidized by the Netherlands Foundation for Medical Research (FUNGO).

REFERENCES

1. Amato, D., Cowan, D. H., and McCulloch, E. A. Separation of immunocompetent cells from human and mouse hemopoietic cell suspensions by velocity sedimentation. *Blood, 39*:472, 1972.

2. Boak, J. L., and Wilson, R. E. Modification of the graft-versus-host syndrome by anti-lymphocyte serum treatment of the donor. *Clin. Exp. Immunol., 3*:795, 1968.

3. Cantor, H., and Asofsky, R. Synergy among lymphoid cells mediating the graft-versus-host response. III. Evidence for interaction between two types of thymus-derived cells. *J. Exp. Med., 135*:764, 1972.

4. Chen, M. G., Price, G. B., and Makinodan, T. Incidence of delayed mortality (secondary disease) in allogeneic radiation chimeras receiving bone marrow from aged mice. *J. Immunol., 108*:1370, 1972.

5. Dicke, K. A., and Van Bekkum, D. W. Allogeneic bone marrow transplantation after elimination of immunocompetent cells by means of density gradient centrifugation. *Transplant Proc., III* (1):666, 1971.

6. Goedbloed, J. F., and Vos, O. Influences on the incidence of secondary disease in radiation chimeras: Thymectomy and tolerance. *Transplantation, 3*:603, 1965.

7. Jooste, S. V., Lance, E. M., Levey, R. H., Medawar, P. B., Ruszkiewicz, M., Sharman, R., and Taub, R. N. Notes on the preparation and assay of anti-lymphocytic serum for use in mice. *Immunology, 15*:697, 1968.

8. Lance, E. M., Medawar, P. B., and Taub, R. N. Antilymphocyte serum. In Dixon, F. J., and Kunkel, H. G., eds., *Advances in Immunology.* New York: Academic Press, 1973, p. 1.

9. Ledney, G. D., and Van Bekkum, D. W. Secondary disease in irradiated mice grafted with allogeneic bone marrow from anti lymphocyte serum treated donors. *J. Nat. Cancer Inst., 42*:633, 1969.

10. Löwenberg, B., De Zeeuw, H. M. C., Dicke, K. A., and Van Bekkum, D. W. Nature of the delayed graft-versus-host reactivity of fetal liver cell transplants in mice. *J. Nat. Cancer Inst., 58*:959, 1977.

11. Lydyard, P. M., and Ivanyi, J. Chimaerism of immunocompetent cells in allogeneic bone marrow-reconstituted lethally irradiated chickens. *Transplantation, 20*:155, 1975.

12. Miller, J. F. A. P. Effect of thymectomy in adult life on immunological competence. *Nature (London.) 208*:1336, 1965.

13. Norin, A. J., and Emeson, E. E. Effects of restoring lethally irradiated mice with anti Thy-1.2 treated bone marrow: graft-vs-host, host-vs-graft, and mitogen reactivity. *J. Immunol., 120*:754, 1978.

14. Osmond, D. G., and Nakatsui, T. Graft versus host activity of lymphoid tissue and bone marrow cells following X-irradiation and bone marrow transfusion in normal and thymectomized rats. *Anat. Rec., 172*:377, 1972.

15. Simmons, R. L., Wolf, S. M., Chandler, J. G., and Nastuk, W. L. Effect of allogeneic bone marrow on lethally irradiated thymectomized mice. *Proc. Soc. Exp. Med. 120*:81, 1965.

16. Sprent, J., Von Boehmer, H., and Nabholz, M. Association of immunity and tolerance to host H-2 determinants in irradiated F₁ hybrid mice restored with bone marrow cells from one parental strain. *J. Exp. Med., 142*:321, 1975.

17. Thierfielder, S., and Rodt, H. Antilymphocytic antibodies and marrow transplantation. V. Suppression of secondary disease by host-versus-θ-graft reaction. *Transplanation, 23*:87, 1977.

18. Van Bekkum, D. W., Löwenberg, B., and Vriesendorp, H. M. Bone marrow transplantation. In Jirsch, D. W., ed., *Immunological Engineering,* Lancaster, England: Falcon House, 1978, p. 179.

19. Van Bekkum, D. W., Van Putten, L. M., and De Vries, M. J. Anti-host reactivity and tolerance of the graft in relation to secondary disease in radiation chimeras. *Ann. N.Y. Acad. Sci., 99*:550, 1962.

20. Van der Ham, A. C., Benner, R., and Vos, O. Mobilization of B and T lymphocytes and haemopoietic stem cells by polymethacrylic acid and dextran sulphate. *Cell. Tissue Kinet., 10*:387, 1977.

21. Van Putten, L. M. Thymectomy: effect on secondary disease in radiation chimeras. *Science, 145*:935, 1964.

22. Wolters, E. A. J., and Benner, R. Immunobiology of the graft versus host reaction. I. Symptoms of graft versus host disease in mice are preceded by delayed type hypersensitivity to host histocompatibility antigens. *Transplantation* (In press).

23. Yoshida, Y., and Osmond, D. G. Graft-versus-host activity of rat bone marrow, marrow fractions, and lymphoid tissues quantitated by a popliteal lymph node weight assay. *Transplantation, 12*:121, 1971.

PART V

Bone Marrow Transplantation

M. M. Bortin

22

Factors Influencing Success and Failure of Human Marrow Transplantation: A Review from the International Bone Marrow Transplant Registry

Mortimer M. Bortin and
Alfred A. Rimm for the Advisory
Committee of the International
Bone Marrow Transplant
Registry

It was more than 25 years ago that Lorenz et al. (12) first reported successful marrow transplantation in laboratory animals. Between 1957 and 1967 more than 200 human marrow allografts were attempted, but none of the patients had more than transient benefit (2). Improvements in histocompatibility matching of donor and recipient led to the first successful marrow transplants by Bach et al. (1), Gatti et al. (8), and de Koning et al. (11) in 1968. As a result, the International Bone Marrow Transplant Registry (IBMTR) was founded in 1970 as a division of the American College of Surgeons/National Institutes of Health Organ Transplant Registry. Shortly thereafter, it was thought that a central registry of all bone marrow transplants would be useful to summarize experience and progress in the field. By 1974, it was apparent that a rapidly increasing number of marrow transplant teams were successfully performing marrow transplants. Because each marrow transplant patient yields a wealth of information relevant to transplantation immunology, cancer immunology, immunogenetics, and hematology, the IBMTR Advisory Committee elected to request comprehensive pre- and post-transplant data from all the marrow transplant teams in the world. It was their belief that it would be possible to uncover important relationships and to find the factors that influence success and failure of clinical marrow transplantation by analyzing pooled data and thereby accelerating achievement in the field. In 1976, with the dissolution of the American College of Surgeons Organ Transplant Registry, the IBMTR became a free-standing agency based in Milwaukee, Wisconsin and supported by the U.S. Department of Health, Education and Welfare. Today, almost all bone marrow transplant teams throughout the world participate in this cooperative venture and report their bone marrow transplant experience to the IBMTR.

The cumulative accession rates for patients and transplant teams reporting to the IBMTR are plotted in Figure 22.1. The distribution of patients according to disease category is presented in Table 22.1. A life table analysis indicating the probability for survival following transplantation of marrow from major histocompatibility complex (MHC) compatible donors for patients with severe aplastic anemia (3), severe combined immunodeficiency disease (SCID) (4), and end-stage acute myeloblastic leukemia (AML) in relapse (5) is shown in Figure 22.2. For several reasons, caution should be exercised in interpreting these survival curves. A general problem with interpretation of data from any registry is that there may have been selective reporting of cases. Further, survival data from teams with little experience and modest facilities have been pooled with survival data from teams with demonstrated expertise, elaborate facilities, and sophisticated laboratory backup.

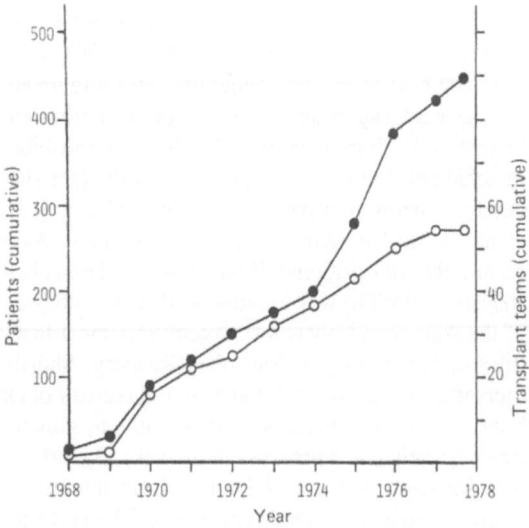

FIGURE 22.1. Cumulative number of patients (●——●) reported and teams (○——○) reporting to the International Bone Marrow Transplant Registry (to 31 August 1978).

TABLE 22.1 Distribution of Patients According to Disease Category

DISEASE CATEGORY	PATIENTS	TRANSPLANTS
Leukemia, Lymphoma, and other malignant diseases	169	193
Aplasia and Hemoglobinopathies	168	214
Immunodeficiency diseases	125	186
	462	593

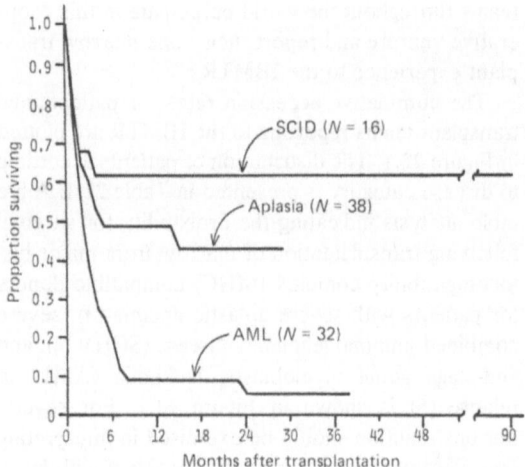

FIGURE 22.2. Life table survival curves for patients treated with marrow transplants from HLA compatible donors. Data for severe aplastic anemia from 1976 report (3); severe combined immunodeficiency disease (SCID) from 1977 report (4); and end-stage acute myeloblastic leukemia (AML) from 1978 report (5).

Finally, survival data accumulated in the late 1960s and 1970s have been combined with more recent results. Nonetheless, the curves are representative of overall survival data from almost all bone marrow transplant teams in the world during the past 8 to 10 years.

Summarized in this chapter are the most important findings uncovered to date by analyses performed by the IBMTR. Subsequent chapters of this section contain reports from six of the foremost bone marrow transplant teams in the world in which are described the innovative approaches they are using in attempts to resolve some of the remaining problems associated with marrow transplantation.

SEVERE APLASTIC ANEMIA

An analysis of HLA-compatible marrow transplantation in patients with severe aplastic anemia revealed a number of factors which affected the prognosis (3):

1. Survival with a functioning graft (apparent cure) was significantly higher among patients transplanted within 3 months of diagnosis in comparison with patients transplanted more than 9 months after diagnosis.
2. Patients who received under 15 pretransplant transfusions had a significantly higher survival rate than patients who received over 15.
3. Patients under 21 years of age at the time of transplantation had a significantly higher survival rate than older patients.
4. Patients who had infections at the time of transplantation had a survival rate as high as patients who did not have infections.

It was noted that almost all deaths occurred within the first 4 months following marrow transplantation; patients who survived beyond 4 months appeared to have an excellent probability for long-term survival (Figure 22.2). Second transplants for severe aplastic anemia were attempted in 16 patients and succeeded in four. Among patients who died, the most frequent primary and/or contributory causes of death were sepsis (73%), aplasia (61%), and graft versus host disease (GvHD) (21%).

SEVERE COMBINED IMMUNODEFICIENCY DISEASE

In 1977, an analysis of bone marrow and fetal tissue transplantation for 69 patients with SCID disclosed a number of factors that affected prognosis (4):

1. The highest survival rates occurred in patients who were transplanted with marrow from HLA genotypically identical donors (Figure 22.2).

2. Patients who received marrow from donors who (a) were HLA phenotypically identical; (b) were HLA-A and/or B incompatible, but mixed leukocyte culture (MLC) compatible; or (c) received fetal tissue from HLA incompatible donors had significantly higher 6-month survival rates than those patients who received marrow from MLC incompatible donors. Transplantation of marrow from MLC incompatible donors was invariably unsuccessful.

3. Male patients who received transplants from female donors had significantly more intense GvHD and lower 6-month survival rates than female patients who received transplants from male donors or patients of either sex who received transplants from a member of the same sex.

4. Patients under 6 months of age at the time of transplantation had a lower incidence of severe GvHD than older patients.

5. Although 80% of the patients had infections at the time of transplantation, their survival rate was as high as the survival rate in patients who did not have infections.

6. Multiple transplants were attempted in 18 patients and were successful in six.

Patients who survived at least 4 months following transplantation of marrow or fetal tissue had an excellent probability for long-term survival with improvement or apparent cure of their disease. Among the patients who died, the most frequent primary and/or contributory causes of death were infection (85%) and GvHD (23%).

END-STAGE ACUTE MYELOBLASTIC LEUKEMIA

A 1978 analysis of bone marrow transplantation for patients with end-stage AML in relapse revealed a number of factors that affected prognosis (5):

1. Patients had about a 10% probability for 20-month survival following high-dose chemoradiotherapy plus HLA compatible marrow transplantation.

2. Patients under 21 years of age had a significantly higher 6-month survival rate than patients over 30 years of age.

3. Patients transplanted within 8 months of diagnosis had a significantly higher survival rate than patients transplanted later in the course of the disease.

4. Patients free of overt infection at the time of transplantation had an increased incidence of lethal idiopathic interstitial pneumonitis.

The primary and/or contributory causes of death in end-stage AML patients treated with marrow transplants from MLC compatible donors were infection (67%), GvHD (58%), recurrent leukemia (30%), and interstitial pneumonitis (29%). There was no evidence of a graft-versus-leukemia effect following transplantation of MHC compatible marrow cells. The true incidence of recurrent leukemia would probably be higher than the 30% figure cited, if competing causes of death were eliminated; the earliest appearance of recurrent leukemia was more than 3 months post-transplant, and more than 60% of the patients were dead of other causes by 3 months (Figure 22.2).

DISCUSSION

After more than 20 years of clinical trial, it is apparent that only now is the great therapeutic potential of allogeneic marrow transplantation for the successful treatment of heretofore incurable diseases *beginning* to be realized. Analyses performed by the IBMTR have helped to identify a number of factors that appear to influence success and failure of marrow transplantation. Formidable problems require resolution before marrow transplantation can be applied as a more or less routine clinical procedure. Briefly discussed below are the main causes for failure and several of the strategies presently under investigation to eliminate them.

Infection

Data from the IBMTR and reports of others point up the fact that during the past 10 years, infection was directly or indirectly responsible for approximately 75% of the deaths following MHC compatible marrow transplantation in patients with severe aplastic anemia, SCID, and AML. The causes for infection following marrow transplantation are well known and include (a) diminished or absent immune function because of the underlying disease; (b) the presence of pretransplant infection; (c) injury to the immune apparatus and to granulocyte function as a consequence of pretransplant immunosuppressive treatments to achieve engraftment or post-transplant immunosuppressive treatments to prevent and/or treat GvHD; (d) the immunosuppressive effects of GvHD; and (e) the long interval between transplantation and *restitutio ad integrum* of immune function. Strategies aimed at reducing the risk of fatal infection following allogeneic marrow transplantation are under intensive study by marrow transplant teams throughout the world and include (a) pretransplant antibiotic treatment for bacterial decontamination of the gastrointestinal tract; (b) strict isolation of the patient in a protected environment; (c) liberal use of granulocyte transfusions prophylactically or at the first sign of fever during the post-transplant, granulocytopenic period; (d) administration of polyvalent convalescent serum for patients with serologic or biopsy evidence of cytomegalovirus infection; and (e) administration *in vivo* of B- and T-cell amplifying agents in an effort to accelerate recovery of immunocompetence. It is anticipated that implementation of at least several of these approaches will result in a substantial reduction in the incidence of lethal infection with resultant improvement in survival rates. In addition, survival rates should improve with the development of new antibiotics and with improvements in supportive care and microbiologic monitoring of marrow transplant patients.

Sustained Engraftment

Almost always, MHC compatible marrow transplantation for patients with SCID and AML has resulted in permanent engraftment (4,5). But in patients with severe aplastic anemia, no engraftment or unsustained engraftment were common causes of failure prior to 1976, accounting for 61% of the deaths (3). As a result, a number of transplant teams intensified immunosuppressive conditioning regimens for those aplasia patients who were thought to have a high risk of graft rejection (9,10). Although this approach accomplished the goal of sustained engraftement, it was not associated with concomitant improvement in survival rates because of increased toxicity and worsening of GvH reactions. In the chapters that follow, Graze and members of the UCLA Bone Marrow Transplant Team (9) and Kersey and members of the University of Minnesota Bone Marrow Transplant Team (10) present early results of two innovative and promising approaches to the dilemma of engraftment without toxicity in patients with severe aplastic anemia.

Graft-versus-Host Disease

Despite the use of donor-recipient pairs who are identical at HLA on the basis of serotyping and MLC data, GvHD remains the major cause of failure in allogeneic bone marrow transplantation (3–5). In a subsequent chapter in this section, Sanders and the Seattle Bone Marrow Transplant Team (15) report that the incidence and severity of GvHD in leukemic patients was similar irrespective of whether the patients were in remission or in relapse at the time of transplantation. Rodt and the members of the Munich Bone Marrow Transplant Team (14) describe their initial results testing antibody treatment of marrow *in vitro* for prevention of GvHD in experimental animals and in humans. Graze et al. (9) and Kersey et al. (10) report the influence of new pretransplant immunosuppressive conditioning regimens on GvHD in aplasia patients.

Recurrent Leukemia

Analyses of MHC compatible bone marrow transplantation for patients with end-stage AML transplanted in relapse disclosed that only a very small proportion of the patients are likely to have long-term, disease-free survival (5,7,16). Recent results reported by the Seattle Bone Marrow Transplant Team in this volume (15) confirm the earlier analyses; only one-fifteenth (7%) of their end-stage AML patients transplanted between March, 1976 and December, 1977 are currently alive. Nonetheless, an element of hope is offered by the fact that a small number of patients with end-stage acute leukemia can be saved when all other treatments have failed (5,7,16) and by the fact that higher survival rates were reported by the IBMTR when marrow transplantation was undertaken early rather than late in the course of the disease (5). Sanders et al. (15) report impressive early results in 43 patients with acute leukemia who received marrow transplants from MHC compatible donors at a time when their primary disease was in remission. The problem of recurrent leukemia following marrow transplantation is addressed also by Lotsova and her colleagues in Houston (13), who describe the results of their tests of natural killer cellular cytotoxicity in normal individuals and AML patients. They discuss the possible importance of natural killer cells in marrow transplant patients both from the point of view of their graft-versus-leukemia and their GvH reactivities.

Other Problems

In the chapters that follow, Graze et al. (9), Kersey et al. (10), and Sanders et al. (15) describe factors that may have influenced the incidence and severity of interstitial pneumonitis. Kersey et al. (10) describe the initial results of their attempts to transplant marrow from MHC incompatible donors. Finally, Braine and the members of the Johns Hopkins Bone Marrow Transplant Team (6) describe an innovative approach aimed at prevention of problems associated with ABO incompatible bone marrow transplantation.

SUMMARY

Reviewed in this chapter are a number of factors that significantly affect the outcome of clinical marrow transplantation. The conclusions presented here were based upon analyses conducted by the IBMTR. In the chapters that follow are reports from seven leading bone marrow transplant teams in which they describe their new approaches to improve the results of allogeneic marrow transplantation.

ACKNOWLEDGMENTS

Tenth Report from the International Bone Marrow Transplant Registry.

Advisory Committee Members: Fritz H. Bach, M.D. (Chairman); Dirk W. van Bekkum, M.D., Ph.D.; Mortimer M. Bortin, M.D.; Robert P. Gale, M.D.; Eliane Gluckman, M.D.; Robert A. Good, M.D., Ph.D.; Walter H. Hitzig, M.D.; Georges Mathé, M.D.; Alfred A. Rimm, Ph.D.; John J. van Rood, M.D.; George W. Santos, M.D.; and Kenneth W. Sell, M.D., Ph.D. The data and analyses presented here were approved by the members of the Advisory Committee; opinions expressed are those of the authors and do not necessarily reflect those of the Advisory Committee.

The research on which this publication is based was performed pursuant to Contracts NO1-HB-6-2963 and NO1-HB-6-2964 from the National Heart, Lung, and Blood Institute; National Cancer Institute, and the Bureau of Community Health Services, Health Services Administration, Public Health Service, U.S. Department of Health, Education and Welfare.

Institutions contributing patient data for this report were Academisch Ziekenhuis; Leiden; Albany Medical Center, Albany, New York; M.D. Anderson Tumor Institute, Houston; Baltimore City Hospitals, Baltimore; Blegdamshospitalet, Copenhagen; Children's Hospital, Philadelphia; Cleveland Clinic, Cleveland; III Clinica Medica, Rome; Clinique Infantile Universitaire, Lausanne; Duke University Medical Center, Durham, North Carolina; Hahnemann Medical College, Philadelphia; Hôpital Eduoard Herriot, Lyon; Hôpital des Enfants Malades, Paris; Hôpital Saint Louis, Paris; Institut de Cancerologie et d'Immunogenetique, Villejuif; Kantonspital, Basel; Kantonspital, Zurich; Kinderklinik der Universitat Munchen, Munich; Massachusetts General Hospital, Boston; Memorial Sloan-Kettering Institute, New York; Michigan State University, East Lansing, Michigan; Mount Sinai Medical Center, Milwaukee; National Cancer Institute, Bethesda; New York Hospital, New York; North Carolina Memorial Hospital, Chapel Hill, North Carolina; Oak Ridge Associated Universities, Oak Ridge, Tennessee; Ontario Cancer Institute, Toronto; Ospedale Regionale, Genova; Princess Margaret Hospital, Toronto; Royal Marsden Hospital, London; Roswell Park Memorial Institute, Buffalo; Semmelweis University, Budapest; Universitat Ulm, Ulm; Universitatsklinik, Basel; University of Alabama, Huntsville, Alabama; University of California Cancer Institute, San Francisco; University of California Center for Health Sciences, Los Angeles; University of Chicago, Chicago; University of Cincinnati Medical Center, Cincinnati; University of Colorado Medical School, Denver; University of Wisconsin Center for Health Sciences, Madison; and Westminster Medical School, London.

REFERENCES

1. Bach, F. H., Albertini, R. J., Joo, P., Anderson, J. J., and Bortin, M. M. Bone marrow transplantation in a patient with Wiskott-Aldrich syndrome. *Lancet, II*:1364–6, 1968.
2. Bortin, M. M. A compendium of reported human bone marrow transplants. *Transplantation, 9*:571–7, 1970.
3. Bortin, M. M., and Rimm, A. A., for the Advisory Committee of the International Bone Marrow Transplant Registry. Bone marrow transplantation from histocompatible, allogeneic donors for aplastic anemia. *JAMA, 236*:1131–5, 1976.
4. Bortin, M. M., and Rimm, A. A., for the Advisory Committee of the International Bone Marrow Transplant Registry. Severe combined immunodeficiency disease: Characterization of the disease and results of transplantation. *JAMA, 238*:591, 1977.
5. Bortin, M. M., and Rimm, A. A., for the Advisory Committee of the International Bone Marrow Transplant Registry. Bone marrow trasnplantation for acute myeloblastic leukemia. *JAMA, 240*:1245, 1978.
6. Braine, H. G., Sensenbrenner, L. L., and the Johns Hopkins Bone Marrow Transplant Team. Erythrocyte incompatible bone marrow transplants. In Baum, S. J., and Ledney, G. D., eds., *Experimental Hematology Today 1979*. New York: Springer-Verlag.
7. Gale, R. P., Feig, S., Ho, W. G., Falk, P. et al. Bone marrow transplantation in acute leukemia. *Lancet, II*:1197, 1977.
8. Gatti, R. A., Allen, H. D., Meuwissen, H. J., Hong, R., and Good, R. A. Immunological reconstitution of sex-linked lymphopenic immunological deficiency. *Lancet, II*:1366, 1968.
9. Graze, P. R., Ho, W., and Gale, R. P., for the UCLA Bone Marrow Transplant Team. Bone marrow transplantation for aplastic anemia: Conditioning with cyclophosphamide plus low dose total body irradiation. In Baum, S. J., and Ledney, G. D., eds., *Experimental Hematology Today 1979*. New York: Springer-Verlag, this volume.
10. Kersey, J., Levitt, S., Ramsay, N., Krivit, W., Nesbit, M., Coccia, P., and Tim, T. Absence of rejection of human allogeneic marrow grafts as a result of immunosuppression with total lymphoid irradiation (TLI) and cyclophosphamide. In Baum, S. J., and Ledney, G. D., eds., *Experimental Hematology Today 1979*. New York: Springer-Verlag, this volume.
11. de Koning, J., Dooren, L. J., van Bekkum, D. W., van Rood, J. J., Dicke, K. A., and Radl, J. Transplantation

Bone Marrow Transplantation

of bone marrow cells and foetal thymus in an infant with lymphopenic immunological deficiency. *Lancet I*:1223, 1969.

12. Lorenz, E., Uphoff, D., Reid, T. R., and Shelton, E. Modification of irradiation injury in mice and guinea pigs by bone marrow injections. *J. Nat. Cancer Inst., 12*:197, 1951.

13. Lotsova, E., McCredie, K. B., Dicke, K. A., and Freireich, E. J. Natural killer cells in man: Their possible involvement in leukemia and bone marrow transplantation. In Baum, S. J., and Ledney, G. D., eds., *Experimental Hematology Today 1979*. New York: Springer-Verlag, this volume.

14. Rodt, H., Netzel, B., Kolb, H. J., Rieder, I., Janka, G., Belohradsky, B., Hass, R. J., and Thierfelder, S. Antibody treatment of marrow grafts *in vitro:* A principle for prevention of GVH disease. In Baum, S. J., and Ledney, G. D., eds., *Experimental Hematology Today 1979*. New York: Springer-Verlag, this volume.

15. Sanders, J. E., for the Seattle Marrow Transplant Team. Allogeneic marrow transplantation for patients with acute leukemia. In Baum, S. J., and Ledney, G. D., eds., *Experimental Hematology Today 1979*. New York: Springer-Verlag, this volume.

16. Thomas, E. D., Buckner, C. D., Banaji, M. et al. One hundred patients with acute leukemia treated by chemotherapy, total body irradiation, and allogeneic marrow transplantation. *Blood, 49*:511, 1977.

23

Combined Cyclophosphamide-Total Lymphoid Irradiation Compared to Other Forms of Immunosuppression for Human Marrow Transplantation

John H. Kersey, William Krivit, Mark E. Nesbit, Norma K. C. Ramsay, Peter F. Coccia, Seymour H. Levitt, and Tae H. Kim

Severe aplastic anemia is a generally fatal disease that may be successfully treated by cellular reconstitution using allogeneic bone marrow (3,11,12). The post-grafting period, however, is often associated with a variety of problems including graft rejection or graft versus host disease (GvHD) (2,10). In this report, results are presented of allogeneic marrow transplantation for severe aplastic anemia at the University of Minnesota in the period 1974–1978. In that period, several immunosuppressive programs were utilized in succession in an attempt to improve overall success while reducing the incidence of graft rejection and GvHD. The earliest patients were treated with cyclophosphamide (CY) alone or in combination with procarbazine (PZ), antithymocyte globulin (ATG), and 6-mercaptopurine (6MP). In a second protocol, patients have received total-body irradiation (TBI) in combination with PZ and ATG. Patients recently have received a combination of CY plus total lymphoid irradiation (TLI). This most recent protocol was developed to reduce morbidity associated with TBI, as well as to take advantage of the possibility that TLI might be associated with less severe GvHD, as suggested by recent animal data (9). Using the newer radiation methodologies, we transplanted mismatched marrow in three cases, and the results of these cases are summarized.

PATIENTS AND METHODS

All patients in this study had severe aplastic anemia meeting the criteria of the International Aplastic Anemia Study Group (Table 23.1). Patient M-026 had exposure to solvents as a possible etiology, M-035 had Fanconi's anemia, and M-055 had post-hepatitic aplasia. The remaining patients had idiopathic aplasia (Table 23.2). All patients had been multiply transfused prior to transplantation, but none with family member donors. The HLA genotypes of donor and recipient are shown in Table 23.1. Additional patient characteristics are summarized in Table 23.2. Relative response indices (RRI) are also shown. Sensitization was determined using complement-dependent cytotoxicity in the standard Johnson assay (4). Cellular cytotoxicity utilized recipient effectors and donor targets in a chromium release assay. Results were determined at 4 and 24 hr.

As shown in Table 23.1, patients received immunosuppression in varying combinations. The three major groups are (a) CY alone or in combination with other chemical agents, (b) TBI in combination with PZ and ATG, and (c) CY + TLI. Characteristics of these three combination programs are shown in Table 23.3. Total-body irradiation was administered at a midplane total dose of 750 rad at 26 rad/min. This method has been described previously (6).

TABLE 23.1 Minnesota Allogeneic Marrow Transplants for Severe Aplastic Anemia, August 1978

ID #	AGE	SEX	DONOR	RECIPIENT HLA	DONOR HLA	MLC-RRi[a] DRx	RDx	SENSITIZATION SERUM (Cytotoxicity)	CELLS (Cr Release)	IMMUNO-SUPPRESSION	STOMA-TITIS	INTERSTITIAL PNEUMONITIS	GRAFT REJECTION	ACUTE GvHD	CHRONIC GvHd	OTHER COMPLICATIONS	SURVIVAL (days)
M-021	17	M	Sister	A1-B8 / A11-BW22	A1-B8 / A11-BW22	0	2	Neg	ND	CY	No	No	Yes	No	No	None	
			Sister	A1-B8 / A11-BW22	A1-B* / A11-BW22	0	2	ND	ND	CY	No	No	No	Yes	Yes		731
M-023	10	M	Sister	AW19-B12 / AW19-BW14	AW19-B12 / AW19-BW14	0	0	Neg.	ND	CY	No	No	No	Yes	No	None	>1,259
M-025	16	F	Sister	A3-B27 / A10-BX	A3-B27 / A10-BX	0	7	Pos.	ND	CY	No	No	Yes	No	No	Candida sepsis	235
M-026	6	F	Brother	A1-B8 / AW32-BW14	A1-B8 / AW32-BW14	0	0	Neg.	ND	CY	No	No	Yes	Yes	No	None	
			Brother	AW32-BW14	A1-B8 / AW32-BW14					TBI + CY	Yes	Yes	No	Yes	Yes	None	161
M-033	5	F	Brother	A2-BW15 / A2-BW15	A2-BW15 / A2-BW15	0	6	Neg.	ND	CY	No	No	Yes	No	No	None	
			Brother	A2-BW15	A2-BW15					CY			Yes	No	No	None	
			Brother	A2-BW15	A2-BW15					Cy + ATG + PZ	No	Yes	No	Yes	Yes	None	400
M-035	12	M	Sister	A28-B35 / A29-B12	A28-B35 / A29-B12	0	3	Neg.	ND	CY 6MP	No	No	No	Yes	NE	None	35
M-036	15	F	Brother	A1-B8 / A26-B40	A1-B8 / A26-B40	0	0	Neg.	ND	CY + 6MP	No	No	No	No	Yes	None	>804

Patient	Age	Sex	Relationship	HLA	HLA						Treatment					Complication	RRI[a]
M-039	15	F	Sister	A3-B7 / A9-B40	A3-B7 / A9-B40	0	0	Neg	Neg	No	CY	No	No	No	No	None	>691
M-040	17	M	Sister	A3-B7 / A1-B8	A3-B7 / A1-B8	0	2	Pos.	Neg.	No	CY + 6MP	No	Yes	No	No	None	>664
			Sister	A3-B7 / A1-B8	A3-B7 / A1-B8					No	CY + ATG + PZ	No	No	Yes	Yes	None	
M-042	16	F	Sister	A2-BW15 / A3-BW16	A2-BW15 / A3-BW16	0	8	Neg.	ND	No	CY + 6MP	No	<u>Yes</u>	No	No	Candida sepsis	93
M-046	11	F	Father	A26-B7 / A1-B8	AX-B7 / A1-B8	8	16	Neg.	Neg.	Yes	TBI + PZ + ATG	No	No	Yes	Yes	None	>467
M-048	8	M	Sister	A24-BX / A30-B13	A24-Bx / A30-B13	2	2	Neg.	ND	Yes	TBI + PZ + ATG	No	No	No	No	None	>458
M-054	14	F	Sister	A3-B7 / A3-B40	A3-B7 / A3-B40	0	0	Neg.	Neg.	No	CY TLI	No	No	No	No	None	>269
M-055	11	M	Sister	A1-B8 / A11-B8	A1-B8 / A11-B8	0	0	Neg.	ND	No	CY TLI	No	No	No	No	None	>242
M-056	10	F	Sister	A2-B12 / A2-B39	A2-B27 / A2-B39	3	0	Pos.	Neg.	No	CY TLI	No	No	No	Yes	None	>201
M-057	18	M	Brother	A23-B7 / A31-B45	A23-B7 / A31-B45	0	2	Neg.	ND	NE[b]	CY + TLI	NE	NE	NE	NE	Candida sepsis	11
M-062	14	F	Brother	A1-B51 / A2-B18	A1-B51 / A2-B18	0	1	ND	Neg.	No	CY + TLI	No	No	No	No	None	>117
M-065	10	M	Half-brother	A1-B14 / A29-B12	A1-B14 / A11-B17	16	38	Pos.	Neg.	No	CY TLI	No	<u>Yes</u>	Yes	NE	Klebsiella sepsis	58
M-066	8	M	Sister	A24-B35 / A29-B44	A24-B35 / A29-B44	1	0	Neg.	ND	No	CY + TLI	NE	Ne	No	NE	None	>20

[a]RRI, relative response index.

[b]NE, not evaluable.

TABLE 23.2 Summary of Patient Characteristics

$N = 19$

Disease: Severe aplastic anemia (idiopathic, 16; hepatitis, 1; Fanconi, 1; solvents, 1)
Age: 5 to 18 years (median, 11)
Sex: Male, 9; female, 10
Donors: HLA-MLC matched siblings, 16 (same sex, 5; opposite sex, 11)
  Haploidentical siblings, 2
  Father, 1
Transfused prior to transplant: 19/19
Sensitized to donor (RRI > 2, or positive crossmatch): 8/19

TABLE 23.3 Summary of Three Immunosuppressive Programs

1. *Cyclophosphamide* (CY), 50 mg/kg daily × 4 either alone or in combination with
   6-mercaptopurine (6MP), 500 mg/m daily × 5
   Antithymocyte globulin (ATG), 15 mg/kg daily × 3
   Procarbazine (PZ), 15 mg/Kg/day × 3
2. *Total-body irradiation* (TBI), 750 rad at 26 rad/min, using the linear accelerator, in combination with ATG and PZ
3. *Cyclophosphamide + Total Lymphoid Irradiation* (TLI)

Day	Regimen
−6	CY, 50 mg/kg
−5	CY, 50 mg/kg
−4	CY, 50 mg/kg
−3	CY, 50 mg/kg
−2	Rest
−1	TLI (750 rad total, 26 rad/min)
0	Transplant marrow

Total lymphoid irradiation was to a field including the mantle and inverted Y as used for therapy in Hodgkin's disease. The spleen was included in the field. Irradiation was split between anterior and posterior ports. Individual lung blocks were made for each patient.

RESULTS

Results of marrow transplantation in these 19 consecutive patients are reported in Table 23.1 and summarized in Tables 23.4 and 23.5. As shown in Table 23.4, overall results indicate that 11 of 19 patients are living, at post-transplant times ranging from >20 to >1,259 days. Median survival is 458 days in living patients. Factors contributing to the death of eight patients are underlined in Table 23.1, and the causes of death are summarized in Table 23.4. Results shown in Table 23.5 are presented to show major events, e.g., survival, evidence of "rejection" of marrow (i.e., positive evidence of engraftment as determined by marrow biopsy and increased periph-

TABLE 23.4 Overall Results of Allogeneic Marrow Transplantation

Patients living: 11/19 alive (>20 − >1,259 days; median, 458 days)
Patients dead: 8/19

Primary causes of death	Number
GvHD	3
Rejection + infection	3
Interstitial pneumonitis	1
Candida sepsis	1

eral counts followed by graft failure), evidence of GvHD (grade II or greater using the Seattle criteria), interstitial pneumonitis (diffuse nonbacterial), and stomatitis (mouth ulceration of greater than minor degree). In Table 23,5, the major groups of (a) Cy, (b) TBI, and (c) CY + TLI are shown, with results for HLA-MLC matched sibling donors and non-

TABLE 23.5 Three Immunosuppressive Programs: Major Events

	CY	TBI	CY + TLI
	HLA-MLC Matched Sibling Donors		
Living	4/10 (>664–>1259 days)	1/2 (>458 days)	4/5 (>20, >117, >242, >269 days)
Rejection	7/14 (6/10 first courses)	0/2	0/3
GvHD	7/14	1/2	0/4
Interstitial pneumonitis	1/14	1/2	0/3
Severe stomatitis	0/14	2/2	0/5
	Non-HLA-MLC Matched Donors		
Living	—	1/1 (>467 days)	1/2 (>201 days)
Rejection	—	0/1	1/2
GvHD	—	1/1	1/2
Interstitial pneumonitis	—	0/1	0/2
Severe stomatitis	—	1/1	0/2

HLA-MLC matched donors. The latter group includes patients M-046 (TBI), M-056 (CY + TLI), and M-065 (CY + TLI). Results summarized in Table 23.5 suggest that survival has improved with the development of the CY-TLI protocol (and perhaps with other factors, since this is the most recent protocol). Of some interest to us are patients receiving non-HLA-MLC matched donor marrow. As indicated in Table 23.5, one of these patients, M-046, is alive at >467 days and has recovered from grade II GvHD. This patient was prepared for transplant using TBI + PZ + ATG; her case is to be reported in more detail (Warkentin et al., in preparation).

The CY-TLI combination for immunosuppression was very well tolerated with no significant morbidity that could be directly attributed to this protocol. No stomatitis, acute parotitis, or interstitial pneumonitis was noted, no cardiac toxicity was observed, and there were no other significant complications from the procedure. Overall, the patients appeared to tolerate the CY-TLI combination as well as they had previously tolerated CY alone and significantly better than they had tolerated TBI. As indicated in Table 23.5, graft rejection was not observed in the three evaluable patients receiving CY + TLI followed by HLA-MLC matched sibling marrow, but was observed in one of two patients who received marrow from a mismatched sibling.

Graft versus host disease was not observed in the cases receiving CY + TLI followed by HLA-MLC matched sibling donor marrow. But GvHD was observed in the patient receiving CY + TLI followed by nonmatched marrow. Patient M-056 has chronic GvHD, grade III, involving the skin and the bowel. This patient is currently receiving prednisone 60 mg/ m^2 every other day and azothioprine 1.5 mg/kg daily. Patient M-065 was difficult to evaluate for GvHD because of the transient graft; in this patient skin rash and fever were noted, but no diarrhea was observed. A skin biopsy consistent with early GvHD was recorded.

DISCUSSION

Currently, graft rejection and GvHD remain as major obstacles to the widespread application of bone marrow transplantation in humans. Graft rejection is a significant problem in patients with aplastic anemia prepared for marrow transplantation using CY alone (12) or, as described in this series, CY in combination with other agents. In the Minnesota experience, six of ten patients prepared for marrow transplantation with CY alone or CY in combination with other chemical agents rejected their marrow. Because of this high rejection rate (3), several transplant centers in addition to Minnesota have utilized

TBI at a total dose of 750 to 1,000 rad (2,10). In Minnesota and in other centers, such protocols have significantly reduced graft rejection, but have been associated with increased morbidity, including stomatitis, and mortality, especially with interstitial pneumonitis (7).

Graft versus host disease has also been a frequent problem in patients treated with all previous protocols, including CY alone or with other agents, or TBI, despite the use of HLA-MLC matched sibling donors. Incidence of acute GvHD approximates 50%, and chronic GvHD approximates 20 to 30% (1,5,8). Of interest are recent studies by Slavin et al. (9) that indicate that mice treated with TLI, given in a fractionated form at 200 rad × 17 doses, did not reject marrow or develop GvHD despite transplantation across the major histocompatibility barriers of the mouse.

With this background and our previous experience, we developed a combination CY-TLI protocol for preparation of patients with aplastic anemia and other nonmalignant disorders for marrow transplantation. One objective was to reduce graft rejection without permitting the development of pulmonary and mucosal toxicity associated with TBI. A second objective was to attempt to reduce GvHD by utilizing a form of TLI. In order to compress the treatment protocol into a short period to minimize discomfort to patients who are frequently acutely ill, the protocol provides a single dose of TLI to be combined with high dose CY.

Despite the preliminary results of the CY-TLI protocol, several observations can be made. First, in the seven patients with aplastic anemia so treated, the morbidity of the combined procedure was no greater than that observed previously in ten patients receiving CY alone or CY plus other agents (e.g., PZ or 6MP). The absence of stomatitis, when compared to TBI both in patients with aplasia and in patients with leukemia (6), was impressive, since patients were free of mouth ulcers, did not lose their appetites, and were able to eat throughout the transplantation course. Second, although patient numbers are still very small, interstitial pneumonitis has not been observed in these patients.

Patients receiving CY + TLI followed by HLA-MLC matched marrow did not reject their marrow, despite having been previously transfused. Of the two patients receiving CY + TLI followed by mismatched marrow, one (M-056) did not reject the marrow and is engrafted at >201 days despite a positive crossmatch at the time of transplantation. Patient M-065 appeared to have rejected his graft in that peripheral counts peaked at 950 and marrow cellularity was markedly improved, followed by a period of rapidly falling leukocytes and loss of cells from the bone marrow.

183

The patients receiving CY + TLI followed by HLA-MLC matched marrow have not shown evidence of GvHD, and we are encouraged by these preliminary results. It is clear, however, that this CY-TLI protocol does not prevent GvHD in patients receiving non-HLA-MLC matched marrow, since GvHD has been observed in patients receiving CY + TLI and subsequently, the mismatched marrow. Whether the incidence or severity of GvHD will be different from that experienced in previous protocols using CY alone, CY in combination with other agents, or TBI will require a significantly larger number of patients. Such studies are currently underway.

ACKNOWLEDGMENTS

The authors acknowledge the outstanding Bone Marrow Unit Nursing Team led by Ms. Sharon Roell and Ms. Jeanette Mefford, as well as the editorial assistance of Ms. Susan Perry.

This work was supported in part by grants from the United States Public Health Service (CA-21737, CA-15548, HL-07145, CA-07306) and by the American Cancer Society (#340).

REFERENCES

1. Fenyk, J. R., Smith, C. M., Warkentin, P. E., Krivit, W., Goltz, R. W., Neely, J. E., Nesbit, M. E., Ramsay, N. K. C., Coccia, P. F., and Kersey, J. H. Sclerodermatous graft-versus-host disease limited to an area of measles exanthem. *Lancet I*:472, 1978.
2. Gale, R. P., Cahan, M., Fritchen, J. H., Opelz, G., and Cline, M. J. Pretransplant lymphocytotoxins and bone-marrow graft rejection. *Lancet I*:170, 1978.
3. Gluckman, E., Devergie, A., Marty, M., Bussel, A., Rottembourg, J., Dausset, J., and Bernard, J. Allogeneic bone marrow transplantation in aplastic anemia—Report of 25 cases. *Transplant. Proc., 10*:141, 1978.
4. Johnson, A. H., Rossen, R. D., and Bulter, W. T. Detection of allo-antibodies using a sensitive antiglobulin microcytotoxicity test: Identification of low levels of preformed antibodies in accelerated allograft rejection. *Tissue Antigens, 2*:215, 1972.
5. Kersey, J. H., Meuwissen, H. J., and Good, R. A. Graft versus host reactions following transplantation of allogeneic hematopoietic cells. *Human Pathol., 2*:389, 1971.
6. Kim, T. H., Kersey, J. H., Sewchand, W., Nesbit, M. E., Krivit, W., and Levitt, S. H. Total body irradiation with a high dose rate linear accelerator for bone marrow transplantation in aplastic anemia and neoplastic disease. *Radiology, 122*:523, 1977.
7. Kraemer, K. G., Neiman, P. E., Reeves, W. C., and Thomas, E. D. Prophylactic adenine arabinoside following marrow transplantation. *Transplant Proc., 10*:237, 1978.
8. Shulman, H. M., Sale, G. E., Lerner, K. G., Barker, E. A., Weiden, P. L., Sullivan, K., Gallucci, B., Thomas, E. D., and Storb, R. Chronic cutaneous graft-versus-host disease in man. *Am. J. Pathol., 91*:545, 1978.
9. Slavin, S., Fuks, Z., Kaplan, H. S., and Strober, S. Transplantation of allogeneic bone marrow without graft-versus-host disease using total lymphoid irradiation. *J. Exp. Med., 147*:963, 1978.
10. Storb, R., Prentice, R. L., and Thomas, E. D. Treatment of aplastic anemia by marrow transplantation from HLA identlical siblings: Prognostic factors associated with graft versus host disease and survival. *J. Clin. Invest., 59*:625, 1977.
11. Storb, R., Thomas, E. D., Weiden, P. L., Buckner, C. D., Clift, R. A., Fefer, A., Goodell, B. W., Johnson, F. L., Neiman, P. E., Sanders, J. E., and Singer, J. One-hundred-ten patients with aplastic anemia (AA) treated by marrow transplantation in Seattle. *Transplant. Proc., 10*:135, 1978.
12. Thomas, E. D., Storb, R., Clift, R. A., Fefer, A., Johnson, F. L., Neiman, P. E., Lerner, K. G., Glucksberg, H., and Buckner, C. D. Bone-marrow transplantation. *New Eng. J. Med., 292*:832, 895, 1975.

24

Bone Marrow Transplantation for Aplastic Anemia: Conditioning with Cyclophosphamide plus Low-Dose Total-Body Irradiation

The UCLA Bone Marrow Transplantation Team

Bone marrow transplantation is now recognized as an effective form of therapy in selected cases of severe aplastic anemia. Recent reports suggest that nearly one-half of patients with severe aplastic anemia who receive a marrow transplant from a histocompatible sibling will become long-term survivors with restored hemopoietic function (1,2,18,19,27).

Despite the improved survival achieved with bone marrow transplantation in this disease, important clinical problems remain. Graft failure or graft rejection is still a major cause of treatment failure despite the use of histocompatible donor-recipient pairs (10,18,19,21,27,29). Immunologic factors are probably important in most cases of bone marrow graft failures. Storb and coworkers have shown in dogs that sensitization against minor histocompatibility antigens by blood transfusion can result in failure of engraftment that can be overcome by intensive immunosuppression (15,17,31). These investigators have also reported a correlation between successful engraftment and bone marrow dose and the results of certain *in vitro* assays of anti-donor immunity in man (10,21,29). Although better prognostic tests might permit the ·exclusion of patients likely to reject a marrow graft, this would limit the number of patients who might benefit from a marrow transplant. On the other hand, more effective immunosuppressive conditioning regimens might provide a means of overcoming immunologic barriers to successful engraftment. In this report, we compare the results of intensive immunosuppressive conditioning with cyclophosphamide and total-body irradiation to those achieved with cyclophosphamide conditioning alone in a series of 35 consecutive patients with severe aplastic anemia.

MATERIALS AND METHODS

Patients

Between August, 1973 and June, 1978, 35 patients with severe aplastic anemia received a bone marrow transplant from a HLA-identical sibling. Two patients were conditioned with total-body irradiation (TBI) alone, and were excluded from the analysis. Criteria for the diagnosis of severe aplastic anemia included granulocyte count equal to or under 0.5×10^9/liter; platelet count equal to or under 20×10^9/liter; corrected reticulocyte count equal to or under 1.0%; two or more bone marrow biopsy specimens indicating moderate to severe aplasia with more than 70% nonmyeloid cells (9,10,15,17,21,27,29,31).

Eleven patients with severe aplastic anemia were conditioned with cyclophosphamide alone (CY); 9 patients were conditioned with CY followed by 1,000 rad total-body irradiation (CY-TBI-1); and 13 patients were conditioned with CY followed by 300 rad total-body irradiation (CY-TBI-2). The three

TABLE 24.1 Conditioning Regimens for Allogeneic Bone Marrow Transplantation[a]

REGIMEN[b]	CYCLOPHOSPHAMIDE Dose	Schedule	TOTAL-BODY IRRADIATION Dose	Schedule
CY	50 mg/kg	D-5,D-4, D-3,D-2	None	—
CY-TBI-1	60 mg/kg	D-4,D-3	1,000 rad	D-1
CY-TBI-2	50 mg/kg	D-5,D-4, D-3,D-2	300 rad	D-1

[a]Bone marrow transplant on day 0 in all cases.
[b]CY: cyclophosphamide; TBI: total-body irradiation.

treatment regimens are presented in Table 24.1. Clinical data for the three treatment groups are presented in Table 24.2.

Histocompatibility

Marrow donors were selected on the basis of HLA and mixed lymphocyte culture (MLC) testing of available siblings, performed by standard techniques (11,13). Plasma exchange was performed pre-transplant for one patient in the CY-TBI-1 group and one patient in the CY-TBI-2 group with major ABO mismatches (A → O), as described (6).

Bone Marrow Transplantation

Informed consent, approved by the UCLA Human Subject Protection Committee, was obtained from all patients. Bone marrow was obtained from donors under anesthesia and given to the recipients as an intravenous infusion (24).

Methotrexate was given until day 100 following transplantation to prevent or modify graft versus host disease (GvHD) (21). Engraftment of transplanted bone marrow was documented by measuring the increase in bone marrow and peripheral blood cells; by analysis of erythrocyte and leukocyte antigens and isoenzymes; and by chromosome analysis (14).

All patients were maintained in reverse isolation, and received oral nonabsorbable antibiotics until granulocyte counts rose to over 0.5×10^9/liter. Granulocyte transfusions were given to patients who remained febrile after 72 to 96 hr of antibiotic therapy. Platelet transfusions were given to maintain a platelet count over 20×10^9/liter. Blood products given post-transplant were irradiated with 1,500 rad prior to infusion.

Graft Rejection

Failure of engraftment was defined as a granulocyte count under 0.05×10^9/liter, a platelet count under 10×10^9/liter, a reticulocyte count under 0.1%, absence of hemopoietic precursors or sponta-

TABLE 24.2 Clinical Characteristics of Marrow Transplant Recipients

	CY	CY-TBI-1	CY-TBI-2
Number	11	9	13
Age (years)[a]	23(11–48)	21(5–56)	18(3–38)
Etiology			
Unknown	7	6	9
Hepatitis	4	2	1
Drug (?)	0	1	3
Previous therapy			
Androgens	5	5	6
Steroids	8	4	5
Interval from diagnosis to transplant (days)[a]	67(22–52)	71(20–229)	70(15–255)[b]
Cell dose received (× 10/kg)[a]	2.5(0.1–3.7)	3.8(1.2–8.0)	3.6(1.3–6.0)

[a]Numbers indicate mean (range).
[b]Two patients with intervals of 2720 days and 3650 days were excluded.

neously dividing donor cells in the marrow aspirate or biopsy in patients surviving 21 days or longer following transplantation. Graft rejection was defined as the development of marrow aplasia in patients with prior evidence of engraftment. Engrafted patients surviving 28 days or longer were considered at risk for graft rejection.

Graft-versus-Host Disease

The diagnosis of GvHD was based on the presence of dermatitis, hepatitis, and/or diarrhea in engrafted patients. Skin biopsy specimens were examined by light and electron microscopy and by immunofluorescent staining. Criteria for diagnosis and staging of GvHD have been reported (25). Patients with Stage II GvHD were treated with intravenous corticosteroids. Patients with grafts surviving 30 days or longer post-transplant were considered at risk for GvHD.

Interstitial Pneumonitis

The diagnosis of interstitial pneumonitis was based on typical radiologic findings and significant hypoxemia. Evaluation included fiberoptic bronchoscopy with brushings and washings, transbronchial biopsy, and open-lung biopsy. Specimens were examined with Gomori-methenamine-silver stain to identify pneumocystis. The diagnosis of viral pneumonitis required recovery of the agent from pulmonary tissue, identification of intranuclear inclusions typical of cytomegalovirus (CMV), or a more than twofold rise in antibody titer temporally related to the pneumonia. Patients surviving 40 days or more post-transplant were considered at risk for interstitial pneumonitis.

Data Analysis

Data were evaluated using the Chi-square analysis (19). Survival curves were prepared from a life table analysis and survival rate (4).

RESULTS

Clinical data for the patients included in each treatment group are presented in Table 24.2. Criteria for the selection of patients for bone marrow transplantation, and the supportive care they received, were similar for the three treatment groups. Age, sex, previous therapy, ABO group, the interval between diagnosis and bone marrow transplantation, and pretransplant hematologic parameters were evaluated as possible risk factors for subsequent outcome (1). These parameters were similar for all groups. The number of patients who had received minimal (fewer than 10) blood transfusions before transplantation were similar for CY and CY-TBI-1 patients (one of eleven and two of nine, respectively). More patients who received CY-TBI-2 had minimal blood transfusions (five of thirteen), but this group also included two patients with a prolonged interval between diagnosis and transplantation. Virtually all patients presented with hemorrhage, with severe thrombocytopenia and neutropenia. Although the mean marrow cell dose received was lower in patients receiving CY, the proportion of patients who received less than 3.0×10^8 cells/kg was similar for all treatment groups ($p > .3$).

Graft Rejection

The incidence of graft rejection in the three treatment groups is presented in Table 24.3. One patient in each group died early and thus could not be evaluated. Although the graft rejection rate was 40% for patients treated with CY, graft rejection was not observed in any of the 20 evaluable patients treated with CY and TBI. This difference is statistically significant ($p < .03$).

Graft-versus-Host Disease

Clinically significant GvHD (Grade II) was observed in two of six patients who received CY, six

TABLE 24.3 Graft Rejection

REGIMEN	PATIENTS AT RISK	GRAFT REJECTION (%)
CY	10	4 (40)
CY-TBI-1	8	0
CY-TBI-2	12	0
Total	30	4 (13)

TABLE 24.4 Graft-versus-Host Disease

REGIMEN	PATIENTS AT RISK	CLINICAL GRADE				
		0	I	II	III	IV
CY	6	3	1	0	2	0
CY-TBI-1	8	1	1	1	3	2
CY-TBI-2	12	2	6	1	1	2

of eight patients who received CY-TBI-1, and four of 12 patients who received CY-TBI-2 (Table 24.4).

Interstitial Pneumonitis

Interstitial pneumonitis was significantly less frequent in patients who received CY-TBI-2 than in patients treated with CY or CY-TBI-1 (Table 24.5). There was no difference in incidence between patients treated with CY or CY-TBI-1. It should be noted, however, that several patients who received CY-TBI-2 may still be at risk for developing interstitial pneumonitis. Idiopathic interstitial pneumonitis was most frequent in patients who received CY-TBI-1, whereas patients who received CY or CY-TBI-2 and subsequently developed interstitial pneumonitis were more likely to have an infectious (viral) etiology.

Survival

Actuarial survival data for the three treatment groups are presented in Figures 24.1 and 24.2. Fourteen patients are currently alive with normal hemopoietic function 46 days to 5 years post-transplant. Actuarial one-year survival for all patients is 43%. Survival for patients who received either CY or CY-TBI-1 was similar despite the absence of graft rejection in the latter group. Median survival for patients who received CY-TBI-2, however, has not yet been reached, with nine of 13 patients alive 46 to 493 days post-transplant. Graft rejection was a major cause of death in patients receiving CY, whereas interstitial pneumonitis (three patients) and GvHD (two patients) were major causes of death in patients receiving CY-TBI-1.

TABLE 24.5 Interstitial Pneumonitis

REGIMEN	PATIENTS AT RISK		IP (%)
	CY	CY-TBI-1	CY-TBI-2
Patients at risk	7	8	12
IP (%)	5 (70%)	7 (87%)	4 (33%)
Fatal (%)	1 (20%)	3 (43%)	2 (50%)
Etiology			
CMV	3	1	2
Other virus	0	1	2
Pneumocystis	1	0	0
Idiopathic	1	5	0

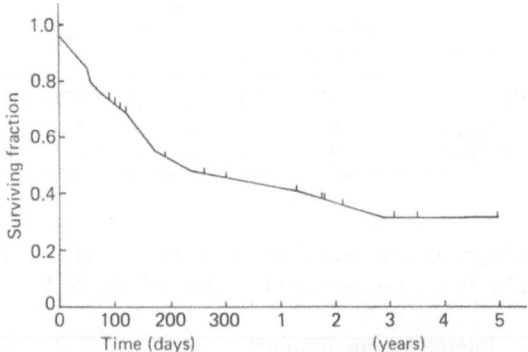

FIGURE 24.1. Actuarial survival for 33 patients with severe aplastic anemia treated with bone marrow transplantation. Day 0, day of marrow transplantation.

DISCUSSION

In contrast to the extremely poor prognosis of patients with severe aplastic anemia treated by conventional approaches (3,8,9,34), 35 to 40% of patients treated with allogeneic bone marrow transplantation become long-term survivors with complete restoration of hemopoietic function (1,2,18,19,-27). It has been recently shown that survival of marrow transplant recipients is superior to that of conventionally treated patients with severe aplastic anemia (2,27) and that transplantation should be considered early in the course of disease (2).

Despite encouraging results, graft rejection has been a major cause of treatment failure in previous series, and has occurred in 30 to 40% of transplant recipients despite complete HLA matching (2,18,19,-23,27). Patients with acute leukemia, on the other hand, who have been conditioned for marrow transplantation with more intensive cytotoxic therapy,

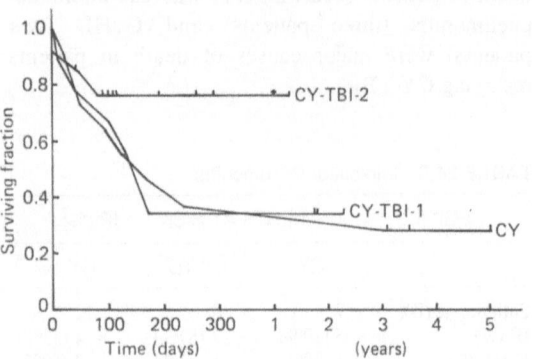

FIGURE 24.2. Actuarial survival for patients with severe aplastic anemia treated with Cy, CY-TBI-1, or CY-TBI-2. Day 0, day of marrow transplantation. Surviving patients are indicated; one late death (+) in the CY-TBI-2 group is also indicated.

rarely fail to engraft (25,26,28). Although graft rejection occurs despite histocompatibility matching, it is thought to be immunologically mediated. Four of our initial series of 11 patients treated with CY rejected their grafts. A retrospective analysis failed to indicate any identifiable prognostic factors. In an attempt to reduce the graft rejection rate in patients with aplastic anemia, we next employed a more intensive cytotoxic and immunosuppressive regimen, consisting of CY and 1,000 rad TBI. This program, CY-TBI-1, appeared to be effective in preventing graft rejection, but overall survival was not improved largely, because of an increase in the incidence of severe GvHD and interstitial pneumonitis.

We next reduced the dose of TBI to 300 rad. None of the 12 evaluable patients with anemia conditioned with this regimen (CY-TBI-2) rejected their grafts. Thus, the incidence of graft rejection was significantly reduced ($p < 0.03$) in these patients. Moreover, the incidence of severe GvHD was significantly lower than in the group treated with CY-TBI-1. Although several of the patients treated with CY-TBI-2 are still at risk to develop interstitial pneumonitis, the incidence of interstitial pneumonitis does appear to be similar to the group treated with CY alone, and significantly lower than the incidence in patients receiving CY-TBI-1 ($p = .025$).

A number of observations have suggested that an immune response causes a large proportion of graft failures in aplastic anemia patients. Canine marrow transplant experiments and clinical studies in man have demonstrated an association between prior sensitization of the recipient to non-HLA transplantation antigens and subsequent graft rejection (15,-16,31,33). The nature of the antigens involved in immunization and presumably graft rejection is unknown, although they seem to be independent of HLA and ABO and sex (21,30). Recently, a multifactorial regression analysis performed on a large series of marrow transplant patients indicated that the presence of recipient sensitization against donor cells (cell-mediated immunity), and a low marrow cell dose each correlated with graft rejection (21). *In vitro* assays of antidonor cell-mediated immunity can be correlated with graft outcome in marrow graft recipients and have been used to identify sensitized patients likely to reject a marrow graft (10,21,29,30).

Results in animal model systems have suggested that immunization against the marrow graft by prior blood transfusions can be overcome by intensifying immunosuppressive regimens (5,17). Recent results suggested that this approach might be successful in man. Ascensao and coworkers described a patient whose bone marrow cells suppressed normal marrow myeloid colony formation *in vitro*. This suppressor effect was eliminated by treatment of the patient marrow with anti-thymocyte globulin and complement prior to culture (32). Moreover, there have been occasionally successful second or third marrow grafts after additional immunosuppression in patients who rejected the first grafts (33).

Graft versus host disease is also a major post-transplant complication in marrow allograft recipients. More than one-half of aplastic anemia patients develop clinically detectable acute GvHD (33). Although GvHD developed more frequently in patients receiving TBI than in those receiving CY alone, the difference was not significant. There was however, a greater tendency for patients receiving CY-TBI-1 to develop severe GvHD than in patients receiving CY-TBI-2.

The incidence of interstitial pneumonitis in allogeneic marrow transplant conditioned with CY is approximately 50% (34). Although the lungs have a limited radiation tolerance, this disease obviously is not directly related to pulmonary radiation toxicity. Nevertheless, fewer patients who received CY-TBI-2 have interstitial pneumonitis than patients who received CY-TBI-1. Patients with severe GvHD are much more likely to develop interstitial pneumonitis than are patients with no or mild GvHD. The lower incidence of both diseases in the CY-TBI-1 patients may reflect a more tolerable systemic effect of low dose TBI in addition to the lower pulmonary radiation exposure.

The mechanism by which low dose TBI in combination with CY prevents graft rejection in marrow transplant recipients is not known. Nor is the effect of TBI on the subsequent development of GvHD understood. It is likely, however, that effective immunosuppression plays a role in ensuring engraftment in patients treated with this regimen. Thus, CY-TBI-2 was not only effective but also well tolerated, without any increased incidence of other life-threatening sequelae post-transplant.

Although the number of evaluated patients is still small, our results with CY-TBI-2 are encouraging. Conditioning with 200 mg/kg CY plus low-dose TBI may assure engraftment in most patients with severe aplastic anemia receiving a marrow transplant from a histocompatible sibling.

SUMMARY

Bone marrow transplantation is increasingly used in the treatment of patients with severe aplastic anemia. Approximately one-half of the patients so treated become long-term survivors. Graft rejection has been recognized as a major cause of treatment failure in 30 to 40% of patients. Several studies have suggested that an immunologic mechanism for allo-graft rejection may be responsible. To approach this problem, we added total-body irradiation (TBI) to cyclophosphamide (200 mg/kg). One group of patients received cyclophosphamide plus 300 rad TBI (CY-TBI-2). Results were compared to two similar groups of patients: one group received cyclophosphamide plus 1,000 rad TBI (CY-TBI-1), and the other received cyclophosphamide alone (CY). Eight of nine patients (mean age, 19 years) who received CY-TBI-1 were evaluable. Their average marrow cell dose was 3.8×10^8/kg. All demonstrated prompt engraftment without subsequent rejection. Three patients are alive 404 to 727 days post-transplant. Interstitial pneumonitis developed in seven of eight patients and was fatal in three. Graft-versus-host disease (GvHD) developed in six of eight patients and was fatal in two. Five of six patients (mean age, 20 years) who received CY-TBI-2 are evaluable. Their average marrow cell dose was 3.6×10^8/kg. All five patients engrafted promptly without subsequent rejection. Four are currently alive 149 to 336 days post-transplant, but interstitial pneumonitis developed in one patient. Graft-versus-host disease developed in four of five patients and was fatal in one. Graft rejection occurred in four of 11 patients who received CY alone, whereas no patient who received CY-TBI rejected the graft. Interstitial pneumonitis developed in eight of 13 patients at risk who received CY-TBI and was fatal in four. Graft-versus-host disease grade 2 developed in three of six CY patients and seven of 13 CY-TBI patients and was especially severe in those who received CY-TBI-1. It was shown that CY-TBI provides sufficient immunosuppression to prevent rejection of marrow allografts. The addition of 300 rad TBI to CY is effective without significantly increasing the risk of interstitial pneumonitis or GvHD in allograft recipients. The data suggest that graft rejection can be overcome by the addition of low-dose TBI to cyclophosphamide.

ACKNOWLEDGMENTS

The UCLA Bone Marrow Transplantation Team includes Peter R. Graze, Stephen Feig, Alan Tesler, Winston Ho, Lowell S. Young, Drew Winston, Robert Sparkes, Gregory Sarna, Mary Territo, Gerhard Opelz, David W. Golde, Melinda Cahan, Nancy Lyddane, John L. Fahey, Martin J. Cline, and Robert P. Gale.

Supported by grants CA23175, CA12800, CA15688, and RR00865 from the National Cancer Institute and the U.S. Public Health Service. Dr. Graze is a Junior Faculty Clinical Fellow of the American Cancer Society. Dr. Gale is a Scholar of the Leukemia Society of America.

REFERENCES

1. Advisory Committee of the Bone Marrow Transplant Registry. Bone marrow transplantation from histoincompatible allogeneic donors with [sic] aplastic anemia. *JAMA, 236*:1131, 1976.

2. Camitta, B. M., Thomas, E. D., Nathan, D. G., Santos, G., Gordon-Smith, E. C., Gale, R. P., Rappeport, J. M., Storb, R. Severe aplastic anemia: A prospective study of the effect of early marrow transplantation on acute mortality. *Blood, 48*:63, 1976.

3. Davis, S., and Rubin, A. D. Treatment and prognosis in aplastic anemia. *Lancet, 1*:871, 1972.

4. Dixon, W. J. BMD Biomedical Computer Programs, X-series suppl. Berkeley, 1969.

5. Elfenbein, G. J., Anderson, P. N., Klein, D. L., Schacter, B. Z., Santos, G. W. Difficulties in predicting bone-marrow graft rejection in patients with aplastic anemia. *Transplant. Proc., 10*:441, 1978.

6. Gale, R. P., Feig, S. A., Ho, W., Falk, P., Rippee, C., Sparkes, R. S. ABO blood group system and bone marrow transplantation. *Blood, 50*:185, 1977.

7. Goyette, D., Mickey, M. R. *Chi Square Probabilities for 2 × 2 Tables*. Technical Reprt. No. 15. Health Sciences Computing Facility, UCLA, 1975.

8. Lewis, S. M. Course and prognosis in aplastic anemia. *Br. Med. J., 1*:1027, 1975.

9. Lynch, R. E., Williams, D. M., Reading, J. C., and Cartwright, G. E. The prognosis in aplastic anemia. *Blood, 45*:517, 1975.

10. Mickelson, E. M., Fefer, A., Storb, R., and Thomas, E. D. Correlation of the relative response index with marrow graft rejection in patients with aplastic anemia. *Transplantation, 22*:294, 1976.

11. Mittal, K. K., Mickey, M. R., Singal, D. P., and Terasaki, P. I. Serotyping for homotransplantation XVIII. Refinement of microdroplet lymphocyte cytotoxicity test. *Transplantation, 6*:913, 1968.

12. Neiman, P. E., Reeves, W., Ray, G., Flournoy, N., Lerner, K. G., Sale, G. E., and Thomas, E. D. A prospective analysis of interstitial pneumonia and opportunistic viral infection among recipients of allogeneic bone marrow grafts. *J. Infect. Dis., 136*:754, 1977.

13. Sengar, D. P. S., and Terasaki, P. I. A semimicromixed lymphocyte culture test. *Transplantation, 11*:260, 1971.

14. Sparkes, M. L., Crist, M. L., Sparkes, R. S., Gale, R. P., Feig, S. R., and UCLA Bone Marrow Transplant Team. Gene markers in human bone marrow transplantation. *Vox Sang., 33*:202, 1977.

15. Storb, R., Epstein, R. B., Rudolph, R. H., and Thomas, E. D. The effect of prior transfusion on marrow grafts between histocompatible canine siblings. *J. Immunol., 105*:627, 1970.

16. Storb, R., Rudolph, R. H., Graham, T. C., and Thomas, E. D. The influence of transfusions from unrelated donors upon marrow grafts between histocompatible canine siblings. *J. Immunol., 107*:409, 1971.

17. Storb, R., Floersheim, G. L., Weiden, P. L., Graham, T. C., Kolb, H. J., Lerner, K. G., Schroeder, M. L., and Thomas, E. D. Effect of prior blood transfusions on marrow grafts: Abrogation of sensitization by procarbazine and antithymocyte serum. *J. Immunol., 112*:1508, 1974.

18. Storb, R., Thomas, E. D., Buckner, C. D., Clift, R. A., Johnson, F. L., Fefer, A., Glucksberg, H., Giblett, E. R., Lerner, K. G., and Neiman, P. E. Allogeneic marrow grafting for treatment of aplastic anemia. *Blood, 43*:157, 1974.

19. Storb, R., Thomas, E. D., Weiden, P. L., Buckner, C. D., Clift, R. A., Fefer, A., Fernando, L. P., Giblett, E. R., Goodell, B. W., Johnson, F. L., Lerner, K. G., Neiman, P. E., and Sanders, J. E. Aplastic anemia treated by allogeneic bone marrow transplantation: A report of 49 new cases from Seattle. *Blood, 48*:817, 1976.

20. Storb, R., Thomas, E. D., Buckner, C. D., Clift, R. A., Fefer, A., Fernando, L. P., Giblet, E. R., Johnson, F. L., and Neiman, P. E. Allogeneic marrow grafting for treatment of aplastic anemia: A follow-up on long-term survivors. *Blood, 48*:485, 1976.

21. Storb, R., Prentice, R. L., and Thomas, E. D. Marrow transplantation for treatment of aplastic anemia. An analysis of factors associated with graft rejection. *N. Eng. J. Med., 296*:61, 1977.

22. Storb, R., Prentice, R. L., and Thomas, E. D. Treatment of aplastic anemia by marrow transplantation from HLA identical siblings. Prognostic factors associated with graft versus host disease and survival. *J. Clin. Invest., 59*:625, 1977.

23. Storb, R., Thomas, E. D., Weiden, P. L., Buckner, C. D., Clift, R. A., Fefer, A., Goodell, B. W., Johnson, F. L., Neiman, P. E., Sanders, J. E., and Singer, J. Results of 110 consecutive HLA identical marrow transplants for treatment of severe aplastic anemia. In Baum, S. J., and Ledney, G. D., eds., *Experimental Hematology Today 1978*. New York: Springer-Verlag, 1978.

24. Thomas, E. D., and Storb, R. Technique for human marrow grafting. *Blood, 36*:507, 1970.

25. Thomas, E. D., Storb, R., Clift, R. A., Fefer, A., Johnson, F. L., Neiman, P. E., Lerner, K. G., Glucksberg, H., and Buckner, C. D. Bone marrow transplantation. *N. Eng. J. Med., 292*:832, 895, 1975.

26. Thomas, E. D., Buckner, C. D., Banaji, M., Clift, R. A., Fefer, A., Flournoy, N., Goodell, B. W., Hickman, R. O., Lerner, K. G., Neiman, P. E., Sale, G. E., Sanders, J. E., Singer, J., Stevens, M., Storb, R., and Weiden, P. L. One hundred patients with acute leukemia treated by chemotherapy, total body irradiation, and allogeneic marrow transplantation. *Blood, 49*:511, 1977.

27. UCLA Bone Marrow Transplant Team. Bone-marrow transplantation in severe aplastic anemia. *Lancet II*:921, 1976.

28. UCLA Bone Marrow Transplantation Group. Bone marrow transplantation with intensive combination chemotherapy/radiation therapy (SCARI) in acute leukemia. *Am. Int. Med., 86*:155, 1977.

29. Warren, R. P., Storb, R., Weiden, P. L., Mickelson, E. M., and Thomas, E. D. Direct and antibody-dependent cell-mediated cytoxicity against HLA identical sibling lymphocytes. Correlation with marrow graft rejection. *Transplantation, 22*:631, 1976.

30. Warren, R. P., Storb, R., Weiden, P. L., Su, P. J., and Thomas, E. D. Cytotoxicity against HLA identical sibling lymphocytes: Genetic determination and independent segregation of the antigenic system involved. *Transplant. Proc., 10*:67, 1978.

31. Weiden, P. L., Storb, R., Kolb, H. J., Graham, T. C., Kao, G., and Thomas, E. D. Effect of time on sensitization to hematopoietic grafts by preceding blood transfusions. *Transplantation, 19*:240, 1975.

32. Weiden, P. L., Storb, R., Slichter, S., Warren, R. P., and Sale, G. E. Effect of six weekly transfusions on canine marrow grafts: Tests for sensitization and abrogation of sensitization by procarbazine and antithymocyte serum. *J. Immunol., 117*:143, 1976.

33. Weiden, P. L., Storb, R., Mickelson, E. M., Warren, R. P., and Thomas, E. D. Immune response to transplantation antigens in human marrow graft recipients. *Transplant. Proc., 10*:409, 1978.

34. Williams, D. M., Lynch, R. E., and Cartwright, G. E. Drug-induced aplastic anemia. *Semin. Hematol., 195*:223, 1973.

25

Allogeneic Marrow Transplantation for Acute Leukemia

Jean E. Sanders for the Seattle Marrow Transplant Group

Marrow transplantation for the treatment of leukemia is based on the concept of destroying diseased marrow and replacing it with normal donor marrow. This basic concept, perhaps oversimplified, provides a setting in which "supralethal" chemo-irradiation therapy may be given without regard to marrow toxicity (10). We have previously reported the application of this approach in the treatment of 110 patients with end-stage refractory acute leukemia. Fourteen of these patients are surviving free of disease on no maintenance therapy from 2½ to 8 years following transplantation (3,11).

Traditionally, antileukemic agents and regimens are first evaluated in patients refractory to established therapy and then incorporated earlier in disease management if they show promise. Following the demonstration of the effectiveness of marrow transplantation in patients with far advanced refractory leukemia, we have now evaluated transplantation earlier in leukemia management. This report summarizes the Seattle experience with marrow transplantation performed in patients in marrow remission who have a poor long-term prognosis. We also report the results of transplantation in a concurrent series of patients transplanted for far advanced disease.

MATERIALS AND METHODS

Between March 1976 and December 1977, 89 patients with acute leukemia were transplanted from a sibling matched for HLA-A, -B, and -D loci (10). Forty-six patients transplanted in relapse had received extensive chemotherapy, but were considered refractory to further attempts at remission induction with conventional agents. Forty-three patients were transplanted in remission, but were considered to have a relatively poor probability of long-term survival. Table 25.1 summarizes the pretransplant characteristics of these patients.

All patients were prepared for transplantation with 1,000 rad of total-body irradiation preceded by 60 mg/kg of cyclophosphamide (CY) given on each of two successive days (10). The technique of marrow transplantation has been previously described (9). All patients received intrathecal methotrexate (MTX) 8 and 4 days prior to marrow infusion. Eight patients with acute lymphoblastic leukemia (ALL) in relapse and five patients with acute nonlymphoblastic leukemia (ANL) in relapse received a single dose of 5 mg/kg of dimethyl busulfan intravenously 7 days before marrow infusion and 1 day before the first dose of CY (12). For convenience, day 0 is designated the day of marrow transplantation. Following transplantation, all patients were given MTX to prevent or ameliorate graft-versus-host disease (GvHD)

TABLE 25.1 Pretransplant History

	ACUTE LYMPHOBLASTIC LEUKEMIA		ACUTE NONLYMPHOBLASTIC LEUKEMIA	
	Remission	Relapse	Remission	Relapse
Number of patients	21	31	22	15
Age	9	17	20	22
	(4–28)	(1–30)	(2–47)	(4–50)
Months following diagnosis	33	18	4	13
	(7–99)	(4–152)	(2–23)	(5–33)
Remission or relapse	2	2	1	1
(number)	(1–4)	(0–4)	(1–3)	(1–3)
Central nervous system disease				
Active	0	5	0	2
History of	7	8	0	2

(8,10). Beginning on day 32, the MTX was given intrathecally every other week for central nervous system prophylaxis.

These patients were also entered on other protocols involving prophylactic granulocytes (2), laminar air flow isolation (1), and prophylactic antithymocyte globulin (13) as described elsewhere.

RESULTS

Table 25.2 summarizes the causes of death in 57 of 89 patients. All patients with evidence of residual or recurrent leukemia were considered to have died of leukemia irrespective of the immediate cause of death. Eighty-six patients achieved engraftment. Three patients, all in relapse at the time of transplantation, failed to achieve engraftment: one died on day 11 of cardiac failure, one died on day 14 of infection, and one died on day 10 with persistent leukemia.

Recurrent Leukemia

Leukemia recurred in 33 patients and was the cause of death in 28. Twenty-one of 46 patients transplanted in relapse failed to achieve remission or developed recurrent leukemia at a median time of 86 days (7-245) post-transplant, 17 of 31 with ALL and four of 15 with ANL. One patient with recurrence in the central nervous system and marrow on day 153 is alive and free of disease 6 months following reinstitution of chemotherapy.

Twelve of 43 patients transplanted in complete remission relapsed at a median of 172 days (51 to 281) post-transplant. Nine of 21 patients with ALL relapsed, six in marrow alone and three with only extramedullary recurrence. Five of six patients with marrow relapse are dead and two of three patients with extramedullary relapse are alive and free of disease on maintenance therapy. Two of six patients with ANL transplanted in second or third remissions have relapsed and one of 16 patients transplanted in first remission has relapsed.

Interstitial Pneumonitis

A total of 37 patients developed interstitial pneumonitis, which was the primary cause of death in 23. Twenty-three patients transplanted in relapse developed interstitial pneumonitis, which was the primary cause of death in 14. Four additional patients died of

TABLE 25.2 Causes of Death for Patients Receiving Allogeneic Transplants

	ACUTE LYMPHOBLASTIC LEUKEMIA		ACUTE NONLYMPHOBLASTIC LEUKEMIA		
	Remission	Relapse	Remission	Relapse	Total
Total number of patients	21	31	22	15	89
Infection without GvHD	0	0	0	1	1
Infection with GvHD	0	1	2	1	4
Interstitial pneumonia with grade 0-1 GvHD	3	5	1	6	15
Interstitial pneumonia with grade 2–4 GvHD	0	1	5	2	8
Cardiac failure	0	1	0	0	1
Leukemia	6	16	2	4	28
Alive	12	7	12	1	32

this disease, but they also had associated recurrent leukemia. Fourteen patients transplanted in remission developed interstitial pneumonitis, which was fatal in nine. The median time of onset was 52 days post-transplant.

All but two patients had an open lung biopsy performed to determine the etiology of the pneumonia. Twelve patients had cytomegalovirus pneumonitis, which was the primary cause of death in 10. Five patients had *Pneumocystis carinii* pneumonitis, which was fatal in two. One patient developed a fatal Herpes zoster pneumonitis. No etiology could be identified in 17 cases, eight of whom died.

Graft versus Host Disease

Table 25.3 summarizes the relationship of GvHD to recurrent leukemia, incidence of interstitial pneumonitis, and disease status. Eighty-six patients with sustained engraftment were at risk for the development of GvHD. Sixty-one patients had mild or absent GvHD not requiring specific therapy. Thirty were transplanted in relapse and 31 in remission. Twenty-five developed grade II or greater GvHD and required specific therapy (13).

Survival

Twenty-four of 43 patients transplanted in remission are alive, 12 of 21 with ALL and 12 of 22 with ANL. Twenty are surviving disease free on no maintenance therapy from 8 to 29 months post-transplant. Four patients are surviving following relapse and are in a chemotherapy-induced remission.

Eight of 46 patients transplanted in relapse are surviving, one of 15 with ANL and seven of 31 with ALL. Seven are disease free from 10 to 26 months on no therapy following transplant. One patient is surviving in a chemotherapy-induced and -maintained remission.

Figure 25.1 is a Kaplan-Meier survival curve for patients transplanted in remission or in relapse for ALL or ANL (4). The log rank test for survival data shows the difference between the two curves to be highly significant ($p < .005$) (11). The probability of surviving to 100 days is 0.79 for patients transplanted

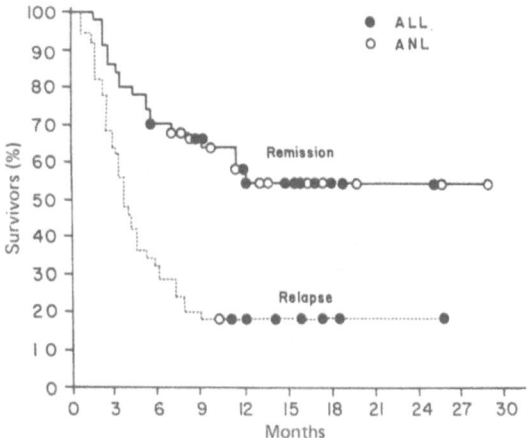

FIGURE 25.1. Kaplan-Meier product estimates of percentage surviving for 43 patients transplanted in remission and 46 patients transplanted in relapse. ALL (●); ANL (○).

in remission and 0.54 for patients transplanted in relapse. Currently, the median survival for patients transplanted in relapse is 4 months. The median survival for remission patients has not yet been reached, but it will be over 12 months.

Twenty of 32 surviving patients are completely well and living a normal life. Seven patients have chronic GvHD of varying degrees of involvement, with two patients being moderately limited in their activities; the other five are able to function almost normally. Five patients are being maintained on systemic chemotherapy and are otherwise well.

DISCUSSION

Advances in combination chemotherapy during the past decade have resulted in considerable improvement in the treatment of ALL with longer relapse-free survival and apparent cure for some children (6). Once the first relapse occurs, however, the prospect for prolonged survival is poor with median survivals of less than 1 year (5). All patients with ANL, adult or child, have a poor outlook with a median duration of first remission of approximately 1 year and few patients survive beyond 3 years (6,14).

We have shown that marrow transplantation may be a curative form of therapy for a small fraction of patients with end-stage acute leukemia (3,11). Eradication of leukemia seems to be due to a total leukemic cell kill with a short intensive course of chemo-irradiation therapy followed by marrow infusion from a suitable donor to rescue the patient from the lethal marrow cell damage. We have also observed that patients who come to transplantation in good clinical condition are much better able to tolerate this intensive approach and have a better

TABLE 25.3 Graft-versus-Host Disease Relationship to Recurrent Leukemia and Incidence of IP in Allogeneic Transplants

	REMISSION		RELAPSE		TOTAL
Grade of GvHD	0–1	> 2	0–1	> 2	
Number of patients	31	12	30	13	86
Recurrent leukemia	10	2	16	4	32
Interstitial pneumonitis (incidence)	8	6	17	6	37
Alive without recurrence	17	3	3	4	27

survival than those patients in poor clinical condition (11).

Encouraged by the demonstration that leukemia could be permanently eradicated in some end-stage refractory patients, it seemed reasonable to apply this form of intensive therapy earlier in the disease course before the leukemia became drug refractory and the patient's clinical condition deteriorated. For those patients with ALL in second or subsequent remission and ANL in first or subsequent remission, where prognosis for long-term disease free survival is known to be poor, we have utilized intensive chemo-irradiation therapy and marrow transplantation (12). The results to date clearly indicate a much improved survival during the first 3 months after grafting for these patients transplanted in remission. Although a median survival for these patients has not been reached, it cannot be less than 12 months. Thus, current survival is now at least as long as could be expected from the best combination chemotherapy. Since, as has been described, end-stage patients who survive relapse free for 2 years following marrow transplantation seem to be cured of their leukemia, it is reasonable to hope that those patients transplanted in remission who have not relapsed by 2 years will have the same opportunity for cure. Followup of over 1 year is needed to confirm this expectation for the patients in this report.

The data presented here also address other major problems associated with allogeneic transplantation. These problems center around the pulmonary opportunistic infections in the immuno-compromised host and GvHD, which may be either acute or chronic. Continued progress is being made toward the prevention of and the more effective treatment of these major transplant complications. The increased survival and the relatively low rate of recurrence of leukemia in the patients transplanted in remission is encouraging.

ACKNOWLEDGMENTS

This investigation was supported by Grant numbers CA18029, CA17117, and CA15704, awarded by the National Cancer Institute, Department of Health, Education and Welfare.

REFERENCES

1. Buckner, C. D., Clift, R. A., Sanders, J. E., Meyers, J. D., Counts, G. W., Farewell, V. T., Thomas, E. D., and the Seattle Marrow Transplant Team. Prospective study of a protective environment in marrow transplant recipients. *Ann. Intern. Med.,* 1978.

2. Clift, R. A., Sanders, J. E., Thomas, E. D., Williams, B., and Buckner, C. D. Granulocyte transfusions for the prevention of infection in patients receiving bone-marrow transplants. *N. Eng. J. Med., 298*:1052, 1978.

3. Fefer, A., Einstein, A. B., Thomas, E. D., Buckner, C. D., Clift, R. A., Glucksberg, H., Neiman, P. E., and Storb, R. Bone-marrow transplantation for hematologic neoplasia in 16 patients with identical twins. *N. Eng. J. Med., 290*:1389, 1974.

4. Kaplan, E. L., and Meier, P. Nonparametric estimation from incomplete observations. *J. Am. Statist. Assoc., 53*:457, 1958.

5. Kung, F. H., Nyhan, W. L., Cuttner, J., Falkson, G., Lanzkowsky, P., Del Duca, V., Nawabi, I. U., Koch, K., Pluess, H., Freeman, A., Burgert, E. O., Leone, L. A., Ruymann, F., Patterson, R. B., Degnan, T., Hakami, N., Pajak, T. F., and Holland, J. Vincristine, prednisone and L-asparaginase in the induction of remission in children with acute lymphoblastic leukemia following relapse. *Cancer, 41*:428, 1978.

6. Mauer, A. M. Treatment of acute leukaemia in children. In Simone, J. V., ed., *Clinics in Haematology,* Vol. 7. Ch. 2. London: Saunders, 1978, p. 245.

7. Peto, R., and Peto, J. Asymptotically efficient rank invariant test procedures (with discussions). *J. R. Statist. Soc. A, 135*:185, 1972.

8. Storb, R., Epstein, R. B., Graham, T. C., and Thomas, E. D. Methotrexate regimens for control of graft-versus-host disease in dogs with allogeneic marrow grafts. *Transplantation, 9*:240, 1970.

9. Thomas, E. D., and Storb, R. Technique for human marrow grafting. *Blood, 36*:507, 1970.

10. Thomas, E. D., Storb, R., Clift, R. A., Fefer, A., Johnson, F. L., Neiman, P. E., Lerner, K. G., Glucksberg, H., and Buckner, C. D. Bone-marrow transplantation. *N. Eng. J. Med., 292*:832, 895, 1975.

11. Thomas, E. D., Buckner, C. D., Banaji, M., Clift, R. A., Fefer, A., Flournoy, N., Goodell, B. W., Hickman, R. O., Lerner, K. G., Neiman, P. E., Sale, G. E., Sanders, J. E., Singer, J., Stevens, M., Storb, R., and Weiden, P. L. One hundred patients with acute leukemia treated by chemotherapy, total body irradiation, and allogeneic marrow transplantation. *Blood, 49*:511, 1977.

12. Thomas, E. D., Buckner, C. D., Fefer, A., Neiman, P. E., and Storb, R. Marrow transplantation in the treatment of acute leukemia. *Adva. Cancer Res., 27*:269, 1978.

13. Weiden, P. L., Doney, K., Storb, R., and Thomas, E. D. Anti-human thymocyte globulin (ATG) for prophylaxis and treatment of graft-versus-host disease in recipients of allogeneic marrow grafts. *Transplant. Proc., 10*:213, 1978.

14. Wiernik, P. H. Treatment of acute leukaemia in adults. In Simone, J. V., ed., *Clinics in Haematology,* Vol. 7, Ch. 3. London: Saunders, 1978, p. 259.

26

Antibody Treatment of Marrow Grafts *in Vitro:* A Principle for Prevention of Graft-versus-Host Disease

H. Rodt, B. Netzel, H. J. Kolb,
G. Janka, I. Rieder,
B. Belohradsky, R. J. Haas,
and S. Thierfelder

Graft-versus-host disease (GvHD) is still a frequent complication in clinical marrow transplantation. It causes severe lesions in several tissues, and it also affects hemopoiesis and the immune response of the recipient. When the major histocompatibility complex (MHC) differ, GvHD is uniformly fatal. A major advance in circumventing GvHD was achieved by selection of histocompatible donors. In 1968, Epstein et al. (4) demonstrated, in dogs, that bone marrow could be grafted successfully from DLA-matched littermates. In man, despite the selection of HLA-matched sibling donors and prophylactic immunosuppressive post-transplant therapy, 50% of the recipients get GvHD and about 25% die from it (27). This mortality was attributed to minor histocompatibility antigens undetected by currently employed *in vitro* test systems. The limited number of compatible siblings and the occurrence of GvHD in HLA-matched combinations led to experimental studies in which GvHD was suppressed by eliminating the immune-reactive cell population of the graft. Advances in cellular immunology have delineated the causal role of thymus-derived (T)-lymphocytes in graft-versus-host (GvH) reactions (8,9,11).

Attempts have been made to reduce T-cells by (a) treatment of the donor, (b) treatment of the recipient after transplantation, or (c) *in vitro* treatment of the donor bone marrow before transplantation (for review, see ref. 18). The last approach has the advantage that it avoids direct treatment of donor and recipient by agents with dangerous side effects. Dicke et al. (3) reported a separation of stem cells from T-cells by a density gradient centrifugation on albumin gradients, which resulted in the suppression of GvHD in mice and in a delay of GvHD in monkeys. Other experimental approaches have been designed to eliminate the GvH-reactive cell populations in the donor marrow by incubating the graft *in vitro* with specific antibody preparations (1,11,17,-29). Trentin and Judd (7) reported suppression of GvHD in mice with spleen-cell-absorbed antithymocyte globulin, and Müller-Ruchholtz et al. (11) reported suppression of GvHD in rats after incubation with a macrophage-absorbed anti-lymphocyte globulin. Immunocompetent lymphocytes may also be removed by the action of anti-idiotypic antibodies, as documented by Binz and Wigzell (1). Investigations by our group showed that, in mice, an *in vitro* treatment of incompatible donor cells with T-cell-specific antibodies before transplantation could suppress an otherwise lethal GvH reaction completely (17,18). The described system is based on the idea of removing GvH-reactive T-lymphocytes by a specific xenogeneic antiserum against T-cells that has been purified from antibodies cross-reacting with hemopoietic stem cells by an extensive absorption procedure.

The present report will summarize the effects of an incubation treatment with anti-T-cell globulin on GvHD in mouse and dog experimental models and show the successful application of this principle to clinical bone marrow transplantation.

MATERIAL AND METHODS

Cell Preparations

Normal and leukemic lymphocytes were separated from peripheral blood and bone marrow by density gradient centrifugation on a Ficoll-Isopaque gradient (2). Granulocytes were prepared by removing erythrocytes through dextran-sedimentation followed by a removal of lymphocytes with Ficoll-Isopaque. Thymocytes were prepared from the thymi of 1- to 3-year-old children undergoing cardiac surgery. The thymus tissue was purified from adhering blood, minced with scissors, and pressed through a wire grid into cold L-15 medium. Large fragments were allowed to settle, and the suspension was then passed through a 5-mm cotton filter. Bone marrow cells used in the test systems were prepared from surgically removed cells and capiti femoris. The marrow was scraped out and suspended in cold TC-199 medium. Large fragments and spongiosa particles were removed by 1 g sedimentation. The supernatant containing the marrow cells was passed through a 5-mm cotton filter.

All cell preparations were washed three times in medium and showed a viability of more than 90%.

Anti-T-Cell Globulin

The preparation of anti-mouse T-cell globulin (ATCG-M) and anti-dog T-cell globulin (ATCG-D) has been described elsewhere (15,18).

The production of an anti-human T-cell globulin (ATCG-H) for application in human marrow transplantation is shown in Figure 26.1. It was prepared from anti-human thymocyte serum. Rabbits were immunized intravenously on day 14, 15, 16, and 23 and were bled 1 week after the last absorption. Sera from at least 10 rabbits were pooled for each preparation. Anti-human thymocyte globulin (ATG-H) was isolated by fractionation of the crude antiserum. For preparation of ATCG-H, the antiserum was absorbed with liver-kidney homogenate, chronic lymphocytic leukemia cells (CLL) (B-cells), and erythrocytes by the following procedures: human kidney and liver were homogenized in a cooled homogenizer (Cenco) and washed with saline until the supernatant was completely clear. The CLL-cells were obtained from patients, treated with an IBM cell-separator, and washed three times in saline without removing the 20% contaminating erythro-

cytes. Absorptions were carried out twice with liver-kidney homogenate and three to five times with CLL cells obtained from different patients in a weight ratio sediment:serum of 1:4. Then the hemagglutination titer of the preparation was tested and removed by additional absorptions with erythrocytes. All absorptions were carried out at 4°C for 30 min. The absorbed antiserum was purified to the globulin fraction by ammonium sulfate precipitation, DEAE cellulose ion-exchange chromatography, and ultracentrifugation. The resulting ATCG-H was reconcentrated to 10 mg/ml. Further details are described elsewhere (16,20). Several precautions were taken before ATCG-H was applied to clinical marrow transplantation. Only tissues and cells negative for Australia antigen were used for the absorptions, and the ATCG-H was controlled for sterility and the absence of pyrogens and hemagglutinins, nephrotoxic reactivity, and general toxicity.

Complement Fixation Test

Complement fixation using human lymphocytes as antigens was performed by a micromethod as described previously (13). Veronal buffered saline (VBS) was used for all dilutions. First, 100 μl of a suspension of target cells (10^7 cells/ml) was added to 100 μl of each antibody dilution, the mixture was then allowed to incubate for 30 min, after which 100 μl of diluted guinea pig complement (GPC') was added. After an additional incubation of 30 min, 100 μl of sensitized sheep red blood cells (SRBC) were added. After 30 min, the reaction mixture was diluted with 0.9 ml of VBS, sedimented in an Eppendorf centrifuge, and examined for hemolysis at 412 nm in a Zeiss PM 6 spectrophotometer equipped for automatic reading. The 50% lysis of sheep red cells was defined as the titer of the antiserum. Sensitization of red cells was performed as described. The GPC' was adjusted to a dilution (about 1:100) that produced 90% hemolysis if the antigen and antibody dilutions were replaced by VBS.

Hemopoietic Toxicity

The hemopoietic toxicity of ATG-H and ATCG-H was measured by culturing human bone marrow cells after incubation in a colony-forming unit culture system or in a diffusion chamber system: 4×10^6 nucleated bone marrow cells were incubated with increasing dilutions of the antibody preparations for 30 min at 4°C. After washing, selected rabbit complement was added and the suspension further incubated for 60 min at 37°C.

The influence of the antibody preparations on colonies in culture (CFU-c) was evaluated after 12 days of cultivation by a modified double layer agar technique (13) originally described by Pike and

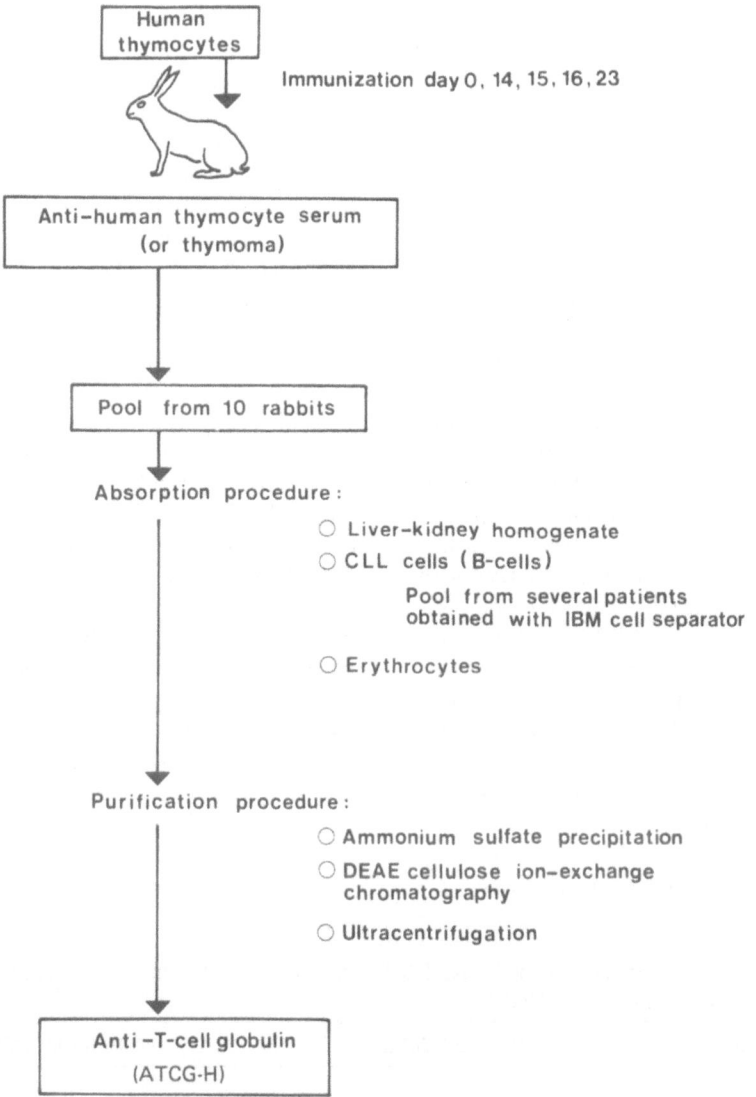

FIGURE 26.1. Production, absorption, and purification of anti-human T-cell globulin (ATCG-H) for application in bone marrow transplantation.

Robinson (14). After gelation at room temperature, the agar dishes were incubated for 12 to 14 days at 37°C in a fully humidified atmosphere continuously flushed with 5% CO_2. Colonies, defined as groups of 50 cells or more, were counted using a Leitz Diavert microscope.

The influence of crude and absorbed antisera on the growth of marrow cells in a diffusion chamber has already been described (16). In brief: aliquots of 5×10^5 nucleated marrow cells were loaded into Millipore chambers and implanted intraperitoneally into male CBA/J mice that had been irradiated with 650 R. On days 3, 6, 9, 13, and 17, the chambers were harvested for each antibody incubation and shaken for 1 hr in 0.5% pronase-buffered medium to

liquify the clot. The resulting cell suspensions were counted to determine the total number of nucleated cells.

Bone Marrow Transplantation

Total-body irradiation of the patient was given 24 hr before transplantation from two opposed ^{60}Cobalt sources at a dose rate of 5.5 R/min and a target distance of 4 m. The technique of bone marrow transplantation in man has been described in detail by Thomas and Storb (28). For incubation treatment the donor marrow was collected in a blood bag, and after the addition of the antiserum, which was diluted in 10 ml of medium TC 199, incubated at 4°C for 30 min with slight, but constant agitation. The marrow infusion was performed immediately after incubation.

RESULTS AND CONCLUSIONS
Effect of Anti-T-Cell Globulin in Animal Models

Characterization of Anti-Mouse T-Cell Globulin and Anti-Dog T-Cell Globulin In 1971, Golub (5) reported that immunization of rabbits with mouse brain resulted in an antiserum directed against mouse thymocytes and T-lymphocytes. This discovery was of particular value because it demonstrated that specific anti-T-cell sera could also be prepared under xenogeneic conditions. Serologic analysis of crude anti-mouse brain sera, however, revealed that they still contain antibodies cross-reacting with several non-T-cell populations, including hemopoietic stem cells. These cross-reactions were abolished (Table 26.1) when the antiserum was absorbed with liver homogenate and plasmacytoma cells and purified to the globulin fraction (ATCG-M). When this ATCG-M was analyzed in several indicator systems, only reactivity with T-lymphocyte populations was detected (17,18). The ATCG-M no longer cross-reacted with hemopoietic cells in the colony forming unit-spleen (CFU-s) test and the colony-forming culture test (CFU-c) and did not suppress the hemopoietic recovery when irradiated recipients were infused with ATCG-M incubated marrow cells.

On the basis of these results, it was possible to produce specific anti-T-cell sera for the dog model (Table 26.1). Anti-dog thymocyte serum was absorbed with liver-kidney homogenate, newborn spleen cells, and erythrocytes and was then purified to globulin. In analogy to the mouse model, this procedure eliminated cross-reactive antibodies with hemopoietic cells present in the unabsorbed antiserum preparations, but still left a high reactivity against dog T-cells. Anti-dog T-cell globulin (ATCG) did not inhibit the growth of CFU-c and an *in vitro* treatment of canine marrow did not interfere with its ability to reconstitute lethally irradiated autologous hosts (7).

Suppression of Graft-versus-Host Disease by in Vitro Treatment with Anti-T-Cell Globulin The effect of ATCG-M on GvHD in the mouse model (Table 26.2) was first tested in a parental → F_1 combination. When C57B1/6 spleen cells were incubated with ATCG-M and then transferred to lethally irradiated (C57B1/6 × CBA) F_1 hybrids, an otherwise lethal GvHD was suppressed completely. Cytogenetic studies (T6 chromosome) and analysis with anti-H-2 sera revealed complete donor cell chimerism of the recipients (9,10). A comparable suppression of GvHD was also obtained in an allogeneic BALB/c → (C57B1/6 × CBA) F_1 combination. Pretransplant conditioning of the recipients, however, had to be improved to combined irradiation and

TABLE 26.1 Characterization of Anti-Mouse T-Cell Globulin (ATCG-M) in the Murine- and Anti-Dog T-Cell Globulin (ATCG-D) in the Canine Model

ATCG	MURINE MODEL	CANINE MODEL
1. Production antigen:	Mouse brain → Rabbits ← Dog thymocytes	
	Anti-mouse brain serum ← Rabbits → Anti-dog thymocyte serum	
Absorption:	Liver, plasmocytoma	Liver-kidney, newborn spleen cells, erythrocytes
Purification:	Ammonium sulfate precipitation, DEAE cellulose ion exchange chromatography, ultrafiltration, ultracentrifugation	
	↓ ATCG-M	↓ ATCG-D
2. Reactivity	• Strong reactions with mouse thymocytes and T-lymphocytes • No cross-reaction with mouse B-lymphocytes and bone marrow cells	• Strong reactions with dog thymocytes and peripheral blood T-lymphocytes • Weaker reaction with dog bone marrow cells
3. Cross-reaction with hemopoietic cells[a]	• No depression of CFU-s and CFU-c after incubation of mouse bone marrow • Normal hemopoietic recovery in syngeneic irradiated recipients of ATCG-incubated marrow	• No depression of CFU-c after incubation of dog bone marrow • Normal hemopoietic recovery of irradiated dogs after transfer of ATCG-incubated autologous marrow

[a]Cross-reaction with hemopoietic cells was evaluated at concentrations of ATCG still highly reactive against T-lymphocytes.

TABLE 26.2 Effect of ATCG on GvHD After Marrow Transfer to Lethally Irradiated Incompatible Recipients

COMBINATION	MURINE MODEL	CANINE MODEL
1. Combination with GvH-incompatibility (one way)	Parental marrow[a] into F_1-hybrids	DLA-homozygous marrow into DLA-heterozygous littermates
1.1 Anti-GvHD effect[b] and chimerism	• Complete suppression of GvHD in over 95%[f] • Total donor cell chimaeras	• Complete suppression of GvHD in 40%, total donor cell chimeras • Delay of GvHD in 40% • No engraftment in 20%
1.2 Immunocompetence of radiation chimeras	• Rejection of third-party skin grafts • Immune reactivity against SRBC	• Donor cells react in the MLC against third-party lymphocytes
1.3 Tolerance of donor cells against recipient's tissues	• Tolerance of the donor cells against the recipient type H-2 antigens in transfer experiments	U.I.[c]
2. Allogeneic combination[d] (two way)	Allogeneic marrow into H-2 incompatible recipients[e]	n.d.[f]
1.1 Anti-GvHD effect[b] and chimerism	• Complete suppression of GvHD in over 90%[f] • Total donor cell chimeras after increased treatment of the recipients (irradiation + ATG)	
1.2 Immunocompetence of radiation chimeras	U.I.	
1.3 Tolerance of donor cells against recipient's tissues	U.I.	

[a]Investigated in the combination C57/BL/6 → (C57/BL6 × CBA) F_1.
[b]GvHD assays in mice were performed with spleen marrow.
[c]U.I., under investigation.
[d]Transfer of allogeneic marrow in H-2 compatible combinations (CBA → AKR) yielded results comparable to 1.1, 1.2, and 1.3.
[e]Investigated in the combination BalbC → (C57/BL/6 × CBA) F_1.
[f]n.d., not done.

ATG treatment in order to obtain a long-lasting donor cell chimerism.

The parental → F_1 combination was studied to determine whether recipients were able to restore their immunocompetence in the post-transplant phase. After transfusion of T-cell-deprived marrow, an immune reconstitution can only be induced by cooperation of the H-2 non-identical different recipient thymus with the donor stem cells by maturation to T-lymphocytes. Reconstitution of immune reactivity in the former chimeras was established by rejection of third-party skin grafts as well by an immune response against sheep erythrocytes (19). These observations have also been confirmed by other authors in comparable combinations (23). The new T-cell population in the chimeras is tolerant for the tissues of the recipient. This could be shown by Thierfelder et al. by transfer of unincubated chimeric spleen cells to a second F_1 host (26). No GvHD occurred after the adoptive transfer of the cells.

The restoration of immune reactivity in complete allogeneic combinations is still controversal. Zinkernagel et al. (30) reported the absence of H-2-restricted T-cell functions in allogeneic bone marrow chimeras.

Table 26.2 also illustrates the effect of an incubation treatment with ATCG-D in the canine model. The DLA-homozygous donor marrow was transplanted to lethally irradiated DLA-heterozygous littermate recipients that shared one haplotype with the donor. Transplantation of unincubated marrow resulted in 100% mortality. After *in vitro* treatment of the donor marrow with ATCG-D, four out of 10 dogs survived, with hemopoietic recovery with no sign of GvHD; in four out of 10 dogs, the course of GvHD was significantly delayed; and two dogs died without sustained hemopoietic engraftment. The surviving dogs showed complete chimerism established cytogenetically or by DLA typing. Lymphocytes of the chimeras reacted in the MLC against third-party cells, which may indicate a recovery of immunocompetence in the transplanted recipients (7).

Clinical Application of Anti-Human T-Cell Globulin

Specificity of Anti-Human T-Cell Globulin The results of the animal experiments suggested an *in vitro* application for suppression of GvHD in man, provided that specific anti-human T-cell sera could be produced. Xenogeneic unabsorbed antisera recognize a broad spectrum of antigens not only expressed on T-lymphocytes but also present on many other cells including stem cells of the same species (10,21). Extensive absorption and purification procedures are required to remove antibodies against nonspecific determinants and to isolate a specific anti-T-cell globulin (ATCG-H) suitable for application in human marrow transplantation (Figure 26.1).

The antibody specificities of unabsorbed ATG-H and absorbed ATCG-D were compared in the complement fixation test, which indicates the binding of antibodies to cells in a semiquantitative manner (Table 26.3). Both antiserum preparations were tested against normal cells, thymocytes, peripheral blood lymphocytes, bone marrow cells, granulocytes, and lymphatic leukemia cells (E$^+$ALL, Ig$^+$-CLL) of the B- or T-cell type. Unabsorbed ATG-H did not show detectable specificity for T-lymphocytes. It also cross-reacted with the B-lymphocyte populations and even with granulocytes. The antibody titers of different pools of unabsorbed ATG-H ranged between 1:512 and 1:2,048. After absorption, the resulting ATCG-H did not cross-react with B-lymphocyte populations (CLL) and granulocytes. It still reacted strongly against T-lymphocyte with such cell populations as thymocytes (over 90% T-cells), peripheral blood lymphocytes (about 70% T-cells), and T-ALL (over 90% T-cells). The antibody titer

against bone marrow cells was diminished to 1:32. The residual reactivity may be caused in part by contaminating T-lymphocytes of the blood.

Hemopoietic Toxicity of Anti-Human T-Cell Globulin The toxicity of ATG-H and ATCG-H against hemopoietic cells was evaluated in colony-forming unit tests (CFU-c) and diffusion chamber tests. In order to relate hemopoietic toxicity to anti-T-cell activity, all the globulins were adjusted to the same anti-T cell activity prior to testing.

Figure 26.2 shows the effect on colony growth of incubated human bone marrow cells in serial dilutions of unabsorbed ATG-H or absorbed ATCG-H in the presence of complement. Unabsorbed ATG-H was found to be highly toxic; at low antibody dilutions, the number of colony-forming cells was almost completely reduced. When marrow cells were incubated with absorbed ATCG-H, no reduction of CFU-c was observed, compared with marrow cells incubated in normal rabbit globulins. The arrow indicates the dilution of ATCG-H, which was later used for incubation treatment in clinical marrow transplantation. This dilution (1:100) still guarantees an optimal reactivity against T-lymphocytes in the complement fixation test and no depression of CFU-c in the bone marrow preparations.

Figure 26.3 shows the proliferation of bone marrow cells in diffusion chambers after incubation with crude ATG-H and absorbed anti-T-cell globulin and complement. For this assay, which was performed in cooperation with Dr. Körbling and Dr. Fliedner, University of Ulm, an anti-T-cell globulin preparation, which was also absorbed with lymphoblastoid cell lines, was used. Incubation with crude ATG led to decreasing growth curves until day 13 of implantation. No cell differentiation was found. Growth curves and cell differentiation after incubation with anti-T-cell globulin and complement at a titer of 1:20 did not differ from controls incubated without antiserum.

From the results with both stem cell assays we conclude that only highly absorbed ATCG-H is effective for successful incubation of human marrow. As the number of marrow cells is often suboptimal for marrow engraftment, unabsorbed antibody-preparations of anti-thymocyte serum may cause too high a loss of stem cells at concentrations that are necessary to eliminate T-cells effectively.

Bone Marrow Treatment with Anti-T-Cell Globulin Before ATCG was used clinical bone marrow transplantation, it was tested for the absence of several side effects. The antiserum was shown to be sterile, and without pyrogens, haemagglutinins,

TABLE 26.3 Antibody Titers of Unabsorbed ATG-H and Absorbed ATCG-H Against Various Human Cell Populations in the Complement Fixation Test

| CELLS[a] | COMPLEMENT FIXATION TITER[b] OF | |
	Unabsorbed ATG-H	Absorbed ATCG-H
Thymocytes	1:2,048	1:1,024
Peripheral blood lymphocytes	1:1,024	1:512
Bone marrow cells	1:1,024	1:32
Granulocytes	1:512	Neg.
Acute lymphoblastic leukemia (E-rosette$^+$, T-cell Type)	1:1,024	1:512
Chronic lymphoblastic leukemia (Membrane Ig$^+$, B-cell type)	1:1,024	Neg.

[a]Cell concentration 10/ml.
[b]50% hemolysis.

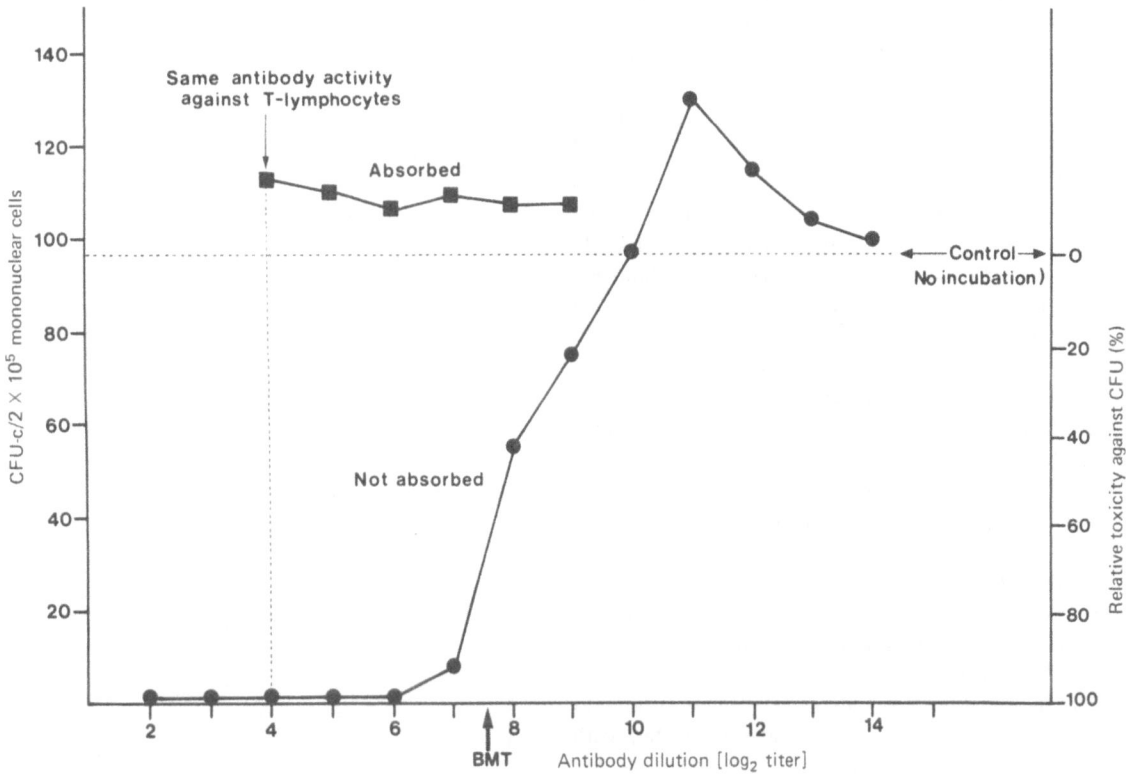

FIGURE 26.2. Effect of unabsorbed ATG-H (●——●) and absorbed ATCG-H (■——■) on the recovery of CFU-c after incubation of human bone marrow cells at different antibody dilutions and complement (control = incubation without antiserum, CFU-c:97). Both antisera were adjusted to the same activity against T-lymphocytes prior to the test.

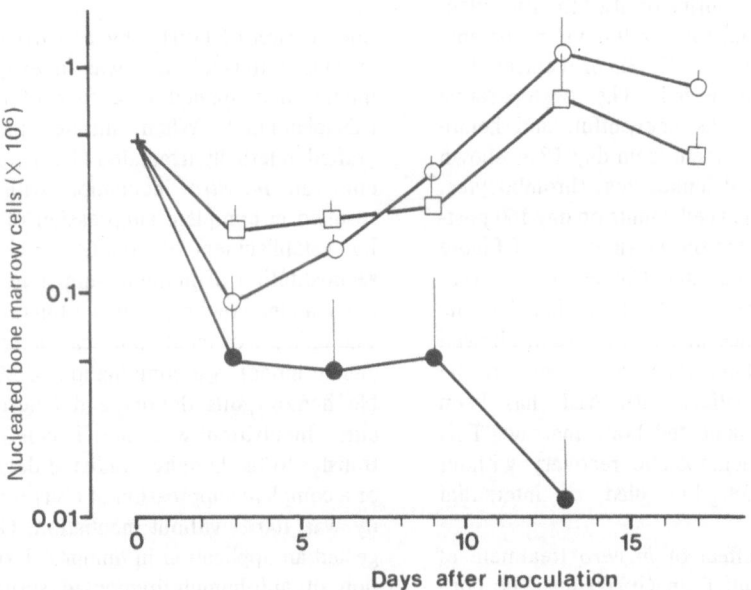

FIGURE 26.3. Proliferation of bone marrow cells in diffusion chambers after incubation with crude ATG (●——●) or absorbed anti-T cell globulin (○——○) or no antiserum as control (■——■), upon addition of complement (antibody concentration 1:20).

nephrotoxic antibodies, and cross-reacting antibodies with plasmaproteins. There was also no activation of complement by complexes and no general toxicity.

In the following case, a marrow incubation treatment is described. An 11-year-old girl with a second relapse of common acute lymphoblastic leukemia was prepared for bone marrow transplantation. The patient received a pre-transplant course of BCNU (200 mg/m² on days -13, -12), cytosine arabinoside (200 mg/m² on days -12 to -8), cyclophosphamide (1.8 g/m² on days -7, -6), and a total-body irradiation of 1,000 rad 24 hr before transplantation. An 8-year-old HLA-identical and MLC-negative brother was available as donor. From investigations by Kolb et al. (6) in dogs, and from clinical reports of Storb et al. (24), we know that sex-different combinations show a significantly higher incidence of GvH reactions. To exclude this possibility, incubation treatment of the donor marrow was performed before transfer. The marrow was first separated from erythrocytes and concentrated to 15% of the initial volume by a technique described by Netzel et al. (12). The removed erythrocytes were transfused back to the donor. Then the marrow was incubated with ATCG-H in a final concentration of 1:200 at 4°C for 30 min. The marrow-containing blood bag was gently agitated. The procedures are shown schematically in Figure 26.4. The antibody concentration used corresponds to a GvHD-suppressive dose in the canine model. Immediately after incubation, the bone marrow was transfused to the recipient in an amount of 3.2×10^8/kg body weight. There was no sign of any irritation of the patient during or after the transplantation procedure. So far (over 240 days' post-transplantation), no symptoms of GvHD or recurrent leukemia have been observed. The postoperative course (Figure 26.5) was uneventful, and hemopoietic recovery was evident from day 15 as shown by a rise in peripheral leukocytes, thrombocytes, and reticulocytes. Blood cell counts on day 160 post-transplantation are listed on the right side of Figure 26.5 and appear almost normal. Chimerism was complete in this patient, as indicated by a change in the karyotype. A difference in the Rh system allowed the identification of donor-type erythrocytes. In the meantime, another patient with ALL has been treated with ATCG-incubated bone marrow. This patient also had a hemopoietic recovery without GvHD until day 38, but died of interstitial pneumonia.

The suppressive effect of *in vitro* treatment of bone marrow with anti-T on GvHD must be confirmed in other patients. In view of the encouraging animal experiments, a bone marrow incubation with ATCG may help to prevent GvHD in man.

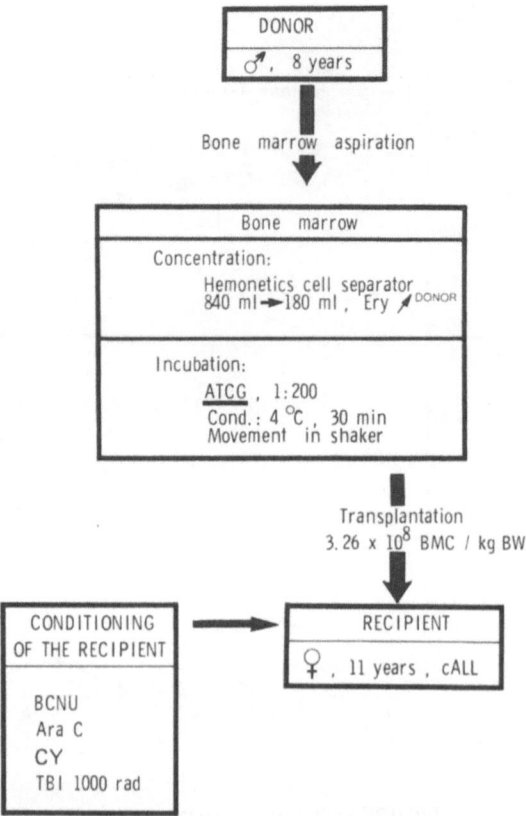

FIGURE 26.4. Incubation treatment of donor marrow with ATCG-H in clinical marrow transplantation.

SUMMARY

Suppression of GvHD by *in vitro* incubation with specific anti-T-cell sera was investigated in animal models and applied in a case of clinical marrow transplantation. When murine spleen cells were grafted in lethally irradiated H-2 incompatible recipients, an *in vitro* incubation with such antisera resulted in complete suppression of GvHD mortality, establishment of donor-type hemopoiesis, and reconstitution of immunocompetence against third-party antigens with persisting tolerance to the host. The efficacy of incubation was demonstrated in the canine model in a combination of MHC-incompatible homozygous donors and heterozygous recipients. Incubation with anti-T cell globulin before transfer to the lethally irradiated dogs led to a delay or a complete suppression of GvH reactions; mortality was 100% without incubation. Our results suggested an application in humans. Extensive absorption of anti-human-thymocyte serum revealed an anti-human-T-cell globulin that is specifically reactive against T-cells and not inhibitory to human CFU-c and bone marrow growth in diffusion cham-

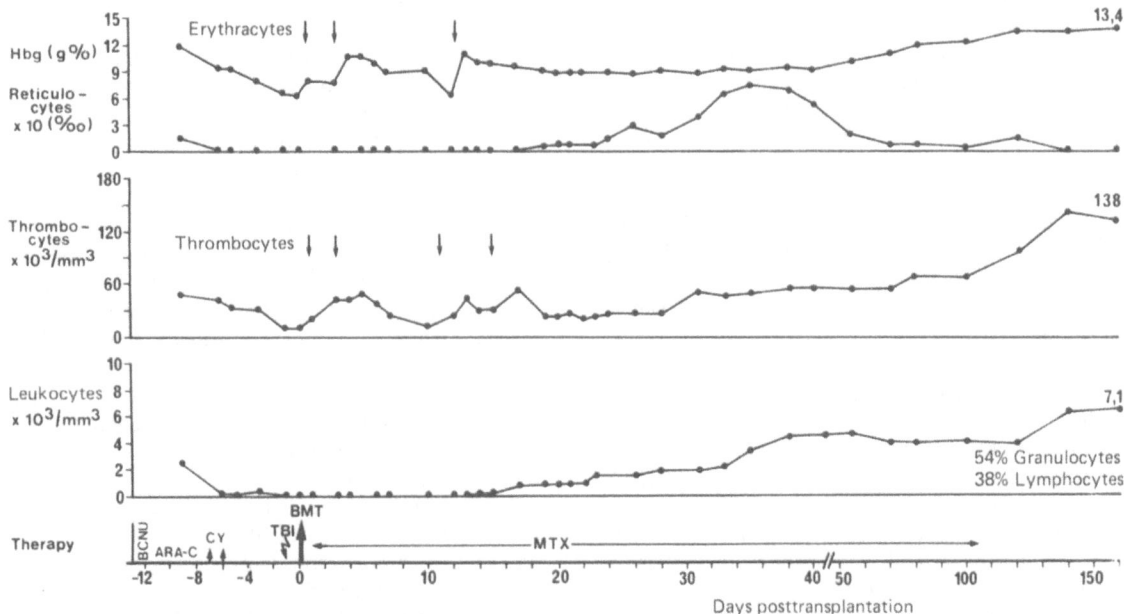

FIGURE 26.5. Clinical course of patient B. V. after transplantation with ATCG-H incubated bone marrow cells (↓, blood and platelet transfusions).

bers. This antibody preparation was applied to clinical bone marrow transplantation in a patient with ALL. The HLA-identical and MLC-negative but sex-different donor marrow was incubated with anti-human T-cells and transferred to the recipient, who had been conditioned with an antiproliferative regimen and total-body irradiation. The patient tolerated the incubated marrow without side effects and had an uneventful hemopoietic engraftment and recovery. So far no symptoms of GvHD have been observed.

ACKNOWLEDGMENTS

The authors are indebted to the staff of the Abteilung Immunologie, Institut für Hämatologie, for skillful technical assistance. The clinical transplantation was performed in cooperation with the "Münchener Arbeitsgruppe für Knochenmark-transplantation."

This study was supported by EURATOM-GSF 089-72-I BIAD and research Grant SFB 37-E3.

REFERENCES

1. Binz, H., and Wigzell, H. Antigen-binding, idiotyric T lymphocyte receptors. *Top. Immunbiol.*, 7:111, 1977.
2. Böyum, A. Separation of leukocytes from blood and bone marrow. *Scand. J. Clin. Lab. Invest.*, 21(Suppl.):97, 1968.
3. Dicke, K. A., van Hooft, J. I. M., and van Bekkum, D. W. The selective elimination of immunologically competent cells from bone marrow and lymphatic cell mixtures. *Transplantation, 6*:562, 1968.
4. Epstein, R., Storb, R., Ragle, H., and Thomas, E. D. Cytotoxic typing antisera for marrow grafting in littermate days. *Transplantation, 6*:45, 1968.
5. Golub, E. S. Brain-associated θ-antigen: Reactivity of rabbit anti-mouse brain with lymphoid cells. *Cell. Immunol., 2*:353, 1971.
6. Kolb, H. J., Rieder, I., Grosse-Wilde, H., Scholz, S., Schäffer, E., and Kolb, H. Graft-versus-host disease following marrow grafts from DLA-matched canine littermates. *Transplant. Proc.* (in press).
7. Kolb, H. J., Rieder, I., Rodt, H., Netzel, B., Grosse-Wilde, H., Scholz, S., Schäffer, E., Kolb, H., and Thierfelder, S. Anti lymphocytic antibodies and marrow transplantation. IV. Graft-versus-host tolerance in DLA-incompatible dogs following "in vitro" treatment of bone marrow with absorbed anti-thymocyte globulin. *Transplantation* (in press).
8. Miller, J. A. F. P. Effect of thymectomy in adult mice on immunological responsiveness. *Nature (London), 208*:1337, 1965.
9. Miller, J. A. F. P., Doak, S. M. A., and Cross, A. M. Role of the thymus in recovery of the immune mechanism in the irradiated adult mouse. *Proc. Soc. Exp. Biol. Med., 112*:785, 1963.
10. Mookerjee, B. K., Azzolina, L., and Poultor, L. Interaction of anti-thymocyte serum with hemopoietic stem cells. I. Effects *in vitro* and *in vivo. J. Immunol., 112*:822, 1974.
11. Müller-Ruchholtz, W., Wottge, H.-U., and Müller-Hermelink, H. K. Bone marrow transplantation in rats across strong histocompatibility barriers by selective

elimination of lymphoid cells in donor marrow. *Transplant. Proc.*, 8:537, 1976.

12. Netzel, B., Haas, R. J., Janka, G. E., and Thierfelder, S. Viability of stem cells (CFU-c) after long term cryopreservation of bone marrow cells from normal adults and children with acute lymphoblastic leukemia in remission. In Rainer, H., ed., *Cell Separation and Cryobiology.* Stuttgart-New York: Schaffauer, 1978.

13. Netzel, B., Rodt, H., Hoffmann-Fezer, G., Thiel, E., and Thierfelder, S. The effect of crude and differently absorbed anti-human T cell globulin on granulocytic and erythropoietic colony-formation. *Exp. Hematol.*, 6:410, 1978.

14. Pike, B. L., and Robinson, W. A. Human bone marrow colony growth in agar gel. *J. Cell. Physiol.*, 76:77, 1970.

15. Rodt, H., Kolb, H. J., Netzel, B., Rider, I., Janka, G., Belohradsky, B., Haas, R. J., and Thierfelder, S. GvHD-suppression by incubation of marrow grafts with anti-T cell globulin: Effect in the canine model and application to clinical bone marrow transplantation. *Transplant. Proc.* (in press).

16. Rodt, H., Netzel, B., Niethammer, D., Körbling, M., Götze, D., Kolb, H. J., Thiel, E., Haas, R. J., Fliedner, T. M., and Thierfelder, S. Specific absorbed antithymocyte globulin for incubation treatment in human marrow transplantation. *Transplant. Proc.*, 9:187, 1977.

17. Rodt, H., Thierfelder, S., and Eulitz, M. Suppression of acute secondary disease by heterologous anti-brain serum. *Blut, 25*:385, 1972.

18. Rodt, H., Thierfelder, S., and Eulitz, M. Anti-lymphocytic antibodies and marrow transplantation. III. Effect of heterologous anti-brain antibodies an acute secondary disease in mice. *Eur. J. Immunol., 4*:25, 1974.

19. Rodt, H., Thierfelder, S., and Götze, D. Cooperation between donor cells and H-2 incompatible recipient thymus after suppression of secondary disease with anti-T cell globulin. *Exp. Hematol., 2*:299, 1974.

20. Rodt, H., Thierfelder, S., Thiel, E., Götze, D., Netzel, B., Huhn, D., and Eulitz, M. Identification and quantification of human T-cell antigen by antisera purified from antibodies crossreacting with hemopoietic progenitors and other blood cells. *Immunogenetics, 2*:411, 1975.

21. Schlesinger, M., and Galili, U. Antigenic differences between T and B lymphocytes in man. *Israel J. Med. Sci., 10*:715, 1974.

22. Simmonsen, M. Graft-versus-host reactions. Their natural history and applicability as tools of research. In Kallos, P., Byron, H., and Waksman, S., eds., *Progress in Allergy,* Vol. 6 Basel-New York: Karger, 1962.

23. Sprent, J., v. Böhmer, H., and Nabholz, M. Association of immunity and tolerance to host H-2 determinants in irradiated F_1 hybrid mice reconstituted with bone marrow cells from one parental strain. *J. Exp. Med., 142*:321, 1975.

24. Storb, R., Weiden, P. L., Prentice, R., Buckner, C. D., Clift, R. A., Einstein, A. B., Fefer, A., Johnson, F. L., Lerner, K. G., Neiman, P. E., Sanders, J. E., and Thomas, E. D. Aplastic anemia (AA) treated by allogeneic marrow transplantation: The Seattle experience. *Transplant. Proc., 9*:181, 1977.

25. Thierfelder, S. Experimental bone marrow transplantation. *Blut, 30*:1, 1975.

26. Thierfelder, S., Rodt, H., Netzel, B., and v. Rössler, R. Studies on chimaeric tolerance induced with anti-T cell globulin. *Exp. Hematol., 2*:299, 1974.

27. Thomas, E. D., Buckner, C. D., Banayi, M., Clift, R. A., Fefer, A., Flournoy, N., Goodell, B. W., Hickman, R. O., Lerner, G. K., Neiman, P. E., Sale, G. E., Sanders, J. E., Singer, J., Stevens, M., Storb, R., and Weiden, P. L. 100 patients with acute leukemia treated by chemotherapy, total body irradiation, and allogeneic marrow transplantation. *Blood, 49*:511, 1977.

28. Thomas, E. D., and Storb, R. Technique for human marrow grafting. *Blood, 36*:507, 1970.

29. Trentin, J. J., and Judd, K. P. Prevention of acute graft-versus-host (GvH) mortality with spleen-absorbed anti-thymocyte globulin (ATG). *Transplant. Proc., 7*:865, 1975.

30. Zinkernagel, R. M., Callahan, G. N., Althage, A., Cooper, S., Klein, P. A., and Klein, J. On the thymus in the differentiation of "H-2 self recognition" by T cells: Evidence for dual recognition?. *J. Exp. Med., 147*:882, 1978.

27

Natural Killer Cells in Man: Their Possible Involvement in Leukemia and Bone Marrow Transplantation

Eva Lotzová, K. B. McCredie,
L. Muesse, K. A. Dicke,
and E. J. Freireich

Over the past few years, there has been an increasing polarization of interest and research on natural cell-mediated cytotoxicity directed against leukemias, lymphomas, and several other types of tumors. The phenomenon of natural cytotoxicity has been described and most extensively studied in rodents, especially in mice (5,10,12,14,18,20), and man (9,13,15,16,17,19). The cells mediating natural cytotoxicity were designated natural killer (NK) cells to emphasize their natural occurrence and killing capacity. These cells appear to be distinct from other cytotoxic lymphocytes and macrophages since they do not possess easily detectable T- and B-lymphocyte markers and are nonadherent and nonphagocytic (6,11,13,16). Human NK cells and at least some murine NK cells carry Fc receptors (7,13,16), although the Fc receptors on murine NK cells are of low avidity (7). Even though some workers have suggested that NK cells are members of a T-cell lineage, (7-8), the data to support this suggestion are still missing. Thus, the precise nature of NK cells remains to be determined.

Human as well as murine NK cells exhibit a certain degree of selectivity in their cytotoxic activity against target cells. Nevertheless, the nature of the antigens that trigger such activity is still obscure. It has been suggested that murine NK cells react against structures associated with C-type viruses (5). But no correlation between expression of MuLV antigens and lytic susceptibility was found in the various Moloney lymphoma lines studied (1). Since NK cells react to a variety of syngeneic, allogeneic and xenogeneic tumors of various histologic types, it is reasonable to assume that their reactivity is not directed to a single antigen but to a rather large spectrum of antigenic determinants. We have found recently that the cells that cause the rejection of bone marrow grafts express characteristics of NK cells (12). This observation suggests that NK cells may also recognize and react to the cell-surface determinants on bone marrow cells. This suggestion is in accord with our observation (Lotzová, unpublished observations) and observations by others (7) that show that NK cells are cytotoxic to bone marrow cells *in vitro*.

Although the NK cell etiology and their possible significance remains obscure, and it is perhaps too early to be dogmatic on the matter, there is evidence that these cells may be involved in at least three important biologic functions: (a) immunosurveillance to tumors, (b) resistance to already-established tumors, and (c) rejection of bone marrow transplants (see Figure 27.1). Actually there is a precedent for the first two suggested functions, arising from experimental animal studies. First, in the murine system, there appears to be a positive correlation between NK cell activities *in vitro* and tumor resistance lev-

207

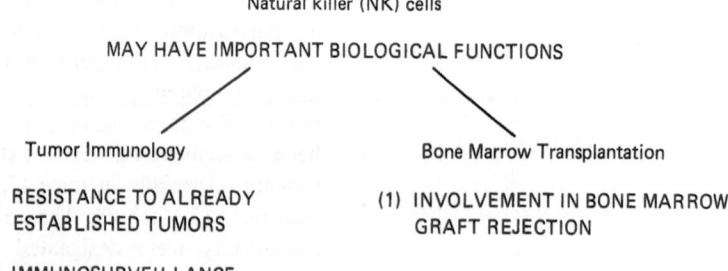

Natural killer (NK) cells

MAY HAVE IMPORTANT BIOLOGICAL FUNCTIONS

Tumor Immunology

(1) RESISTANCE TO ALREADY
ESTABLISHED TUMORS

(2) IMMUNOSURVEILLANCE
TO TUMORS

Bone Marrow Transplantation

(1) INVOLVEMENT IN BONE MARROW
GRAFT REJECTION

FIGURE 27.1. Schematic presentation of the possible biologic functions of natural killer cells.

els *in vivo*, e.g., high leukemic mouse strains express no or very low NK cell cytotoxicity *in vitro*, and vice versa, low leukemic mouse strains express very high NK cell activity *in vitro* (20). Second, the fact that athymic mice (which express high NK cell activity) do not experience a higher incidence of malignancies than conventional mice despite the lack of T-cell-mediated immunity is compatible with the protective role of NK cell effector mechanism. Also, their spontaneous occurrence with no requirement for priming and their capacity to recognize and kill malignant cells, make NK cells almost perfect candidates for "immunosurveillors." The implication of NK cell activity in the rejection of bone marrow grafts is also supported by a few facts, such as their striking similarity to bone marrow effector cells and their reactivity *in vitro* to bone marrow cell antigens.

The above-mentioned facts, for fairly obvious reasons, triggered our interest in NK cells and prompted us to study various parameters of human NK cells, especially with regard to their involvement in resistance to leukemia.

MATERIALS AND METHODS

Target Cells

The cultured human lymphoblastoid cell line CEM, with T-cell characteristics, was used in all our experiments as a target. The line was kindly provided by Dr. Evan Hersh (from our Department) in 1977. It has been maintained as a continuous culture in our laboratory in RPMI 1640 medium supplemented with 10% fetal calf serum (FCS), HEPES buffer, and antibiotics.

For cytotoxicity studies, ten million CEM cells in 0.5 ml of RPMI 1640 medium with 10% FCS were labeled with 100 μCi radioactive sodium chromate (^{51}Cr, Amersham/Searle Corporation, Arlington Heights, Ill.) and incubated for 30 min at 37°C. The labeled cells were washed three times in 25 ml of the supplemented RPMI. The concentration of target cells was adjusted to 4×10^5/ml.

Effector Cells

Peripheral blood lymphocytes were separated from heparinized whole blood obtained from healthy volunteer donors or from patients on a Ficoll-Hypaque gradient as described by Boyum (2). Bone marrow cells were obtained from three groups of donors: (a) healthy individuals serving as bone marrow donors in allogeneic bone marrow transplantation, (b) non-hemopoietic cancer patients in remission, (c) acute myeloid leukemia patients in relapse, and (d) acute myeloid leukemia patients in remission. The first two groups of individuals are classified as normal donors. The bone marrow cells were collected from the iliac crest.

Bone marrow cells were separated from erythrocytes either by Ficoll-Hypaque gradient or by centrifugation and subsequent buffy coat collection. The latter method allowed higher bone marrow cell recovery and thus was used in most experiments. The NK cell cytotoxicity was comparable with either of these separation techniques. In most of the experiments, the target-to-effector cell (T:E) ratio was 1:50; in some experiments 1:100 and 1:200 T:E ratios were employed.

Spleens were kindly provided by Dr. Barry Kahan, Department of Surgery, The University of Texas Medical School. The spleens did not show any anomalies and were removed prophylactically in preparation for kidney transplantation. Spleen cell suspension was prepared in Hank's balanced salt solution.

The discontinuous albumin gradient technique of Dicke et al. (4) was used for fractionation of NK cells. Cryopreservation of bone marrow cells was performed as described previously (3); 10% dimethyl sulfoxide was used as cryoprotective agent; and the cells were stored at −192°C.

Cytotoxicity Test

One-tenth of a milliliter of the desired concentration of effector cells was plated in quadruplicate into the wells of a flat-bottomed microtiter plate (Falcon

Microtest II, Scientific Products, Dallas, Tex.) followed by the addition of 0.05 ml of target cells. The plates were incubated for 16 hr at 37°C in 5% CO_2 humidified atmosphere. After incubation, 0.1 ml of medium was added to each well and the plates were centrifuged at 1,000 rpm for 10 min. A 50-μl aliquot was removed from each well and counted in an Auto-Gamma scintillation spectrometer. The percentage of ^{51}Cr release was determined according to the formula:

$$\% \text{ of } ^{51}\text{Cr release} = \frac{5 \times (\text{Radioactivity of 50 } \mu\text{l of supernatant})}{\text{Total radioactivity}} \times 100$$

The results were expressed as percentage of cytotoxicity:

$$\text{Cytotoxicity } (\%) = \frac{\text{Experimental release } (\%) - \text{Spontaneous release } (\%)}{\text{Maximum release } (\%) - \text{Spontaneous release } (\%)} \times 100$$

Spontaneous release of ^{51}Cr was determined by incubating the target cells in medium alone; it ranged from 16 to 23%. Maximum release of ^{51}Cr was determined after freezing and thawing of the tumor cells three times, and ranged from 83 to 95%.

Statistical Analysis

The differences between individual groups studied were evaluated statistically with a Student's t-test, and the probability (p) was calculated.

RESULTS

Distribution of Natural Killer Cells in Peripheral Blood, Spleen, and Bone Marrow of Normal Donors

In the first series of experiments, peripheral blood from 45, spleens from nine, and bone marrow from 13 different normal donors were tested for NK cell-mediated cytotoxicity against human T-lymphoblastoid cell line, CEM, in the ^{51}Cr cytotoxicity assay. The T:E cell ratio was 1:50. This ratio was found to give optimal results with all three tissues. When the T:E ratio was increased to 1:100 or 1:200 there was either no increase or only a slight increase in cytotoxicity (see Figure 27.3). Therefore, the ratio of 1:50 was adhered to in all the studies reported here.

It can be seen from Figure 27.2 that the NK cells residing in the spleen expressed the highest degree of cytotoxicity to the CEM target cell (mean ± SE was 56.8% ± 5.8%). Next in efficiency were the peripheral blood NK cells, expressing about 1.8 times less cytotoxicity than the splenic NK cell (mean ± SE was 31.2% ± 1.2%). The bone marrow NK cells were least efficient; their cytotoxicity was around

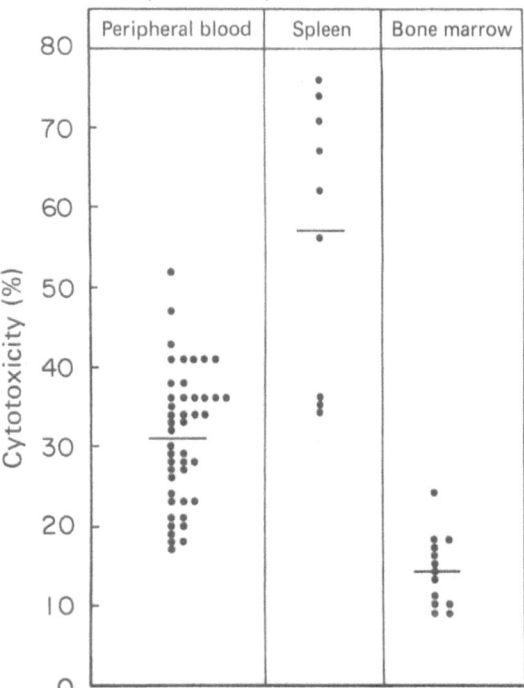

FIGURE 27.2. Natural killer cell activity in various tissues of normal donors. The human T-lymphoblastoid cell line was used as a target; the target-to-effector cell ratio was 1:50. The cultures were incubated for 16 hr.

two and four times lower than the NK cell cytotoxicity of peripheral blood and spleen, respectively (mean ± SE was 14.1% ± 1.2%). From these studies it can be concluded that NK cells are present in all the three tissues tested, but their concentration varies with the tissue.

Natural Killer Cell Activity in the Peripheral Blood and Bone Marrow of Patients with Acute Myeloid Leukemia

It was mentioned in the Introduction that NK cells may represent an important component in resistance and immunosurveillance to malignancies, especially the leukemias. Thus, it was of utmost interest to determine NK cell activity in patients with acute myeloid leukemia (AML) and in AML patients in remission. Normal donors served as controls in these experiments. Peripheral blood of 45 normal donors, 12 AML patients, and 11 AML patients in remission was tested. The results of these studies are shown in Table 27.1. It can be seen that there was a significant decrease in peripheral blood NK cell activity of AML patients with active disease and of AML patients in remission. As illustrated in Figure 27.3, the low NK cell activity in AML

TABLE 27.1 Natural Killer Cell-Mediated Cytotoxicity in Bone Marrow and Peripheral Blood of Normal Donors and AML Patients

SOURCE OF EFFECTOR CELLS	CYTOTOXICITY (%)[a] MEAN ± SE (N)[b]	p VALUE[c]
Peripheral blood		
Normal donors	31.3 ± 1.2 (45)	
AML patients	15.3 ± 3.1 (12)	<.01
AML patients in remission	24.2 ± 0.1 (11)	<.02
Bone marrow		
Normal donors	14.1 ± 1.2 (13)	
AML patients	6.0 ± 0.9 (7)	<.001
AML patients in remission	15.7 ± 1.9 (11)	<.5

[a]The human CEM T-lymphoblastoid cell line was used as a target; target-to-effector cell ratio was 1:50. The cultures were incubated for 16 hr.

[b]N numbers of individuals.

[c]The difference between normal donors and patients was evaluated statistically with a Student's t- test, and the probability (p) was calculated.

patients cannot be explained by a dilution effect caused by a much larger number of blast cells present in the peripheral blood, since the increase in the T:E cell ratio to 1:100 or 1:200 was not accompanied by an increase in NK cell activities. The low NK cell activity could not be explained by the presence of suppressor cells or blast cells that would interfere with NK functions either, since no inhibition of NK cell cytotoxicity was found when mixtures of normal and leukemic individuals were tested (these data are not shown).

As indicated in the lower part of Table 27.1, NK

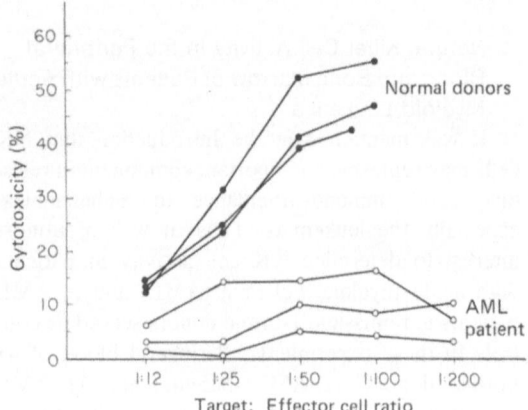

FIGURE 27.3. Evaluation of natural killer T-cell activity in the peripheral blood of three normal donors and three AML patients with regard to different target-to-effector cell ratios. The human T-lymphoblastoid cell line was used as a target. The cultures were incubated for 16 hr.

cell activity in bone marrow of AML patients with active disease were also significantly lower than those of normal individuals. Similar to peripheral blood, an increase in T:E cell ratio to 1:100 or 1:200 was not accompanied by higher bone marrow NK cell activity. Actually, when cytotoxicity was determined with a 1:200 T:E ratio, the cytotoxicity of NK cell was always lower than when the 1:50 ratio was used. In contrast to peripheral blood of AML patients in remission where the NK cell cytotoxicity was significantly lower than that of peripheral blood of normal donors, NK cells in bone marrow of AML patients in remission did not show less activity in comparison to normal donors.

Fractionation of Human Bone Marrow Natural Killer Cells by the Discontinuous Albumin Gradient Method

In the previous two series of experiments, it was shown that the activity of NK cells in bone marrow is rather low. If NK cells are actually involved in resistance to leukemia, and bone marrow transplantation is used as therapy for leukemia patients, then the fact that NK cells express low activity in bone marrow is very distressing. No doubt there are always some resideual leukemic cells in the bone marrow recipient, in spite of chemotherapy and irradiation. Moreover, in the case of autologous bone marrow transplantation, there are always some residual leukemic cells in the remission marrow. Therefore, it would be highly desirable to supply the patients with maximal numbers of NK cells. These reasons stimulated our search for techniques to "enrich" NK cells. The discontinuous albumin gradient technique appears to be a promising approach to NK cell concentration. Employing this technique, we have studied unfractionated and fractionated bone marrow cells of six AML patients in remission. As can be seen from Table 27.2, there was a signifi-

TABLE 27.2 Natural Killer Cell Activities of Unfractionated and Discontinuous Albumin Gradient Fractionated Bone Marrow Cells

	CYTOTOXICITY (%)[a]				
		Fractions			
DONOR[b]	Unfractionated	1 and 2	3	4	5
1	14	65	10	4	1
2	13	58	17	0	3
3	14	44	21	12	2
4	10	61	29	11	4
5	15	46	12	11	0
6	16	54	36	8	2

[a]The human CEM T-lymphoblastoid cell line was used as a target; target-to-effector cell ratio was 1:50. The cultures were incubated 16 hr.

[b]Bone marrow cells from AML patients in remission were used in these studies.

cant enrichment of NK cells in fractions 1 and 2 (around three to six times) in all bone marrow samples tested. In fraction 3, the NK cell activities were comparable to those of unfractionated marrow, with the exception of donor 4 and 6 in whom NK cell activity doubled in comparison to unfractionated marrow. Low NK activities were found in fraction 4, and practically no NK cell activity was found in fraction 5.

Activities of Cryopreserved and Piperazinedione-Treated Natural Killer Cells

When autologous bone marrow transplantation is performed, the bone marrow cells from leukemic patients in remission are collected and frozen until the patient experiences a crisis, at which time the bone marrow cells are thawed and transplanted back to the patient. Thus, it was important to determine whether cryopreserved bone marrow NK cells were as active as fresh bone marrow NK cells. Eleven samples of fresh bone marrow and five samples of cryopreserved bone marrow of AML patients in remission were tested for NK cell activity. The data in Table 27.3 clearly indicate that cryopreserved NK cells had 2.5 times less activity than fresh NK cells.

Piperazinedione and total-body irradiation is one of the cytoreductive regimens used before bone marrow transplantation to eradicate the recipients' residual leukemic cells. It is known that NK cells are relatively radioresistant, but there is no information on the effect of piperazinedione on these cells. This information is critical to the understanding of NK cell involvement in reccurrence of leukemia. Therefore, in another series of experiments, we studied the peripheral blood of seven prospective bone marrow recipients before and after piperazinedione treatment. Two doses of 25 mg/m^2 of piperazinedione were given 5 and 6 days before NK cell assay. The results of these studies are illustrated in Table 27.4. It is obvious that NK cell activity was totally

TABLE 27.3 Activities of Cryopreserved and Fresh Bone Marrow Natural Killer Cells of AML Patients in Remission

TYPE OF BONE MARROW CELLS	CYTOTOXICITY (%)[a] MEAN ± SE (N)[b]	VALUE[d]
Fresh	15.7 ± 1.9 (11)	
Cryopreserved[c]	6.1 ± 1.1 (5)	<.01

[a]The human CEM T-lymphoblastoid cell line was used as a target; target-to-effector cell ratio was 1:50. The cultures were incubated for 16 hr.

[b]N, numbers of individuals.

[c]Bone marrow cells in 10% dimethylsulfoxide solution were stored in liquid nitrogen (−192°C).

[d]Difference between the activities of fresh and cryopreserved bone marrow NK cells was evaluated statistically with a Student's t test and the probability (p) was calculated.

TABLE 27.4 Effect of Piperazinedione on Peripheral Blood Natural Killer Cells of Acute Myeloid Leukemia Patients

PATIENT	UNTREATED[a] Cytotoxicity (%)	PIPERAZINEDIONE[b] TREATED
1	17	0
2	14	0
3	12	1
4	29	3
5	14	0
6	18	3
7	10	0

[a]Piperazinedione was given in two doses of 25 mg/m^2, 5 and 6 days before NK cell assay.

[b]The human CEM T-lymphoblastoid cell line was used as a target; target-to-effector cell ratio was 1:50. The cultures were incubated for 16 hr.

depleted after two treatments with piperazinedione. It can be concluded from these studies that leukemic recipients most likely do not possess any NK cell activities after the piperazinedione regimen and thus are solely dependent on the NK cells from the infused bone marrow.

DISCUSSION

Although the etiology and the biologic significance of NK cells remains obscure, there is evidence that these cells may be involved in resistance to and immunosurveillance of malignancies. Besides this important function in tumor immunity, NK cells may also play a role in bone marrow transplantation (12). These two possibilities strongly justify further research on NK cells.

In the studies on tissue distribution of NK cells, we found that NK cell cytotoxicity against T-lymphoblastoid cell line, CEM, was expressed in the spleen, peripheral blood, and bone marrow of normal individuals. But the degree of NK cell cytotoxicity varied with regard to the tissue, being highest in the spleen and lowest in the bone marrow. The AML patients with active disease expressed significantly lower NK cell activity in both peripheral blood and bone marrow. Lower NK cell activity of AML patients cannot be explained by a dilution effect caused by a larger number of blast cells in the bone marrow and peripheral blood, since the increase of target-to-effector cell ratio to 1:100 or even to 1:200 was not accompanied by an increase in NK cell activity. The low NK cell activity could not be explained by the interference of blast cells with NK cell cytotoxicity, or by the presence of suppressor cells in AML patients either, since we did not detect any inhibition of NK cell cytotoxicity when cells of normal individuals and AML patients were mixed. The significantly lower NK cell activity in AML patients is in accordance with the involvement of NK cells in resistance to leukemia. Further studies, however, must be done to determine the cause of this NK cell anomaly in AML patients and to understand its mechanism.

If one expects NK cells to be an important factor in resistance to the leukemias, then our observation that NK cells activity is low in fresh bone marrow and that there are virtually no NK cells in cryopreserved bone marrow is distressing, specifically, in the view of bone marrow transplantation as a therapy in leukemia. It would be advantageous to provide the leukemic patients not only with hemopoietic stem cells but also with maximum numbers of NK cells so as to strengthen their defense to leukemia. This would be especially useful in view of the fact that the cytoreductive regimen used in the clinic before bone marrow transplantation includes piperazinedione, which has been shown in our studies to destroy NK cell activity. This problem could be avoided, however, by employing the techniques allowing NK cell "enrichment." We have shown in this presentation that the discontinuous albumin gradient technique was such an efficient technique.

Even if it is optimistic, it is not unjustified to speculate that future NK cell therapy, in the form of transfusions, may be beneficial for patients with malignancies. Since the major histocompatibility complex homology is not required for NK cell function, even allogeneic NK cell therapy may be possible. But the function of NK cells in the graft-versus-host reaction must be precisely determined first.

SUMMARY

Natural killer (NK) cells may be an important cell type in resistance to and immunosurveillance of malignancies as well as in the rejection of bone marrow grafts. Therefore, we have studied several parameters of NK cell-mediated cytotoxicity. We have found that spleens of normal individuals had expressed the highest NK cell activity, next in efficiency were the peripheral blood, followed by the bone marrow. Our AML patients showed significantly lower NK cell activity in both peripheral blood and bone marrow, which suggested an involvement of NK cells in resistance to leukemia. It was demonstrated that NK cell function was sensitive to piperazinedione treatment and to cryopreservation, and that NK cells could be concentrated by the discontinuous albumin gradient technique. The highest NK cell activity was found in fraction 1 and 2, and NK cell activity in these fractions was three to six times higher than that of unfractionated bone marrow. Other fractions showed either comparable or lower NK cell activity than unfractionated bone marrow.

ACKNOWLEDGMENTS

The authors wish to express their gratitude to Dr. D. Verma for collecting some of the human peripheral blood and bone marrow samples, to Dr. Axel Zander for many helpful discussions and for providing peripheral blood and bone marrow samples, to S. Thompson for performing the discontinuous albumin gradient technique, and to Rachel Rojas for excellent assistance in preparation of this manuscript.

This work was supported by Grant CA-21062 from the National Cancer Institute.

REFERENCES

1. Becker, S. E., Fenyo, E. M., and Klein, E. The "natural killer" cell in the mouse does not require H-2 homology and is not directed against type or group-specific antigens of murine C-viral proteins. *Eur. J. Immunol., 6*:882, 1976.

2. Boyum, A. Separation of leukocytes from blood and bone marrow. *Scand. J. Clin. Lab. Invest., 21* (Suppl. 97):77, 1968.

3. Dicke, K. A., Lotzová, E., Spitzer, G., and McCredie, K. B. Immunobiology of bone marrow transplantation. *Semin. Hematol., 15*:263, 1978.

4. Dicke, K. A., van Noord, M. J., Maat, B., Schaefer, V. W., and van Bekkum, D. W. Attempts at morphological identification of the hemopoietic stem cell in primates and rodents. In Wolstenholme, G. E. W., and O'Connor, M., eds., *Haemopoietic Stem Cells: Ciba Foundation Symposium 13 (New Series)*. Amsterdam: Elsevier, 1973, p. 47.

5. Herberman, R. B., Nunn, M. E., and Lavrin, D. H. Natural cytotoxic reactivity of mouse lymphoid cells against syngeneic and allogeneic tumors. I. Distribution of reactivity and specificity. *Int. J. Cancer, 16*:216, 1975.

6. Herberman, R. B., Nunn, M. E., Holden, H. T., and Lavrin, D. H. Natural cytotoxic reactivity of mouse lymphoid cells against syngeneic and allogeneic tumors. II. Characteristics of effector cells. *Int. J. Cancer, 16*:230, 1951.

7. Herberman, R. B., and Holden, H. T. Natural cell-mediated immunity. *Adv. Cancer Res., 27*:305, 1978.

8. Kaplan, J., and Callewaert, D. M. Expression of human T-lymphocytes antigens by natural killer cells. *J. Nat. Cancer Inst., 60*:961, 1978.

9. Kay, H. D., and Sinkovics, J. G. Cytotoxic lymphocytes from normal donors. *Lancet, II*:296, 1974.

10. Kiessling, R., Klein, E., and Wigzell, H. "Natural"

killer cells in the mouse. I. Cytotoxic cells with specificity for mouse Moloney leukemic cells. Specificity and distribution according to genotype. *Eur. J. Immunol., 5*:112, 1975.

11. Kiessling, R., Klein, E., Pross, H., and Wigzell, H. "Natural" killer cells in the mouse. II. Cytotoxic cells with specificity for mouse Moloney leukemia cells. Characteristics of the killer cells. *Eur. J. Immunol., 5*:117, 1975.

12. Lotzová, E., and Savary, C. A. Possible involvement of natural killer cells in bone marrow graft rejection. *Biomedicine, 27*:341, 1977.

13. Lotzová, E., and McCredie, K. B. Natural killer cells in mice and man and their possible biological significance. *Cancer Immunol. Immunotherap.* 1978 (in press).

14. Nunn, M. E., Djeu, J. Y., Glaser, M., Lavrin, D. H., and Herberman, R. B. Natural cytotoxic reactivity of rat lymphocytes against syngeneic Gross-virus-induced lymphomas. *J. Nat. Cancer, Inst., 56*:393, 1976.

15. Pross, H. F., and Jondal, M. Spontaneous cytotoxic activity as a test of human lymphocyte function. *Lancet, I*:355, 1975.

16. Pross, H. F., and Baines, M. Spontaneous human lymphocyte-mediated cytotoxicity against tumor target cells. VI. A Brief Review. *Cancer Immunol. Immunotherap., 3*:75, 1977.

17. Rosenberg, E. G., McCoy, J. L., Green, S. S., Donelly, F. C., Siwarski, D. H., Levin, P. H., and Herberman, R. B. Destruction of human lymphoid tissue culture cell lines by human peripheral lymphocytes in ^{51}Cr-release cellular cytotoxicity assay. *J. Nat. Cancer Inst., 52*:345, 1974.

18. Shellam, G. R., and Hogg, N. Gross-virus-induced lymphoma in the rat. IV. Cytotoxic cells in normal rats. *Int. J. Cancer, 19*:212, 1977.

19. Takasugi, M., Mickey, M. R., and Terasaki, P. T. Reactivity of lymphocytes from normal persons on cultured tumor cells. *Cancer Res., 33*:2898, 1978.

20. Zarling, J. M. Nowinski, R. C., and Bach, F. H. Lysis of leukemia cells by spleens of normal mice: *Proc. Nat. Acad. Sci. U.S., 72*:2780, 1975.

28

Erythrocyte-Incompatible Bone Marrow Transplants

H. G. Braine,
L. L. Sensenbrenner,
S. K. Wright, G. J. Elfenbein,
P. J. Tutschka, H. Kaizer,
W. B. Bias, and G. W. Santos

Allogeneic bone marrow transplantation of patients with major ABO erythrocyte incompatibility between the donor and recipient has been successfully performed at several centers (5–8). Transplantation in this mismatch situation in which the bone marrow recipient has circulating isohemagglutinins against donor-type erythrocytes is associated with three potential clinical problems: First—the danger of acute hemolysis during infusion of the marrow, which contains 400 to 800 ml of incompatible peripheral blood; second—the possibility that the preformed isohemagglutinins could adversely affect stem cell engraftment resulting in an increased incidence of graft rejection; and third—the production of erythrocyte elements post-transplantation in the presence of incompatible hemagglutinins with resultant chronic hemolysis or delayed recovery of other blood elements.

Two approaches have been used to prepare patients for ABO-incompatible transplantation. The first and most commonly used has been that of antibody reduction (5–8). By intensive plasma exchange, the recipient's isohemagglutinin titer can be significantly reduced. Further reduction is then achieved by antibody absorption with A-substance or donor-type incompatible erythrocytes. This approach, while successful, has several limitations. Exchanges usually involve plasmapheresis of one to two times the patient's circulating blood volume on several consecutive days through an arteriovenous shunt. The patient, at this time, may be exceedingly ill with severe thrombocytopenia and neutropenia and susceptible to serious infection or bleeding. Despite intensive plasmapheresis, low isohemagglutinin titers usually persist and subsequent transfusion of incompatible donor-type erythrocytes carries the risk of an acute hemolytic transfusion reaction. In addition, such massive exchanges require a great amount of time from a skilled plasmapheresis staff and are wasteful of large volumes of frequently scarce blood components.

An alternate solution to the ABO incompatibility problem, antigen reduction, has been the approach employed at this center. With the use of the Haemonetics Model 30® interrupted flow centrifugation system, bone marrow is processed in a manner similar to peripheral blood and the buffy layer, with as little erythrocyte contamination as possible, is prepared for transplantation. No attempt is made to reduce the donor's circulating isohemagglutinins. In this manner, the risk of massive hemolytic transfusion reactions during marrow infusion can be minimized. The marrow, however, must engraft in the presence of circulating isohemagglutinins, which has the potential of increasing the risk of graft rejection or delaying graft recovery. We report here our results in the first 10 patients prepared in this manner.

MATERIALS AND METHODS

Patient Population

Ten consecutive transplant candidates with major ABO incompatibilities were included in the study. Five of the patients had aplastic anemia (AA), four had acute lymphocytic leukemia (ALL), and one patient had chronic myelocytic leukemia in blast crisis (CML-B) (Table 28.1). Patients were between 5 and 28 years of age. Five donors had A antigens on their erythrocytes that were missing from the recipients (A mismatches), and five donors had B antigens on their erythrocytes that were missing from the recipients (B mismatches). Pre-transplantation circulating isohemagglutinins were determined using the microtiter technique of Crawford (3). The results of these studies indicated a range of isohemagglutinin titers of between 1:2 and 1:256 (Table 28.1).

Recipient Preparation

Prior to marrow infusion, all patients were hydrated with normal saline and diuresed to maintain a urine flow of greater than 100 ml/hr. Approximately 30 min before infusion, adult recipients were given 650 mg of Tylenol and 50 mg of Benadryl, both orally. Two hundred and fifty milligrams of Solucortef was given intravenously with 50 to 100 g of 10% mannitol. Marrow was infused without a filter over 30 to 60 min with the vital signs being monitored frequently.

Bone Marrow Procurement

Bone marrow was harvested by multiple needle aspirations from donors under general anesthesia in the general operating rooms. It was suspended in heparinized TC-199.

Removal of Red Blood Cells from Bone Marrow (Table 28.2)

The first seven separation procedures were performed using the Haemonetics 30 cell processor equipped with a 225-ml bowl and standard tubing sets. The techniques used were a modification of that reported by Weiner et al. (9). The bone marrow was transferred first to a 2,000-ml transfer bag. One unit of washed, packed erythrocytes and one unit of fresh frozen plasma, each of which was compatible with both the donor and recipient, were added to the bone marrow suspension. This was done to assure that a erythrocyte mass greater than 200 ml would be present to allow for bowl filling. It also served to dilute the incompatible erythrocytes such that the final erythrocyte contamination would be made up of approximately 50% erythrocytes compatible with the recipient. The 2,000-ml bag was then connected to the Haemonetics machine with the standard harness needle. Bowl filling was accomplished at 80 ml/min. When the buffy coat reached the shoulder of the bowl, the pump speed was slowed to 30 ml/min. The flow was redirected into the collection bag as the buffy coat started across the ledge of the bowl, and 60 ml was collected; the ACD-NIH-A in a ratio of 1:8 was bled into the marrow cell preparation during centrifugation. A total of one to three passes were generally required to process a preparation of marrow in this manner. Overall, there was an 80% recovery of total nucleated marrow cells and 100% recovery of small cells which morphologically were lymphocytes. The average packed erythrocyte contamination in the final infusate was 37 ml (range, 17 to 64 ml). Since approximately one-half of the erythrocytes in this product were of recipient type, the average total incompatible erythrocyte contamination was less than 20 ml.

TABLE 28.1 Patients Receiving ABO-Incompatible Bone Marrow Transplants

PATIENT #	TYPE OF MISMATCH[a]	AGE (YEARS)	SEX	DIAGNOSIS[b]	DONOR BLOOD TYPE[c]	RECIPIENT BLOOD TYPE[c]	INCOMPATIBLE TITER
119	A	19	M	AA	A+	0+	1:256
82	A	26	M	AA	A+	0+	1:32
88	A	5	M	AA	A+	0+	1:16
103	A	13	M	ALL	A+	0+	1:16
57	A	18	M	CML	A+	0+	1:32
76	B	12	F	AA	B−	0−	1:32
117	B	20	M	AA	B+	0+	1:32
87	B	18	M	ALL	B+	0+	1:32
78	B	6	M	ALL	AB+	A−	1:4
118	B	28	M	ALL	AB+	A+	1:2

[a]Antigen on donor erythrocyte, to which recipient has antibody, absent from recipient.
[b]AA, aplastic anemia; ALL, acute lymphocytic leukemia; CML, chronic myelogenous leukemia.
[c]ABO and major Rh type.

TABLE 28.2 Method for Marrowpheresis by Interrupted Flow Centrifugation[a]

METHOD	225-ML BOWL [WEINER ET AL. (9)]	100-ML BOWL [FLIEDNER ET AL. (4)]
Preparation	Packed erythrocytes to allow bowl filling	None
Procedure		
1. Filling rate	80 ml/min	40 ml/min
2. Collection rate	30 ml/min	20 ml/min
3. Collection volume	60 ml/pass	13 ml/pass
4. Number of passes	1–3	2–5
5. Additional anticoagulant	ACD-NIH A (1:8)	ACD-NIH A (1:8)
Results		
1. Number of procedures	7	3
2. Average number of nucleated marrow cells recovered (% of pre-separation number)	1.5×10^{10} (80%)	0.7×10^{10} (34%)
3. Average number of lymphocytes recovered (% of pre-separation number)	0.81×10^{10} (100%)	0.2×10^{10} (34%)
4. Average volume packed erythrocytes (range)	37 ml (17–64 ml)	6 ml (4–9 ml)

[a]Haemonetics Model 30® instrument used.

In order to avoid the addition of priming erythrocytes and to try to decrease the number of contaminating erythrocytes, a method similar to the 100-ml "Pediatric" procedure described by Fliedner et al. (4) was used in the last three cases. The Haemonetics Model 30® was prepared for use with the 100-ml bowl in the manner previously described (1). The filling rate was 40 ml/min with a collection rate of 20 ml/min. Two consecutive 13-ml volumes were collected on each pass. Overall, two to five passes were generally required to process the donor's bone marrow using the smaller bowl. An average of 34% of the total nucleated marrow cells were recovered in the first of the 13 ml volumes along with 34% of the lymphocyte population. There was a contamination with an average of 6 ml of packed incompatible erythrocytes (range: 4 to 9 ml) in this fraction. This initial volume was used for transplantation in these three cases. The second 13-ml volume on each run, which contained an equal number of nucleated cells, was frozen in liquid nitrogen and was thus available for later use if necessary.

Treatment Regimens

All patients undergoing transplantation received the treatment protocol extant at the time for the patient's primary diagnosis and evidence of sensitization to histocompatibility antigens (Table 28.3). Non-sensitized patients with AA received cyclophosphamide (50 mg/kg daily for 4 days). Sensitized patients with AA received total-body irradiation following either procarbazine and anti-thymocyte globulin or cyclophosphamide (50 mg/kg intravenously for 4 days). All patients with ALL received a com-

bination of high-dose cyclophosphamide and total-body irradiation. The one patient with CML-B was prepared with high-dose cytosine arabinoside, daunomycin, and cyclophosphamide.

RESULTS

Bone Marrow Infusion

Administration of bone marrow was well tolerated and renal function was unchanged in all patients. One patient showed a transient gross hemoglobinuria and a second patient manifested transient systolic hypertension.

Engraftment

Prior to transplantation, seven of the 10 patients manifested at least one factor known to be associated with poor engraftment (Table 28.3). Five of the 10 patients were felt to be sensitized to transplantation by prior *in vitro* testing. Five of the 10 patients were sex mismatches. The total nucleated marrow cell dose averaged 224×10^6 cells/kg (range 69 to 551 $\times 10^6$/kg). Overall, six out of the 10 patients demonstrated prolonged donor engraftment and one patient with AA had a transient graft with ultimate recovery of his own bone marrow. The remaining three patients died without evidence of engraftment. Patient 57, with CML-B, died 60 days post-bone marrow transplantation while still aplastic. At autopsy, isolated areas of recovering hemopoiesis were noted in his marrow, but karyotypes could not be obtained. Two patients with AA failed to engraft. One would be considered a high risk for failure, since he manifested *in vitro* sensitization, sex mismatch,

TABLE 28.3 Factors Influencing Engraftment in ABO-Incompatible Transplants

PATIENT NUMBER	TYPE OF MISMATCH[a]	SENSITIZED TO HISTOCOMPATIBILITY ANTIGENS BY IN VITRO TESTS	DONOR/ RECIPIENT SEX MATCH	NUCLEATED MARROW CELLS/KG ($\times 10^{-6}$)	INCOMPATIBLE ABO TITER	PREPARATIVE REGIMEN[b]	BONE MARROW ENGRAFTMENT
119	A	Yes	F/M	69	1:256	PA/TBI	No
82	A	No	F/M	220	1:32	CY	Yes
88	A	Yes	F/M	125	1:16	CY/TBI	Yes
103	A	Yes	M/M	175	1:16	CY/TBI ARA-C/ DNR/	Yes
57	A	No	M/M	171	1:32	CY	?
76	B	No	M/F	158	1:32	CY	No
117	B	No	M/M	206	1:32	CY	No
87	B	Yes	F/M	125	1:32	CY/TBI	Yes
78	B	No	M/M	551	1:4	CY/TBI	Yes
118	B	Yes	M/M	440	1:2	CY/TBI	Yes
		5/10 sensitized	5/10 sex mismatched	Average 224 (range 69–551)			6/10 engrafted

[a]Antigen on Donor erythrocytes, to which recipient is sensitized, absent from recipient.
[b]CY, Cyclophosphamide 50 mg/kg daily $\times$ 4; TBI, total-body irradiation 800 R (either a single exposure or 400 R $\times$ 2).

and received a low dose of nucleated marrow cells. The other patient was not sensitized to transplantation antigen as measured by *in vitro* testing, but was a sex mismatch.

Hemopoietic Recovery (Table 28.4)

Six patients with proven engraftment were eligible for analysis of graft recovery. Granulocyte recovery to greater than 0.8×10^9 circulating polymorphonuclear leukocytes/liter occurred 18 to 35 days post-transplantation. Platelet recovery to greater than 40×10^9/liter was not observed in all patients. Four patients manifested a fairly prompt platelet recovery to greater than 40×10^9/liter on

days 16, 17, 22, and 30 post-transplantation. One patient who died at 42 days post-transplantation had not achieved greater than 40×10^9 platelets/liter. This patient manifested several causes for platelet consumption, including graft-versus-host disease (GvHD) fever, and interstitial pneumonitis. One patient (118) showed an unusual isolated thrombocytopenia, with markedly reduced megakaryocytes in the bone marrow. This patient underwent a second bone marrow transplant from the same donor and achieved a count of greater than 40×10^9 platelets/liter 90 days after his first bone marrow transplant.

All patients manifested a delay in reticulocytosis. The earliest recovery to greater than 1% reticulo-

TABLE 28.4 Hematologic Recovery in Six ABO-Incompatible Transplants Who Demonstrated Engraftment Compared to 32 ABO-Compatible Patients

PATIENT NUMBER	TYPE OF MISMATCH[a]	DAYS POST-TRANSPLANT TO OVER $.8 \times 10^9$ GRANULOCYTES/ LITER	DAYS POST-TRANSPLANT TO OVER 40×10^9 PLATELETS/ LITER	DAYS POST-TRANSPLANT TO OVER 1% RETICULOCYTES
82	A	30	16	26
88	A	23	17	60
103	A	31	30	>39
87	B	35	>42	>42
78	B	18	22	31
118	B	22	>85	60
Average of 32 aplastic anemia patients	0	26	25	21

[a]Antigen on Donor erythrocytes, to which recipient is sensitized, absent from recipient.

TABLE 28.5 Isohemagglutinin Titers Post-ABO-Incompatible Transplantation[a]

WEEK POST-TRANSPLANT	CASE NO. 82	CASE NO. 88	CASE NO. 118
0	1:32	1:16	1:2
1	1:8	1:8	1:1
2		1:4	1:1
3	1:4[b]		
4	1:4[c]	1:8	1:1
5	1:4		
6	1:4	1:2	1:1
7	1:2		
8		1:1	1:2
9		1:1[b,c]	1:2[c]
10		1:1	1:1
11		1:1	1:2
12	0	0	1:2[b]
13	1:1		1:2
14	1:2	0	1:1

[a]Based only on the three patients surviving over 14 weeks.
[b]Week reticulocyte count was first over 1%.
[c]Week donor RBC types was first detected in peripheral blood of recipient.

cytes occurred 26 days post-transplantation. Two patients required 60 days to obtain a similar level. Three patients lived more than 14 weeks post-transplantation. Their isohemagglutinin titers are shown in Table 28.5. All three developed direct Coomb's test immediately following transplantation, which persisted for more than 14 weeks. Reticulocytosis to more than 1% occurred at 3 to 12 weeks' post-transplantation in this group. At approximately the same time, patients began to show detectable donor-type erythrocytes in the peripheral blood. At 14 weeks, all three patient were continuing Coomb's positive, which suggested that a low grade but clinically insignificant chronic hemolysis was taking place. One patient has survived longer than a year and has now become Coomb's negative.

Overall Outcome

Two of the six patients manifesting engraftment are alive 368 and 95 days post-transplantation. One of these patients has chronic debilitating GvHD. The other patient has mild thrombocytopenia. The other four patients died of complications related to their engraftment. All showed GvHD, severe immunodeficiency, and interstitial pneumonitis. One of these four had an unusual transverse myelitis and encephalitis without specific etiology.

DISCUSSION

Transplantation of bone marrow across major ABO erythrocyte cell mismatches has been successfully achieved in many centers (5–8). In one series, 17 out of 18 transplants with major ABO-erythrocyte incompatibility were successfully engrafted using an intensive plasma exchange technique (2). During one to three plasmapheresis sessions, 1 to 7 liters of incompatible antibody-containing plasma were exchanged prior to transplant. While successful, these exchange techniques have several practical limitations. Plasma exchange in children and small adults or under clinically unstable situations are not without their risk of either volume depletion or overload, infection, and bleeding. Similarly, the exchange used in most of the reported techniques involves transfusion of large volumes of ABO-incompatible erythrocytes in order to successfully lower the isohemagglutinin titer. Despite the intense attempts to abolish preformed isohemagglutinins in many situations, low titers have persisted or have returned rapidly, post-transplantation. For these reasons, an antigen reduction approach was adopted to see if an acceptable engraftment could be achieved. In this series, six out of 10 patients were successfully engrafted. Although the series is small, this experience is not dissimilar to that which would be expected in a similar population with high risk for graft failure who did not have major erythrocyte incompatibility. Two patients with *in vitro* evidence of sensitization to transplantation antigen, sex mismatch, each of whom received a total dose of 125 × 10⁶ nucleated marrow cells/kg, were successfully engrafted across isohemagglutinin titers of 1:32 and 1:16, respectively. It is to be noted, however, that only one patient in this series had an isohemagglutinin titer over 1:32 and that patient failed to engraft. Six patients were successfully engrafted with isohemagglutinin titers as high as 1:32.

The kinetics of granulocyte and platelet recovery in this patient population were generally similar to that observed in our larger group of transplant patients with AA not having a major erythrocyte mismatch. All patients, however, demonstrated a chronic, low-grade hemolytic anemia as manifested by a persistently positive direct Coomb's test and a delay in recovery of reticulocytes beyond that seen in other transplanted patients. This hemolysis was clinically insignificant except for creating a slightly greater transfusion requirement. Patients manifesting sustained engraftment and surviving over 14 weeks required no further transfusion support.

Complications following major ABO incompatible transplantation included GvHD, interstitial pneumonitis, and hemorrhagic cystitis, which are those that have been generally found to follow allogeneic transplantation in ABO compatible donor/recipient pairs. No unusual toxicity that could be attributed to the ABO mismatch was encountered.

SUMMARY

Ten patients with either AA or leukemia received bone marrow transplants from donors who possessed major erythrocyte antigens that were absent from the recipients and to which the recipients had detectable antibody levels. The methodology used included the removal of as much donor erythrocyte antigen as possible from the transplanted marrow specimens prior to infusion by the use of a Haemonetics Model 30® cell separator. No significant untoward results, as a result of the incompatible erythrocytes, were seen in any of the 10 patients. Six of the 10 patients achieved prolonged engraftment of donor cells, and one had a transient engraftment. Three had no evidence of engraftment. The recovery of granulocytes and platelet numbers in the peripheral blood of these patients was similar to that seen in other transplant patients. Reticulocyte recovery was delayed in several of the patients, and a positive direct Coomb's test persisted for over 14 weeks in all patients who survived that period of time. The complications of transplantation seen in this group of ABO-incompatible patients included graft-versus-host disease, interstitial pneumonitis, hemorrhagic cystitis, and severe immunodeficiency, which were generally similar in incidence and severity. Thus, it appears that reducing the incompatible erythrocytes from a marrow suspension prior to infusion by the use of a procedure such as that described here is a safe, economical, and feasible approach to major ABO-mismatched bone marrow transplantation.

ACKNOWLEDGMENTS

The authors wish to thank the staffs of the Hemopheresis Laboratory of The Johns Hopkins Oncology Center and the Johns Hopkins Blood Bank for their special expertise and assistance with these studies.

This work was supported by grants CA-06973 and CA-15396 from the National Cancer Institute, NIH, DHEW.

REFERENCES

1. Braine, H. G. and Wright, S. K. Cell fractionation characteristics of the Haemonetics 100-cc Pediatric Bowl: Preliminary experience. Haemonetics Research Institute Advanced Component Service. Boston, June 1978.
2. Buckner, C. D., Clift, R. A., Sanders, J. E., Williams, B., Gray, M., Storb, R., and Thomas, E. D. ABO incompatible marrow transplants. *Transplantation* (in press) 1978.
3. Crawford, M. N., Gottman, F. E., and Gottman, C. A. Microplate system for routine use in blood bank laboratories. *Transfusion, 10*:258, 1970.
4. Fliedner, T. M. personal communication.
5. Gale, R. P., Feig, S., Ho, W., Falk, P., Rippee, C., and Sparkes, R., for The U.C.L.A. Bone Marrow Transplant Team. ABO blood group system and bone marrow transplantation. *Blood, 50*:185, 1977.
6. Graw, R. G., Yankee, R. A., Leventhal, B. G., Rogentine, G. N., Herzig, G. P., Halterman, R. H., Merritt, C. B., Carolla, R. L., Alvegard, T. A., Bull, J. M., McGinnis, M. H., Krueger, G. R. D., Gullion, D. S., Lippman, M. F., Bleyer, W. A., Berard, C. W., Whang-Peng, J., Trapani, R. J., Terasaki, P. I., Steinberg, A. S., Gralnick, H. R., and Henderson, E. S. Bone marrow transplantation in acute leukemia employing cyclophosphamide. *Exp. Hematol., 22*:118, 1972.
7. Santos, G. W., Sensenbrenner, L. L., Anderson, P. N., Burke, P. J., Klein, D. L., Slavin, R. E., Schacter, B., and Borgaonkar, D. S. HL-A-identical marrow transplants in aplastic anemia, acute leukemia and lymphosarcoma employing cyclophosphamide. *Trans. Proc., VIII*:607, 1976.
8. Storb, R., Thomas, E. D., Weiden, P. L., Buckner, C. D., Clift, R. A., Fefer, A., Fernando, L. P., Giblett, E. R., Goodell, B. W., Johnson, F. L., Lerner, K. G., Neiman, P. E., and Sanders, J. E. Aplastic anemia treated by allogeneic bone marrow transplantation: A report on 49 new cases from Seattle. *Blood, 48*:817, 1976.
9. Weiner, R. S., Richman, C. M., and Yankee, R. A. Semicontinuous flow centrifugation for the pheresis of immunocompetent cells and stem cells. *Blood, 49*:391, 1977.

29

The Use of Autologous Bone Marrow Transplantation after High-Dose Cytoreductive Therapy in Various Types of Malignancies in Man

Karel Dicke, Gary Spitzer,
Dharmvir Verma, Axel Zander,
Sharon Thomson,
Kenneth McCredie, L. J. Peters,
and L. Vellekoop

The availability of methods of irradiation and chemical agents to destroy tumor cells has stimulated research in the mechanism and in the kinetics of destruction by such modalities. Extensive experimental data that tumor cell kill follows first-order kinetics—with a given dose of cytotoxic therapy destroying a constant fraction of the tumor cell load and not an absolute number of cells—have been provided by Skipper (31). He demonstrated in the leukemia 1210 model in mice that a reduction of the number of tumor cells from 10^6 to 10^5 (1-log kill) increased the life-span by 2 days, whereas a 99.999% reduction (5-log kill) could cause a 10-day prolongation of survival. In this model, cure of the recipient infused with leukemic cells was only possible when the initial tumor cell burden was small (19,30).

There are a number of clinical studies suggesting that Skipper's data are relevant in man: the higher the dose, the higher the cell kill and the longer the survival. Doubling the dose of theotepa resulted in a fivefold increased response rate in lymphoma (16). Pinkle et al. (25) observed, in childhood leukemia, a delay in onset of and a decrease in central nervous system relapse as well as a prolongation of hematologic remission after high-dose combination chemotherapy as compared with the group of patients treated with only 50% of the dose. Additional evidence suggesting a greater effect after high-dose chemotherapy is the increased response rate in oat cell carcinoma treated with CCNU, cytoxan, and methotrexate in protective environments (4). In these studies, the response rate was compared with results in patients treated with 50% of the dosages of the same drugs (4). Cortes (5) demonstrated an increased response to high dosages of Adriamycin in osteosarcoma. In other studies, it became evident that the response of squamous carcinoma in head and neck to high-dose methotrexate with citrovorum factor rescue was superior to the response to normal-dose methotrexate (20).

The data presented clearly favor the use of maximum drug dosages. Increasing risks of aggravating side effects, which may accompany high-dose chemotherapy, frequently eliminates this advantageous approach. Chemotherapeutic agents, as well as total-body irradiation (TBI), are predominantly toxic to rapidly dividing cells in the bone marrow and in the gastrointestinal tract. In addition, liver, renal, pulmonary, and carditoxicity is often seen after the administration of chemotherapeutic agents. It has been demonstrated that the major toxicity of methotrexate can be reduced or eliminated without affecting its cytotoxicity against the tumor by the addition of citrovorum factor or L-asparaginase (3,17). Thus far, the possibility of using substances that overcome the biochemical mechanism of damage to normal tissue is limited to methotrexate. But there are methods other than biochemical ones to overcome the toxicity of chemotherapeutic agents, such as transplantation of bone marrow cells, provided that the major toxic effect of the drug is on the hemopoietic system. The myelosuppressive effect of TBI in dosages that suppress hemopoiesis can be overcome by the intravenous injection of marrow cells. The possibility of effective rescue by bone marrow cell infusion from identical twins after high-dose chemotherapy (cytoxan 120 mg/kg twice a week), combined with supralethal TBI (1000 rad) for effective control of leukemia, has been reported by Thomas et al. (35). The combination chemotherapy regimen BACT (BCNU, Ara-C, cytoxan, 6-thioguanine) used by Graw and his group to treat leukemia is also severely myelosuppressive and can be overcome by marrow grafts (38). Also, in tumors other than leukemia, such as lymphoma and ovarian cancer, the hematotoxic side effects of high-dose chemotherapy could be reduced by bone marrow transplantation (2,37).

Several years ago, techniques for storage of bone marrow cells were developed using mice and monkeys (27–29). The availability of such methods allows the use of the patient's own bone marrow collected before high-dose cytoreductive therapy. The advantage of the use of autologous bone marrow instead of allogeneic bone marrow is the absence of immunologic reactions complicating the allogeneic post-transplantation period. We developed, in the M.D. Anderson, an autologous bone marrow transplantation program in acute leukemia and several types of solid tumors. In adult acute leukemia, marrow has been collected and stored in remission. At the time of relapse, the patient is treated with a combination of high-dose chemotherapy and TBI prior to bone marrow transplantation. In several patients in remission, before the marrow cells were stored, potential residual leukemic cells were separated from normal hemopoietic cells by running the marrow cells over albumin density gradients (13) In solid tumor patients (oat cell carcinoma of the lung and embryonal carcinoma of the testis and carcinoma of the breast), the bone marrow was stored unfractionated and used after administration of high-dose combination chemotherapy at the time the patients were failing second line or phase I chemotherapy.

PATIENT GROUPS

Leukemia Patients

Eleven patients had AML, eight ALL, and two AUL. The median age was 28, with a range of 18 to 48 years. All except five bone marrows were aspirated during first remission. Patients with ALL and AUL entered the transplantation program in second, third, and fourth relapse; patients with AML in first and second relapse. All patients had received extensive induction, consolidation, maintenance, and late intensification therapy (8). At the time of relapse before bone marrow transplantation, 15 of the patients had received second line chemotherapy, including two who had received investigational chemotherapy, as well, which had failed. Six patients entered the transplantation program without receiving second line chemotherapy. The median time from diagnosis to transplantation was 23 months in AML and 33 months in ALL and AUL. In three patients, the effect of the bone marrow transplantation program could not be evaluated because of early death within 5 days after transplant due to preexisting conditions such a fungemia, liver failure, and *Pseudomonas* septicemia (8). Therefore, 18 patients were evaluable. Eleven of the 21 patients, which was 9 of the 18 evaluable cases, received fractionated marrow. The decision of marrow fractionation before storage depended on the workload of the laboratory.

Solid Tumor Patients

Eight patients with oat cell carcinoma of the lung were entered into the program. Seven of eight patients failed first line chemotherapy (cytoxan, ifosphamide, Adriamycin) (with or without VP-16). One patient entered the program without previous chemotherapy. All patients were ambulatory. Four out of eight patients had extensive disease. In six patients with embryonal cell carcinoma of the testis first line chemotherapy (velban, bleomycin, cis-platinum) had failed and they had extensive bulky disease at the time of bone marrow transplantation.

The two patients with breast cancer had extensive disease and the FAC (5-fluoruracil, Adriamycin, cytoxan) and the CMF (cytoxan, methotrexate, 5-fluoruracil) programs had failed in these patients.

MATERIALS AND METHODS

Preparation of Bone Marrow Cell Suspensions

Cells were collected by multiple bone marrow aspirations under general anesthesia from the posterior iliac crest, and 1,500 to 2,500 ml of bone marrow cells were harvested and suspended in Hank's balanced salt solution with preservative-free heparin. After collection, the nucleated cells were harvested from the buffy coat and prepared for either fractionation over the albumin gradient or for immediate storage according to the method described previously (14).

The Density Gradient Technique For Separation of Hemopoietic Stem Cells From Leukemic Cells

Fractionation was done on the basis of differences in density using bovine serum albumin (BSA, Fracation V, Sigma Chemical Co., St. Louis, Mo.) gradients. The gradient consisted of layers of BSA of different solutions. The stock solution was 35% w/v, having a specific density of 1.1004. This stock solution was prepared in tris-(hydromethyl) amino methane buffer (pH 7.2). (The pH of the 35% stock solution was 5.1.) The osmolarity of the 35% stock solution was 350 mOsm. This osmolarity was the sum of the residual osmolarity in the albumin fraction V, measured by dissolving 1.75 g in 8.3 cc distilled water and multiplying this value by two, and the osmolarity of the tris buffer. The osmolarity of the tris buffer was adjusted on the basis of the residual osmolarity of the albumin. Careful adjustment of the osmolarity is critical, since it determines the amount of water in the cell and thus influences the density of the cell (6).

The BSA solutions used in the gradient were obtained by dilution of the stock solution with a sodium chloride-sodium phosphate buffer (PBS; 0.154 M NaCl; 0.01 M Na phosphate buffer; pH 7.2). The osmolarity of the PBS buffer was 300 mOsm and was maintained. The BSA concentrations (5 ml/solution) were layered in a glass centrifugation tube (diameter 3 cm) using a pipette, starting with 25% at the bottom, then layering successive concentrations of 23 and 21% BSA. The concentrations of albumin determined the specific density. Relatively small variations in concentration caused significant changes in density, which disturbed the reproducibility of the cell distributions. Concentrations were checked routinely by refractometry. Refractometry was preferred to pyknometry, since the former method was simple and faster and needed only 1 drop of albumin/measurement. The values of the different indices of the albumin solutions have been listed in Table 29.1. After layering the 21%, the cells suspended in 17% albumin solution were pipetted on top of the 21%. Following this step, the tube was centrifuged for 30 min at 10°C in an International model PR-2 centrifuge. The cell fraction between the 17 and 21% BSA layers was called fraction 1 + 2—a combination of the fractions 1 and 2, respectively, between 17, 19% BSA and 19, 21% BSA of the "old" gradient (7,9)—between 21% and 23% fraction 3, between 23% and 25% fraction 4 and at the bottom in the 25% BSA solution fraction 5. Each fraction was collected using a pipette and transferred into centrifuge tubes, diluted with Hank's balanced salt solution (1 volume of the fraction to 1 volume Hank's) and centrifuged. The sedimented cells were resuspended in Hank's

TABLE 29.1 Specific Density[a] and Refractive Index[b] of the Various Albumin Concentrations of the Gradient

ALBUMIN CONCENTRATION[c] (W/V)	SPECIFIC DENSITY (G/CM³)	REFRACTIVE INDEX
35	1.1004 ± 0.001[d]	1.4003
25	1.0734 ± 0.0012	1.3815
23	1.0682 ± 0.001	1.3778
21	1.0637 ± 0.0008	1.3740
19	1.0575 ± 0.001	1.3703
17	1.0525 ± 0.0012	1.3665

[a]Estimated by pyknometry at 21°C.
[b]Measured with Abbe refractometer (Zeiss) at 21°C.
[c]35% w/v (fraction V powder Sigma Chemical Co.) in tris (hydroxy methyl)-amino urethane buffer; the other concentrations are prepared from the 35% stock solution by diluting with NaCl phosphate buffer.
[d]2 S value (95%) limit, $N = 3$.

solution. The number of cells collected from each fraction was counted in Turk's solution in a hemocytometer. The eosin exclusion test for cell viability could not be applied to the cells after exposure to albumin, because dead cells do not take up eosin in such suspensions. Following this, the suspension was diluted with Hank's solution to obtain the required cell concentration for culture, storage, or transplantation.

Bone Marrow Storage and Thawing

Hemopoietic cells were stored according to the technique described by Schaefer et al. (29). In short, bone marrow cells were suspended in Hank's balanced salt solution (305 mOsm). After addition of 10% DMSO and 20% calf serum to the cell suspension, the cells were transferred to 2- or 5-cc polypropylene ampoules and cooled at 1°C/min to −40°C, using a cryoson automatic controlled freezer. After rapid cooling from −40 to −80°C, the cells were stored in liquid nitrogen at −192 °C. The cell concentrations during freezing were kept between 20 to 200 × 10⁶ cells/ml. Cell viability was tested immediately after storage and compared with the unfrozen cells using the *in vitro* colony-forming assay for myeloid progenitor cells (CFU-c assay) (see below).

At the time of transplantation, the ampoules were thawed rapidly in a 50°C waterbath. Immediately after thawing, the cells were slowly diluted with Hank's balanced salt solution. Quantities of Hank's balanced salt solution equivalent to 1/50, 1/25, 1/12, etc., of the ampoule volume were added drop by drop until the original volume was diluted tenfold. Between each dilution step, the suspension was carefully mixed for 3 to 5 min. Routinely, DNase (deoxyribonuclease, Sigma 4 mg, 7600 U)/500 cc collected marrow was added to avoid clump formation by extracellular DNA (9). After dilution, the cells were centrifuged, resuspended in Hank's, and filtered through G_2 glass filters (Jena glass, pore size, 40 to 80 μu). Viability of the cells was again tested by the CFU-c assay.

In Vitro Assay for Hemopoietic Stem Cells (CFU-c Assay)

Two hundred thousand cells were suspended in 1 ml of agar medium (0.3% agar, α-MEM, and 15% fetal calf serum) pipetted onto previously prepared human peripheral blood leukocyte (1×10^6 ml) underlayers in agar medium (0.5% agar, α-MEM, and 15% fetal calf serum). In our recent studies, the leukocytes were replaced by placenta-conditioned medium (PCM) to stimulate colony growth as described by Metcalf (22). After gelation, cultures were incubated for 7 days and colonies over 40 to 50 cells were evaluated visually, using an inverted or dissecting microscope (24).

In Vitro Assay for Leukemic Cells (PHA Assay)

Basically the technique consists of two phases: an initial liquid phase of 15 hr at 37°C and a semisolid phase of 7 days' incubation at 37°C (12,33). In the liquid phase, 2×10^6 cells/ml medium (α-MEM + 15% calf serum were cultured in Falcon plastic tissue culture tubes) to which 2 to 4 μg PHA (Welcome)/ml was added. After 15 hr of incubation, the cells were washed twice using Hank's balanced salt solution (305 mOsm) and resuspended in agar medium (final concentration in agar, 0.25% medium + 15% calf serum). After resuspending in agar medium, 1×10^5 cells/ml in each dish were pipetted into Falcon plastic Petri dishes containing 1 ml agar medium to which PCM ranging from 0.1 to 0.25 ml/culture dish was added simultaneously. The cells were plated in Petri dishes containing agar underlayers without PCM. After 7 days of incubation in 7.5% CO_2 gas-controlled humidified incubators at 37°C, the colonies were visible microscopically. These colonies were counted, using an inverted or dissecting microscope. Aggregates containing 40 to 50 cells or more were considered colonies, aggregates containing fewer than 40 to 50 cells clusters.

Conditioning of the Leukemic Patients for Bone Marrow Transplantation

Prior to bone marrow transplantation, the patients were treated with intravenous piperazinedione (NSC 135785), 25 mg/m² administered on days 6 and 5 before TBI. In the first 17 patients, 850 to 950 rad calculated at the mid-abdominal plane were delivered at a dose rate of approximately 12 to 14 rad/min using a 25 MEV linear accelerator. In the last four patients, the dose was decreased to 750 rad. The patients were treated horizontally with one-quarter of the total dose being delivered from the anterior, posterior, right, and left lateral aspects to minimize dose heterogeneity. Within 24 hr after TBI, nucleated cells of fractions 3, 4, and 5 were infused.

Conditioning of the Solid Tumor Patients for Bone Marrow Transplantation

For oat cell carcinoma and embryonal carcinoma of the testis, combinations of cytoxan, BCNU, and VP-16, as listed in Table 29.2, were used. Seventy-two hours after the last dose of cytoxan and VP-16, the marrow cells were infused. For breast cancer, mitomycin (40 mg/m²) was used, and 72 hr later bone marrow cells were infused.

RESULTS

In the leukemia patient population, bone marrow collection occurred at time of remission and in the solid tumor patients, the cells were harvested before high-dose chemotherapy was started. The initial number of marrow cells to be infused to guarantee hemopoietic repopulation in the recipient has been estimated to be 1 to 2×10^8 cells/kg body weight. The total cell loss due to various steps of preparation—buffy coat preparation, gradient centrifugation, storage, thawing and stepwise dilution—is approximately 51% (Table 29.3) so that 2 to 4×10^8 cells/kg body weight should be harvested. In addition, dilution of the bone marrow cell suspension with peripheral blood cells can be up to 50%, so that for a 70-kg recipient, 3 to 4×10^{10} cells should be

TABLE 29.2 The CBV Combination Chemotherapy Program: Drug Dosages[a] in Relation to Normal-Dose Chemotherapy

DRUG	NORMAL DOSE	HIGH-DOSE IN CONJUNCTION WITH TRANSPLANTATION[b]	INCREASE HIGH DOSE/ NORMAL DOSE
Cyclophosphamide	1.0 g/m²	6 g/m²	× 6
BCNU	240 mg/m²	300 mg/m²	1.25
VP-16	250 mg/m²	600 mg	× 2.5

[a]Cyclophosphamide and VP-16 administered over 4 days, BCNU on day 1.
[b]Bone marrow cells infused 2 to 3 days after the last dose of cyclophosphamide and VP-16.

TABLE 29.3 Median Loss of Marrow Cells from Time of Collection to Transplantation in 18 Patients

| | CELL LOSS AS PERCENTAGE OF INITIAL COLLECTION | |
PROCEDURE	Mean	Range
Buffy coat	17	0–35
Fractionation	1	1–4
Storage	32	25–51
Median total loss	51	

collected. In patients who had previously undergone chemotherapy, this number of cells can hardly be obtained in one session, so if possible, the bone marrow was collected twice. The cell yield of the second bone marrow aspiration setting is not different from that of the first, as has been documented in Table 29.4. Also, the CFU-c/10^5 cells plated of the second aspirate does not differ from that of the first. There is, however, a great variation of cell recovery (range 38 to 192%) and of CFU-c recovery (range 41 to 950%) in the second marrow collection expressed as per cent of recovery of the first marrow collection.

Separation between leukemic cells and normal cells using the discontinuous gradient technique was demonstrated in three different experimental settings. In the first experimental setting, the patients' separated bone marrow cells consisted predominantly of normal hemopoietic cells and a small but detectable precentage of leukemic cells. As a monitor assay for normal hemopoietic cells, the CFU-c assay was used and for the leukemic cell population, cytogenetics and electromicroscopy. The results are

TABLE 29.4 Comparison of Cell Numbers /ml Aspirate and CFU-c/10^5 Plated Cells Between Sequential Aspiration in the Same Patient

PATIENT	CELL NUMBERS OF SECOND ASPIRATE (% CELLS IN ASPIRATE I)	CFU-C/10^5 CELLS OF SECOND ASPIRATE (% CFU-C IN ASPIRATE I)	TIME INTERVAL BETWEEN ASPIRATE (DAYS)
RW	192	—	30
VR	150	100	41
BS	130	38	25
CF	109.8	90	21
GA	126	950	33
SA	107	311	7
RW	92	41	37
GA	58.5	165	32
MB	57	140	39
SB	54	72.6	14
NL	46	44	13
GO	38	—	12
Median 100		Median 95	

shown in Table 29.5. Patient DH had an aneuploid leukemic cell clone, and therefore, a cytogenetic analysis was possible. It can be seen in Table 29.5 that in fractions 1 + 2 and 3, 3, respectively, four of the 20 analyzed metaphases revealed the abnormal karyotype, whereas in fractions 4 and 5, this karyotype could not be found. Electronmicroscopic (EM) analysis was positive for leukemic cells in fractions 1 + 2 and 3 and negative in 4 and 5. The nuclear pocket in the cell is characteristic for leukemic cells with abnormal karyotypes (1). The CFU-c population was predominantly in fractions 4 and 5: 64%, so that the majority of this population was deprived of detectable numbers of leukemic cells suggesting separation between the two cell populations. The second experimental setting in which evidence could be provided of separation between normal cells and leukemic cells, was an analysis of the so-called "remission marrow" (with a morphologically undetectable number of leukemic cells) using the PHA assay as a parameter for the leukemic cells. The results were depicted in Table 29.6. An analysis of the fractions revealed a positive PHA assay in fraction 1 + 2, whereas in the other fractions, cells that could give rise to colonies after PHA stimulation could not be demonstrated. Morphologic analysis of the colonies revealed the presence of blast cells, which were morphologically identical to the leukemic cell population before chemotherapy. In fraction 1 + 2, 2% of the normal CFU-c population was present; the majority of CFU-c were found in fractions 3, 4, and 5. The third experimental setting of separation of leukemic cells and normal cells was obtained by comparing, in one patient, the leukemic cell profiles in the gradient of the marrow at the time of relapse and the CFU-c profile of the marrow after achieving remission. Figure 29.1 shows these profiles, and it can be noted that the peak activity of the two populations were in different fractions.

Storage of cells using the technique in Materials and Methods, resulted in no significant loss of CFU-c activity after freezing, whereas the number of bone marrow cells recovered was reduced up to 50% (Table 29.3). Technical simplifications of the step-

TABLE 29.5 Separation Between Normal Stem Cells and Leukemic Cells in Remission Marrow

| | CFU-C (% OF TOTAL) | LEUKEMIC CELL POPULATION | |
		Cytogenetics	EM
Fr 1 + 2	1	4/20	+
Fr 3	35	10/20	+
Fr 4	48	0/20	–
Fr 5	16	0/20	–
Unfractionated	—	2/20	+

TABLE 29.6 Distribution of Leukemic Cell Colonies Using the PHA Assay and of Colonies from Normal Progenitor Cells Using the Robinson Assay in the Gradient of Marrow Cells of a Patient in Remission

| | CFU-C IN ROBINSON ASSAY | | PHA ASSAY | |
	Per 10^5 Plated	Recovery[a] (%)	Leukemic Cell Colonies (10^5 Cells)[b]	Cytogenetic Analysis[c]
Unfractionated	20	100	0	−
Fraction 1 + 2	20	2	10	+
Fraction 3	58	35	0	+
Fraction 4	40	50	0	−
Fraction 5	5	13	0	−

[a]As percentage of total number of CFR-c put on gradient.
[b]Number of colonies in dish containing no leukocyte feeder layer or placenta-conditioned medium.
[c]Twenty metaphases analyzed: +, abnormal karyotype present; −, abnormal karyotype absent.

wise dilution procedure after thawing reduced the loss to 32%, so that the median cell loss from collection of the marrow until transfusion into the patient is 51% (Table 29.1). The simplification consisted of thawing six ampoules in one 250-ml sterile Falcon centrifuge tube instead of thawing 1 ampoule in one 50-ml Falcon centrifuge container, so that much less washing was necessary.

In the transplanted leukemic patient population, the median time elapsing from onset of remission and collection of marrow was 12 months, whereas the median time interval between storage and subsequent relapse of leukemia was 9 months. The median time between the first diagnosis of disease and transplantation was 28 months (Table 29.7). The median storage time of the cells was 15 months. The results obtained in these patients after transplantation have been documented in Table 29.8. It can be noted that 50% of the patients achieved complete remission.

which means a complete hemopoietic recovery (granulocyte count over 1,000/mm^3 and platelet count over 80,000 to 100,000/mm^3 peripheral blood) without evidence of leukemia. The seven of the 16 patients who had evidence of take but who never achieved complete remission died of infection and an acute respiratory distress syndrome. In one patient, leukemia recurred before a complete hematologic recovery took place. Complete remission in one patient lasted for over 13 months, which is a consid-

TABLE 29.7 Timing of Bone Marrow Aspirate

Interval between onset of remission and storage (months)	2–15 (Median: 12)
Interval between storage and subsequent relapse (months)	0.5–30 (Median: 9)
Interval between diagnosis and bone marrow transplantation (months)	17–74 (Median: 28)

TABLE 29.8 Summary of Results of Autologous Bone Marrow Transplantation in Relapsed Adult Acute Leukemia Patients[a]

Number of patients grafted	21
Number of evaluable patients	18
Number of takes	16/18
Number of CR	9/18
Number of CR patients alive	5/9
Survival time in days of patients in CR after transplantation	66, 75, 78, 210 94+, 115+, 123+, 191+, 391+
Recurrence of leukemia	4/18
Time in days of recurrence after transplant	30, 60, 60, 83

[a]Updated, August 1978.

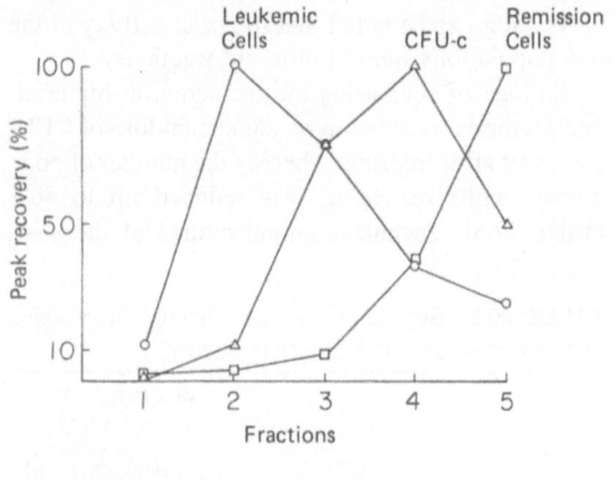

FIGURE 29.1. Density profiles of leukemic cells before treatment and of the CFU-c and bone marrow cells after remission induction. Note a clear-cut difference in the gradient of the peak activity of the leukemic cell population (in fraction 1+2) and the CFU-c population (fraction 4). Leukemic cells in relapse (o——o); CFU-c in remission (Δ——Δ); Hemopoietic cells in remission (□——□).

TABLE 29.9 High-Dose Chemotherapy with Autologous Marrow Rescue in Solid Tumor Patients

TUMOR	MEDIAN AGE	NUMBER OF PATIENTS	CHEMOTHERAPY	RESPONSES
Oat cell carcinoma	58	8	CBV	7/8
Embryonal cell carcinoma of testis	28	6	CBV	3/6
Breast carcinoma	62[a], 48[a]	2	Mitomycin	2/2

[a] Age of individual patient.

erable time, since without bone marrow transplantation therapy the chance of achieving complete remission in that stage of the disease is less than 10%.

Table 29.9 lists the solid tumor patients who were treated with high-dose chemotherapy and autologous marrow rescue. Three types of tumors were tested in this program. The largest series (eight patients) was of oat cell carcinoma of the lung. In seven patients, partial response (six) or complete remission (one) were observed; one patient had no response. Six patients with disseminated embryonal cell carcinoma of the testis resistant to normal-dose chemotherapy were treated with high-dose cytoxan, VP-16, with or without BCNU, followed by autologous marrow infusion. It is still too early to evaluate two of six patients; one patient died shortly after transplantation of an already existing pneumonia so that the results of only three patients can be evaluated here. Those three patients were treated twice with the CBV program, the second course within 2 months of the first course. In two patients, the dose of cytoxan was reduced from 6 to 4.5 g/m² a treatment, and VP-16 from 500 to 375 mg/m²; in one patient cytoxan was reduced from 6 to 3 g/m². The first two patients received, after each course, marrow cells, whereas in the last patient marrow was infused only once. It is remarkable that, in the three patients, partial responses were achieved after the first course and that complete remission was obtained after the second course, which seemed to show that two "reduced" courses are more active than one "full" course.

In breast cancer, a single agent was used. A remarkable regression was observed with 40 mg/m² mitomycin in the two breast cancer patients. This treatment was preceded with mitomycin 20 mg/m² without marrow rescue. One patient is still alive 1 year after mitomycin and marrow rescue with stable disease, although at the time of transplantation the disease was disseminated and progressive.

Table 29.10 shows the hemopoietic recovery of the leukemia patients and of the oat cell patients. In the leukemia patients, recovery started within 3 weeks after transplantation versus 2 weeks in the solid tumor patients. One reason for the difference between the two patient populations may be the presence of residual stem cells in the solid tumor

patients after high-dose chemotherapy, which may play a role in hemopoietic recovery.

DISCUSSION

Autologous marrow rescue after high-dose chemotherapy with or without radiotherapy has been used by various groups (2,18,26). In leukemia, McGovern reported that as early as 1959 he achieved hematologic recovery in three relapsed leukemia patients treated with sublethal TBI and autologous marrow infusion (21). In our patient groups, supralethal dosages of TBI were used in combination with high-dose chemotherapy. Our study was not meant to be used merely to report cases, but to evaluate the role of autologous bone marrow transplantation as a part of anti-leukemic therapy for AML in first and second relapse and ALL and AUL in second, third, or even fourth relapse. At the time of analysis, marrow cells from over 170 leukemia patients have been stored for a period of 3 years. In leukemia, autologous transplantation has three critical elements: survival of stem cells after prolonged periods of storage; elimination of residual leukemic cells from the cell suspension to be grafted; and the cytoreductive program used to treat relapsed leukemia. In 1972, Schaefer et

TABLE 29.10 Hemopoietic Recovery After ABMTR After Chemo-Irradiation in Leukemia and After High-Dose Combination Chemotherapy in Oat Cell Carcinoma

	IN LEUKEMIA[b] (DAYS AFTER TRANSPLANT)		IN OAT CELL CARCINOMA[c] (DAYS AFTER TRANSPLANT)	
	m	range	m	range
Granulocytes[a]				
> 500	22	15–33	14	8–26
>1,000	27	17–42	16	10–27
>1,500	34	21–45	20	13–29
Platelets[a]				
> 20,000	22	20–42	17	7–44
> 50,000	43	24–49	21	9–47
>100,000	55	30–60	23	14–49

[a] Per cubic millimeter peripheral blood.
[b] N = 9.
[c] N = 8.

227

al. published detailed data concerning a storage technique to obtain good survival of stem cells. Extensive transplantation studies in mice and monkeys were done, with no loss of survival of stem cells after freezing (27–29). A correlation was found between transplantation capability and the concentration of CFU-c of the hemopoietic cell suspension in mice and monkeys (10), and since we observed no significant decrease in CFU-c numbers after freezing of the marrow cells of leukemic patients as well as of solid tumor patients, it is not likely that the transplantation potential has been affected. A recent analysis of culture data showed that a significant correlation exists between the number of CFU-c infused and the time of granulocyte recovery. The pattern of hemopoietic recovery with granulocyte levels above 500 and platelet levels above 20,000 by 3 weeks is similar to that reported in the syngeneic bone marrow transplantation study by the Seattle group (15). The transplantation potential of bone marrow stored for up to 30 months in liquid nitrogen was maintained and allowed for full hemopoietic recovery after piperazinedione and TBI.

The cell loss from the collection of marrow until transplantation into the patient is considerable (51%). This can be broken down into a 15% loss due to separation of nucleated cells from erythrocytes and a 35% cell loss due to thawing and stepwise dilution. Simplification of the procedure after freezing may decrease the cell loss, although it cannot be ruled out that freezing damages some cells.

Evidence of separation between leukemic cells and normal cells using the discontinuous albumin gradient technique has been obtained in three experimental settings. In one of the settings, marrow cells were used from patients with low but detectable leukemic cell infiltrates. In these studies, it was evident that the leukemic cell population was predominantly less dense than the normal CFU-c population. These results confirmed the data of Moore et al. (23). In nine patients, fractionated cells were used for transplantation. At this time, no conclusion can be drawn on what effect marrow fractionation has on the relapse rate after transplantation. But there was no significant difference, in terms of hemopoietic recovery, in the patients who received fractionated marrow cells and in those who received unfractionated marrow cells. It is impossible to assess the degree of separation between leukemic cells and normal cells in remission in each individual patient. In one patient described in Results, the PHA assay for leukemic cells was positive in fraction 1 + 2, whereas in the other fractions no PHA-induced leukemic cell colonies could be grown. The PHA assay has been proven to be a valuable assay in CML in the benign phase to predict progression of the disease. In this system, colony formation is specific, which means that only leukemic cells can grow out into colonies; agglutination, however, may hamper the exact quantitation of colony formation in certain patients. Since there are differences in the densities of leukemic cells, it seems appropriate to study the leukemic cells at a time of first presentation and compare the pattern of distribution with that of CFU-c in remission to see if separation is possible. Separation between leukemic cells and normal cells was thus assessed in several patients, and these data confirm the separation pattern as observed in the other experimental settings: the leukemic cell population, in general, is less dense than the CFU-c population. In ALL, marked variations in the density of the leukemic cell population were noted: In four out of eight patients tested, no clear-cut separation could be obtained between leukemic cells and the normal CFU-c. But by changing the osmolarity of the gradient—which is one of the physical parameters influencing the distribution of cells in the gradient—efficient separation was obtained between the two cell populations. This is, for us, a reason to compare in each individual patient the distribution of the leukemic cells before remission induction in the gradient with that of the CFU-c population of remission marrow before fractionation of the actual bone marrow cells used for transplantation so that the most appropriate gradient can be used.

Total-body irradiation or chemotherapy alone used as a cytoreductive regimen in bone marrow transplantation studies has led to limited remissions (36). The combination of chemotherapy and irradiation yielded a higher percentage of long-term surviving patients (36). The actuarial relapse rate of patients treated with the highly toxic SCARI regimen (6-thioguanin, cytoxan, Ara-C, daunorubicin, and irradiation) is lower than the actuarial relapse rate with cytoxan and TBI (34). The percentage of patients surviving more than 2 years, however, is not significantly different (34). For our conditioning regimen, a combination of piperazinedione—a fermentation product of *Streptomyces* with alkylating activity—and TBI was chosen. In all evaluable cases, the size of the leukemia cell population was reduced to morphologicahly undetectable levels, but the period of time after transplantation is too short to draw definite conclusions on its efficiency compared to other regimens. Clinical as well as experimental studies have shown that piperazinedione has good anti-leukemic activity, but its general use in the clinic was abandoned because of its excessive cytotoxicity in the bone marrow. The non-hemopoietic toxicity of this drug is minimal, and restricted to mild nausea and vomiting. The interval of 5 days between piperazinedione and TBI was necessary, since a close proximity of piperazinedione and TBI led to excessive gastrointestinal toxicity in the mouse (11).

Cytoxan, BCNU, and VP-16, used singly, appeared to be extremely active agents in oat cell carcinoma (32). Despite the fact that our patients had been exposed to at least one of those drugs before entering the bone marrow transplantation study, higher doses than normal used in combination showed marked and promising responses. The same holds true for embryonal carcinoma of the testis. In this patient population, we attempted to administer serial courses of high-dose chemotherapy in conjunction with marrow rescue. In the three patients tested, complete remission was achieved only following the second course, which seems to imply that two courses are more effective than one. This phenomenon must be confirmed in additional patients before definite conclusions can be drawn, however. The use of high-dose combination chemotherapy with marrow rescue in patients with solid tumors, such as oat cell carcinoma, embryonal carcinoma of the testis, and breast cancer, may elicit optimal responses and therefore prolong survival, especially when this type of treatment is being used first. Bone marrow collection in those patients should be done before any type of chemotherapy is used, in order to obtain the optimal transplantation capability of the marrow cells collected for transplantation. In this way, hemopoietic toxicity after transplantation can be reduced to minimal levels, thus diminishing the risk of severe morbidity. The relatively short hospitalization time of patients after autologous transplantation is another reason for exploring this therapeutic modality in all its aspects.

SUMMARY

Relapsed adult acute leukemia has been treated with a combination of chemotherapy and total-body irradiation (TBI) in conjunction with autologous marrow transplantation. Marrow cells were collected and stored at the time of remission. Before storage, in several patients, the marrow cells were run over discontinuous albumin gradients to minimize leukemic cell contamination of the cell suspension to be injected. Of the 18 evaluable patients, nine achieved complete remission. The longest remission was 13+ months; the longest time of marrow storage was 30 months. In patients with solid tumors (oat cell carcinoma of the lung, embryonal cell carcinoma of the testis, and disseminated breast cancer) who were resistant to first-line chemotherapy high-dose chemotherapy was used in conjunction with autologous marrow rescue. Hemopoietic recovery was observed within 3 weeks of transplantation. In the majority of patients, the tumor response to chemotherapy was observed. This modality of treatment is extremely promising and should be used in a much

earlier stage of the disease so that bone marrow of better quality can be stored and the clinical condition of the patient is better.

REFERENCES

1. Ahearn, M. J., Trujillo, J. M., Cork, A., Fowler, A., and Hart, J. S. The association of nuclear blebs with aneuploidy in human acute leukemia. *Cancer Res.*, 34:2887, 1974.
2. Applebaum, F. R., Graw, R. G., Herzig, C. P. et al. High dose combination chemotherapy (BACT) followed by bone marrow autografts for the treatment of non-Hodgkin's lymphoma. *Am. Soc. Hematol. 74*:19, 1976.
3. Capizzi, R. Schedule-dependent synergism and antagonism between methotrexate and asparaginase. *Biochem. Pharm. Suppl., 2*:151, 1974..
4. Cohen, M., Creaven, P. J., and Fossiech, B. E., Jr. Intensive chemotherapy of small cell bronchogenic carcinoma. *Cancer Treat. Rept. 61*(3):349, 1977.
5. Cortes, M., Holland, J. F., Wang, J. S. et al. Amputation and adriamycin in primary osteosarcoma. *N.E.S.M., 291*:998, 1974.
6. Dicke, K. A. Bone marrow transplantation after separation by discontinuous albumin density gradient centrifugation. Thesis, Leiden University, 1970.
7. Dicke, K. A., van Hooft, J. I. M., and van Bekkum, D. W. The selective elimination of immunologically competent cells from bone marrow and lymphatic cell mixtures. II. Mouse spleen cell fractionation on a discontinuous albumin gradient. *Transplantation,6*:562, 1968.
8. Dicke, K. A., McCredie, K. B., Spitzer, G., Peters, L., Verma, D. S., Stewart, D., Keating, M. J., and Stevens, E. E. Autologous bone marrow transplantation in patients with adult acute leukemia in relapse. *Transplantation, 26*(3):169, 1978.
9. Dicke, K. A., van Noord, M. J., Maat, B., Schaefer, U. W., and van Bekkum, D. W. Identification of cells in primate bone marrow resembling the hemopoietic stem cell in the mouse. *Blood, 42*:195, 1973.
10. Dicke, K. A., van Noord, M. J., Maat, B., Schaefer, U. W., and van Bekkum, D. W. Attempts at morphological identification of the haemopoietic stem cells in primates and rodents. In *Haemopoietic Stem Cells; CIBA Foundation Symposium No. 13 (new series).* Amsterdam: Elsevier, 1973, p. 47.
11. Dicke, K. A., Scheffers, J. M., Mason, K. A., Knaan, S., Wagemaker, G., and McCredie, K. B. The influence of piperazinedione and total body irradiation on survival of hemopoietic stem cells in vitro in mice. IIIth Annual Conference of the International Society for Experimental Hematology, Trogir, Yugoslavia, September 21–25, 1975.
12. Dicke, K. A., Spitzer, G., and Ahearn, M. J. Colony formation in vitro by leukaemic cells in acute myelogenous leukaemia with phytohaemagglutinin as stimulating factor. *Nature, 259*(5539):129, 1976.
13. Dicke, K. A., Spitzer, G., Zander, A. R., Lanzotti, V. J., Verma, D. S., Peters, L. J., Valdivieso, M., Lot-

zová, E., and McCredie, K. B. Autologous bone marrow transplantation in relapsed adult acute leukemia and solid tumors. *Transplan. Proc.* (in press) 1978.

14. Dicke, K. A., Tridente, G., and van Bekkum, D. W. The selective elimination of immunologically competent cells from bone marrow and lymphocyte cell mixtures. III. *In vitro* test for detection of immunocompetent cells in fractionated mouse spleen cell suspensions and primate bone suspensions. *Transplantation, 8*:422, 1969.

15. Fefer, A., Thomas, E. D., Buckner, C. D., Storb, R., Neiman, P., Glucksberg, H., Clift, R. A., and Lerner, K. G. Marrow transplants in aplastic anemia and leukemia. *Sem. Hematol., 2*:353, 1974.

16. Frei, E. Effect of dose schedule on response. *Cancer Med., 195*:717, 1973.

17. Frei, E., III, Jaffe, M. et al. New approaches to cancer chemotherapy with methotrexate. *N. Eng. J. Med., 292*:846, 1975.

18. Gorin, M. C., Najman, A., and Duhamel, G. Autologous bone marrow transplantation in acute myelocytic leukemia. *Lancet, 1*:1050, 1977.

19. Hall, T. C., and Karren, K. Leukemic agents in experimental systems by varying dose schedules and by combination therapy. In *Proceedings of the Vth International Congress on Chemotherapy. Dose Schedules and Combinations*. Vienna: Verlag der Wiener Me dizinischen Academie, 1968, p. 65.

20. Lefkowitz, E., Papac, R. S., and Bertino, S. R. Head and neck cancer III methotrexate toxicity of 24-hour and protection by leucovorin (RSC-3590) in patients with epidermoid carcinomas. *Cancer Chemo. Rep., 51*:305, 1967.

21. McGovern, J. J., Jr., Russell, P. S., Atkins, K., and Webster, E. W. Treatment of terminal leukemic relapse by total-body irradiation and intravenous infusion of stored autologous bone marrow obtained during remission. *N. Eng. J. Med., 260*:675, 1959.

22. Metcalf, D. *Hemopoietic Colonies*. New York: Springer-Verlag, 1977.

23. Moore, M. A. S., Williams, N., and Metcalf, D. *In vitro* colony formation by normal and leukemic human chemotopoietic cells: interaction between colony forming and colony stimulating cell. *J. Nat. Cancer Inst., 50*:603, 1973.

24. Pike, B. L., and Robinson, W. A. Human bone marrow colony growth in agar-gel. *J. Cell. Physiol., 76*:77, 1970.

25. Pinkel, D., Hernandez, K., Borella, L. et al. Drug dosage and remission duration in childhood lymphocytic leukemia. *Cancer, 27*:247, 1971.

26. Schaefer, U. W. Transplantation of fresh allogeneic and cryopreserved autologous bone marrow (BM) in acute leukemia. *Exp. Hematol., Vol. 5* (Suppl. 2):101, 1977.

27. Schaefer, U. W., and Dicke, K. A. Use of frozen bone marrow cells for restoration of haemopoiesis. *Int. J. Radiat., Biol., 23*:195, 1973.

28. Schaefer, U. W., and Dicke, K. A. In Weiner, R. S., Oldham, R. K., and Schwerzenberg, L., eds., *The Cryopreservation of Normal and Neoplastic Cells*. Proceedings of the International Conference held at l'Institut de Cancerologie et d'Immunogenetique, Villejuiff, France. Paris; Inserm, 1973, p. 83.

29. Schaefer, U. W., Dicke, K. A., and van Bekkum, D. W. Recovery of haemopoiesis in lethally irradiated monkeys by frozen allogeneic bone marrow grafts. *Rev. Eur. Etud. Clin. Biol., 17*:483, 1972.

30. Skipper, H. E., Schabel, F. M., Jr., Mellet, L. B. et al. Implications of biochemical, cytokinetic, pharmacologic and toxocologic relationships in the design of the optimal therapeutic schedules. *Cancer Chemo. Rept., 54*:431, 1970.

31. Skipper, H. E., Schabel, F. M., Jr., and I. Wilcox. Experimental evaluation of potential anti-cancer agents XII on the criteria and kinetics associated with "curability" of experimental leukemia. *Cancer Chemo. Rept., 35*:1, 1964.

32. Spitzer, G. Dicke, K. A., Lanzotti, V., Valdivieso, M., Verma, D. S., Zander, A. R., and McCredie, K. B.: High dose BCNU and BCNU combinations with autologous bone marrow transplantation (ABMT) in adult solid tumors. Proceedings of the 1978 AACR Meeting, April 1–5, 1978.

33. Spitzer, G., Dicke, K. A., Schwartz, M. A., Trujillo, J. M. and McCredie, K. B. Significance of PHA induced clonogenic cells in chronic myeloid leukemia and early acute myeloid leukemia. *Blood Cells, 2*: 149. 1976.

34. The UCLA Bone Marrow Transplantation Team. Bone marrow transplantation in acute leukemia. *Lancet, II*(8050):1197, 1976.

35. Thomas, E. D., Buckner, C. D., Banajii, M. et al. One hundred patients with acute leukemia treated by chemotherapy, total body irradiation and allogeneic marrow transplantation. *Blood, 49*(4):411, 1977.

36. Thomas, E. D., Storb, R., Clift, R. A. et al. Bone marrow transplantation. *N. Eng. J. Med., 292*:832, 1975.

37. Tobias, J. S., Weiner, R. S., Cerillths, C. et al. Autologous bone marrow transplantation following high dose chemotherapy. *Clin. Res., 23*:344, 1975.

38. Ziegler, J. L., Deisseroth, A. B., Applebaum, F. R. et al. Burkitt lymphoma a model for intensive chemotherapy. *Sem. Oncol., 4*:317, 1977.

PART VI

Animal Models of
Clinical Conditions

D. E. Harrison

30

Marrow Allograft Survival in *W/W*v Anemic Mice: Effects on Skin Graft Survival and Effects of Preimmunization

D. E. Harrison
and L. E. Mobraaten

The necessity in marrow transplantation of matching donor and recipient for the major histocompatibility locus has been well established (13,14). It is possible, however, that antigenic disparities determined by other loci are also important in marrow grafting. Thus, the *W*-anemic mouse recipients of marrow grafts offer a valuable method to explore this problem, because their hemopoietic systems can be completely populated (10,11) and their immunopoietic systems mostly populated (5) by grafts of histocompatible normal donor cells. This occurs without irradiation or other treatment of the recipients (reviewed in ref. 12). Thus, the immune responses of *W*-anemic recipients are intact, and they are as effective in skin graft rejections (6) and in life-long responses to sheep erythrocytes and phytohemagglutinin (PHA) (4,9) as are the immune responses of normal mice.

In previous experiments, marrow and skin grafts were compared from many different types of donors, each differing from the *W*-anemic recipients at only a single-locus determined antigenic disparity (7). Although two of the histocompatibility differences studied caused moderately rapid skin graft rejections but failed to prevent marrow graft acceptances, a third histocompatibility difference caused only weak skin graft rejections, but prevented successful marrow grafts (7). These results suggest that certain antigens are strongly immunogenic against skin, but weakly immunogenic against marrow, whereas others are strongly immunogenic against marrow but weakly immunogenic against skin. Several antigenic differences that seemed to have such uneven tissue distributions were studied further to determine whether *W*-anemic mice that had been cured and populated with donor marrow could still reject donor skin grafts, and whether immunization would cause rejection of these marrow grafts.

METHODS AND MATERIALS

Mice

All mice were produced and maintained at the Jackson Laboratory, which is fully accredited by the American Association for Accreditation of Laboratory Animal Care: WB/Re-*W*/+ and C57BL6J-*W*v/ + parents were mated to produce WBB6F1-*W/W*v, *W*v/+, *W*/+, and +/+ offspring. Only *W/W*v mice were used as recipients; they were white with black eyes, sterile, and had a severe macrocytic anemia. Results of skin and marrow grafting did not differ significantly with male and female recipients; both were used and the data were pooled unless otherwise noted. Donor mice congenic with the C57BL/6By (B6) background were produced, and kindly supplied by D. W. Bailey. The background strain-locus differing from B6 (congenic line or strain name) were as follows: B6-*H-2*d (B6.C-*H-2*d or Hw 19); B6-*H-*

25^c (B6.C-H-25^c or Hw 65); B6-H-24^c (B6.C-H-24^c or Hw 54); B6-H-17^c (B6.C-H-17^c or Hw 14); B6-H-$15^c,16^c$ (B6.C-H-$15^c,16^c$ or Hw 13); B6-Ea-2^a [(B6.RIII(76 NS)]. (These are also listed in ref. 7.) Some C57BL10/Sn (B10) donors were also used, as were C67BL/6J donors carrying the T6 and T70 chromosome markers (B6T6, B6T70), which were kindly supplied by E. M. Eicher, who introduced these chromosome translocations onto the B6 background by six to eight successive generations of backcrosses to B6 mice. The B6-W^v/+ and CBA/CaJ-W^x/+ parents produce B6CBAF$_1$-W^x/W^v mice. These were used with CBA/HT6J (CBAT6) and B6CBAT6F$_1$ donors.

Although donors from the different congenic lines are listed as differing from B6 or WBB6F$_1$ recipients at single-named antigenic loci, they may actually differ at several antigenic loci whose combination determines the results of skin and marrow grafting. Three or four different skin grafts were placed on tails of 8- to 10-week-old recipients and scored by Bailey's technique (1). Donor marrow cells (usually 1.0×10^7/recipient) were injected intravenously into warmed recipient mice. Erythrocyte numbers and sizes were measured on a Coulter model ZBI electronic cell counter equipped with a MCT-HCT computer. The W/W^v anemic mice were scored as cured when their erythrocyte numbers increased by 50%, their cellular sizes decreased by 25%, and their hematocrits (packed erythrocyte volume) increased by 20% to values in the normal range. As previously reported (7), there were no partial cures. Either the blood parameters of an animal became normal and remained there for many months, or they did not change significantly. Cures were assessed 3 to 6 months after marrow grafting. X-radiation was obtained from a G. E. Maxitron 250 x-ray machine at 20 amps and 250 kVp with 1 mm Cu and 1 mm Al filtration.

RESULTS

Previous studies have shown that the lymphoid tissues of W-anemic mice cured by grafts of marrow with chromosome markers are mostly but not entirely populated by donor cells (5). Results using B6T6 or B6T70 donors with WBB6F$_1$-W/W^v recipients fell in the same range as results obtained using CBAT6 and B6CBAT6F$_1$ donors with B6CBAF$_1$-W^x/W^v recipients. These data were pooled and are listed in Table 30.1. This information updates our previously reported data. Similar results were obtained with spleen grafts, although more recipient-type cells remained (5). The fact that donor cell repopulation in their immune systems is incomplete

TABLE 30.1 Percentages of Nonerythropoietic Donor Cells in W-Anemic Recipients Cured by Marrow Grafts[a]

TISSUE	NUMBER OF RECIPIENTS	PERCENTAGE OF DONOR CELLS	
		Mean	SE
Thymus	27	95	2
Marrow	31	95	2
Spleen	31	82	3
Lymph nodes	44	57	8
Peyer's patches	16	26	6

[a]For each tissue in each recipient, an average of 27 cells with 40 distinguishable chromosomes was scored. Hemopoiesis was suppressed by injected erythrocytes to give packed erythrocyte volumes of over 60%. Inguinal and mesenteric lymph nodes were pooled. Recipients had been cured for an average of 8 months (3 to 15 month range) when scored.

may explain why B6CBAF$_1$-W^x/W^v recipients of CBAT6 spleen cells do not have graft versus host reactions (Figure 30.1). The immune systems of lethally irradiated B6CBAF$_1$-+/+ recipients are completely populated by donor cells (4,8) and these recipients are rapidly killed by CBAT6 spleen grafts (Figure 30.1).

Results of marrow and skin grafting with the congenic lines described here are summarized in Table 30.2. Antigenic differences caused by the H-2^d and H-25^c alleles completely prevented marrow graft acceptance in W/W^v recipients, and skin was rejected in 2 and 3 weeks. But the H-24^c and H-17^c alleles caused fairly rapid skin graft rejections (4 weeks) but did not prevent marrow graft success. The other antigenic differences described in Table 30.2 caused only weak or nondetectable skin graft rejection, and did not prevent marrow graft success.

After W/W^v mice have been populated by allogenic marrow cell grafts, they will accept skin grafts from the same donors, if the antigen causing skin graft rejection is sufficiently expressed on the cells produced by the marrow graft to cause tolerance. Table 30.3 demonstrates that this was not the case for the antigen determined by H-24^c. The W/W^v mice previously cured by B6-H-24^c donors still rejected B6-H-24^c skin, although the rejection rate may have been delayed. The W/W^v mice previously cured by B6-H-17^c behaved as expected, rejecting skin from other donors, but not B6-H-17^c skin (Table 30.3).

Immunization with spleen cells or with a mixture of thymus and lymph node cells from the same donor affected marrow allografts as indicated in Table 30.4. Immunization did not have any effects on most types of marrow grafts. Thus, B6-H-25^c marrow was still rejected, and B6-H-24^c, B6-H-17^c, B10, and B6-Ea-2^a marrow was still accepted. Immunization, how-

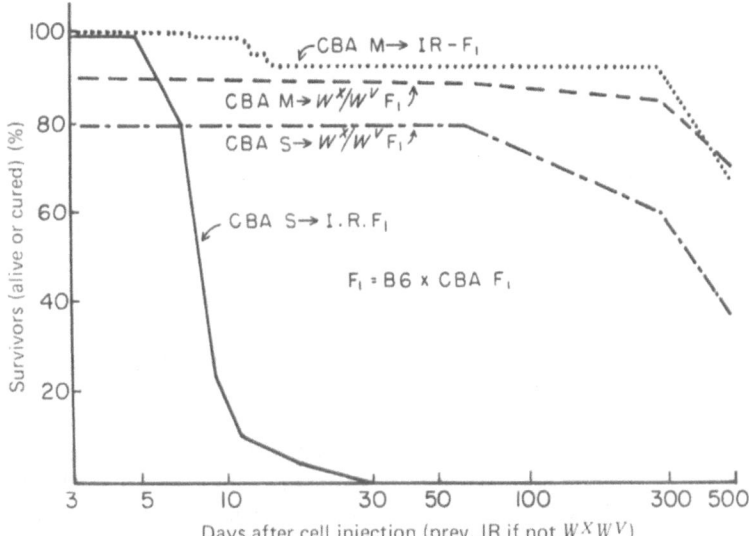

FIGURE 30.1. Recipient (C57BL6 × CBA/Ca)F_1 mice received 1 × 10^7 marrow (CBA M) or 2.0 to 3.0 × 10^7 spleen (CBA S) cells from CBA/HT6J donors. Lethally irradiated, normal B6CBAF$_1$ (IR-F$_1$) recipients could be saved and maintained by parental marrow grafts, but were killed by parental spleen grafts. Parental spleen grafts saved and maintained lethally irradiated CBA/CaJ recipients (3), and this figure shows that parental spleen grafts also cured unirradiated B6CBAF$_1$-W^x/W^v recipients almost as effectively as marrow grafts. Immune systems in such recipients were largely, but not completely populated by donor cells (5).

TABLE 30.2 Survival of Marrow and Skin Allografts in WBB6F$_1$-W/W^v Recipients[a]

DONOR STRAIN	MARROW FAIL/ INJECTED	SKIN (REJECTED/ TRANSPLANTED)	SKIN GRAFT SURVIVAL (WEEKS)		
			Med	Mean	SD
B6-H-2^d	6/6	17/17	2	2.2	0.5
B6-H-25^c	16/16	21/21	3	3.3	1.3
B6-H-24^c	1/33	24/25	4	4.2	2.3
B6-H-17^c	2/20	10/10	4.5	4.4	1.3
B10	1/32	16/17	8	11.6	6.5
B6-H-15^c,16^c	3/17	9/19	>25	—	—
B6-Ea-2^a	2/21	5/12	>25	—	—
B6	1/29	2/8	>25	—	—

[a]Marrow graft results are given as number of mice not cured/number of mice injected with 1.0 × 10^7 marrow cells. Median values of skin graft survival times are derived from all transplants and mean values only from those that were rejected. Skin graft results given as rejected/transplanted = (the number rejected)/(the number transplanted).

TABLE 30.3 Rejection of Skin by W-Anemic Mice Grafted Previously with Marrow of the Skin Donor Type[a]

SKIN DONOR	PRIOR MARROW GRAFT	SKIN REJECTED/ TRANSPLANTED	SKIN SURVIVAL			PERCENTAGE % CURED[b]
			Med	Mean	SD	
B6-H-24^c	B6-H-24^c	5/7	9	8.8	3.1	100 (7)
B6-H-24^c	Other	21/21	5	5.7	3.7	—
B6-H-17^c	B6-H-17^c	0/6	>16	—	—	100 (6)
B6-H-17^c	Other	22/22	7	8.2	3.4	—
B6-H-15^c,16^c	B6-H-15^c,16^c	0/17	>16	—	—	100 (7)
B6-H-15^c,16^c	Other	3/22	>16	—	—	—
B6	B6	0/8	>16	—	—	100 (8)
B6	Other	0/22	>16	—	—	—

[a]Skin survival times in weeks; Med, median of all grafts; mean and standard deviation only of those rejecting in 16 weeks or less. Each WBB6 F$_1$-W/W^v recipient had been populated and cured by a graft of 1 × 10^7 marrow cells injected intravenously 3 to 4 weeks before skin grafting. Each received tail skin grafts of all four types. Donors were female and recipients were male; rejected/transplanted same as Table 30.2.
[b]Figures in parentheses are number of mice cured.

TABLE 30.4 Effect of Immunization with Lymphoid Cells on Marrow Graft Acceptance in *W*-Anemic Mice[a]

| DONOR | % *W/W^v* CURED[b] FOR | | SKIN GRAFT SURVIVAL (WEEKS) |
	Untreated	Immunized	
B6-*H-25^c*	0 (23)	0 (9)	3
B6-*H-24^c*	97 (33)	100 (10)	4
B6-*H-17^c*	90 (20)	86 (7)	4.5
B10	97 (32)	100 (3)	8
B6-*H-15^c,16^c*	82 (17)	0 (12)	>25
B6-*Ea-2^a*	90 (21)	100 (7)	>25
B6	97 (29)	100 (9)	>25

[a]Immunized WBB6 F_1-*W/W^v* recipients received 1×10^7 spleen or lymph node + thymus cells intraperitoneally 2 or 6 weeks before marrow transplants. This dose of spleen cells fails to cure by this route. Median tail skin graft survival times are given for untreated *W/W^v* recipients in different experiments. Results of attempting to cure *W/W^v* recipients are given as the percentage of *W/W^v* mice cured (the number of *W/W^v* recipients of 1×10^7 narrow cells intravenously).

[b]Figures in parentheses are number of mice cured.

ever, completely prevented B6-*H-15^c, 16^c* marrow grafts from curing *W/W^v* recipients.

Further studies with immunization are described in Table 30.5. Lethally irradiated marrow grafts, followed by tail skin grafts, were given to each recipient. Curative marrow grafts containing the *H-24^c* or *H-17^c* antigens were not given until each recipient had rejected those skin grafts. The B6-*H-24^c* donors cured 100% of recipients thus immunized, but only 50% of the recipients were cured using B6-*H-17^c* donors (Table 30.5). Skin grafts containing the *H-15^c,16^c* antigens were not rejected, and neither were the curative marrow grafts. Irradiated marrow cells followed by skin grafts failed to immunize against the *H-15^c,16^c* antigens (Table 30.5), although cells from the spleen or other lymphoid organs did (Table 30.4).

DISCUSSION

These experiments suggest three different tissue distribution patterns for the antigens specified by the *H-24^c*, *H-17^c*, and *H-15^c,16^c* alleles. The *H-24^c* allele specifies a skin antigen that is not expressed sufficiently to produce tolerance on any cells that populate *W/W^v* recipients of B6-*H-24^c* marrow, because such recipients reject B6-*H-24^c* skin (Table 30.3). The rejection rate, however, is delayed, which suggests that a second skin antigen specified by *H-24^c* is expressed on cells from the marrow graft.

Antigens specified on skin by the *H-17^c* allele are expressed in recipients of B6-*H-17^c* marrow, because such recipients no longer reject skin from that donor, although they continue to reject other types of skin graft (Table 30.3). Immunization with B6-*H-17^c* lymphoid cells does not affect the success of marrow grafts (Table 30.4), which suggests that this antigen is not strongly enough expressed on lymphoid cells to induce immunization. But exposure to irradiated B6-*H-17^c* marrow followed by B6-*H-17^c* skin graft rejection cuts the marrow graft success rate in half (Table 30.5). Apparently, the antigen specified by the *H-17^c* allele has some expression on marrow stem cells.

Neither skin nor marrow contains enough antigen specified by the *H-15^c,16^c* alleles for rejection (Tables 30.3 and 30.5); but immunization with B6-*H-15^c,16^c* lymphoid tissue prevents marrow engraftment (Table 30.4). Apparently, this antigen is sufficiently expressed on lymphoid cells to induce immunization, and it is expressed in sufficient concentration on marrow stem cells so that they can be rejected (Table 30.4), although alone they fail to immunize (Table 30.5).

The unique tissue distribution of each of the three antigen types studied suggests that many histocom-

TABLE 30.5 Immunization with Non-H-2 Antigens: Effect on Skin Graft Rejection and Cure of *W*-Anemic Mice[a]

| SKIN DONOR | MARROW GRAFT | SKIN (REJECTED/ TRANSPLANTED) | SKIN SURVIVAL | | | PERCENTAGE CURED[b] |
			Med	Mean	SD	
B6-*H-24^c*	B6-*H-24^c*	8/8	3	4.1	3.4	100 (8)
B6-*H-24^c*	Other	24/24	3	3.0	1.4	—
B6-*H-17^c*	B6-*H-17^c*	8/8	5	6.1	4.0	50 (8)
B6-*H-17^c*	Other	23/23	5	4.9	1.2	—
B6-*H-15^c,16^c*	B6-*H-15^c,16^c*	0/8	>12	—	—	100 (8)
B6-*H-15^c,16^c*	Other	1/24	>12	—	—	—
B6	B6	0/8	>12	—	—	100 (8)
B6	Other	0/24	>12	—	—	—

[a]Skin survival times in weeks: Med. median. Each WBB6 F_1-*W/W^v* recipient was immunized by 1×10^7 marrow cells irradiated *in vitro* with 1,000 R, then skin grafted 2 to 4 weeks later. After graft rejections were complete, each *W/W^v* received 1×10^7 marrow cells of the same donor type used to immunize, in an attempt to cure the anemia. All four types of skin were tail grafted to each recipient. Donors were female and recipients were male. Rejected/Transplanted same as Table 30.2.

[b]Figures in parentheses are number of mice cured.

patibility antigens are expressed to varying degrees in different tissues. This may be important in our understanding of how differentiated tissues develop as well as in planning for their transplantation.

SUMMARY

We have found that W/W^v mice offer unique material for studying the effects of various antigenic disparities on marrow graft success The genetically defective erythroid stem cells that cause the W/W^v anemia are replaced by donor cells in a successful marrow graft. No irradiation or other treatment is necessary and the W/W^v recipient has competent immune responses. Thus, a successful marrow graft across an antigenic barrier demonstrates that the erythropoietic stem cells do not express the foreign antigen in a way that stimulates a lethal graft rejection. Marrow grafts differing at non-$H-2$ antigenic loci have succeeded in several cases, although the same antigenic disparities caused W/W^v mice to reject skin rapidly or to produce strongly cytotoxic and hemagglutinating antibody.

A successful marrow graft populates all the erythropoietic system, and much of the immune system, in a W/W^v recipient. All dividing cells in the thymus become donor type, but significant numbers in Peyer's patches, lymph nodes, and spleen remain recipient type, even many months after the graft. The remaining host-type cells may allow most of the immune system of W-anemic F_1 recipients to be populated by parental donor spleen cells without detectable graft-versus-host (GVH) reactions, whereas the same spleen cells rapidly kill irradiated F_1-recipients.

The W/W^v mice cured and populated by marrow cells from donors differing at the $H-24$ locus continue to reject subsequent skin grafts of that type, although the rejection becomes less rapid. Perhaps skin contains antigens specified by genes at the $H-24$ locus that are not expressed either on marrow cells or on any of their differentiated descendants. But after being cured and populated by marrow cells from donors differing at the $H-17$ locus, W/W^v mice no longer reject skin of that genotype, although they reject other incompatible skin grafts. This tolerance suggests that all the skin antigens specified by genes at the $H-17$ locus are expressed on marrow cells or on their descendants.

Mixtures of histoincompatible thymus and spleen cells injected intraperitoneally into W/W^v mice do not cure their anemia but are able to stimulate immune responses. Immunization of W/W^v recipients with thymus and spleen cells from donors differing at the $H-24$ or $H-17$ loci do not prevent successful marrow grafts of the same type. This suggests two possibilities: (a) The antigens specified by genes at these loci are not expressed on thymus and spleen cells and the immunization is not effective; and (b) the antigens are not expressed on the stem cells, and therefore the stem cells are not damaged by immune responses against the antigens and cure the W/W^v recipients.

Immunization with thymus and spleen prevents acceptance of marrow grafts differing at the $H-15,16$; thus an antigen produced by those loci must be expressed on stem cells, although such marrow grafts cure untreated W/W^v recipients, and $H-15,16$ disparate skin grafts are rejected extremely slowly or not at all by W/W^v mice. Immunization with lethally irradiated marrow does not affect skin graft rejections in these three systems. Skin grafts differing at the $H-24$ and $H-17$ loci are still rejected, whereas those differing at $H-15,16$ are still accepted. After skin grafts have been rejected, however, marrow differing at $H-17$ cures only one-half the W/W^v recipients, which suggests than an antigen produced by this locus is expressed on stem cells. Marrow differing at $H-24$ still cures all recipients; this supports the suggestion that antigens produced by the $H-24$ locus are not expressed on stem cells. Marrow differing at the $H-15,16$ locus also cures the recipients. Antigens produced by those loci apparently are not expressed sufficiently on irradiated marrow or skin to immunize.

ACKNOWLEDGMENTS

We are grateful for the dependable technical assistance of Michael Astle and Evelyn Sargeant.

This work was supported by National Institutes of Health Research Grants HL-16119 from the National Heart and Lung Institute, AG-00594 and AG-00250 from the National Institute on Aging and AI-13130 from the National Institute on Allergy and Infectious Disease.

REFERENCES

1. Bailey, D. W. Histoincompatibility associated with X-chromosome in mice. *Transplantation*, 1:71, 1963.
2. Bailey, D. W. Genetics of histocompatibility in mice. I. New loci and congenic lines. *Immunogenetics*, 2:249, 1975.
3. Harrison, D. W. Avoidance of graft versus host reactions in cured W-anemic mice. *Transplantation*, 22:47, 1976.
4. Harrison, D. E. Genetically defined animals valuable in testing aging of erythroid and lymphoid stem cells and microenvironments. In Bergsma, D., and Harrison, D.

E., eds., *Genetic Effects on Aging*. National Foundation Original Article Series Volume XIV, Number 1, New York: A. R. Liss, 1978, p. 187.

5. Harrison, D. E., and Astle, C. M. Population of lymphoid tissues in cured *W*-anemic mice by donor cells. *Transplantation, 22*:42, 1976.

6. Harrison, D. E., and Cherry, M. Survival of marrow allografts in *W/W^v* anemic mice: effect of disparity at the *Ea-2* locus. *Immunogenetics, 2*:219, 1975.

7. Harrison, D. E., and Doubleday, J. W. Marrow allograft survival in *W/W^v* mice: relation to skin graft sur-survival times. *Immunogenetics, 3*:289, 1976.

8. Micklem, H. S., and Loutit, J. K. *Tissue Grafting and Radiation*. New York: Academic Press, 1966.

9. Mekori, T. and Phillips, R. A. The immune response in mice of genotype *W/W^v* and *Sl/Sl^d*. *Proc. Soc. Exp. Biol. Med., 132*:115, 1969.

10. Murphy, E. D., Harrison, D. W., and Roths, J. Giant granules of beige mice: A quantitative marker for granulocytes in bone marrow transplantation. *Transplantation, 15*:526, 1973.

11. Russell, E. S., and Bernstein, S. E. Proof of whole-cell implant in the therapy of *W*-series anemia. *Arch. Biochem. Biophys., 125*:594, 1968.

12. Russell, E. S. Abnormalities of erythropoiesis associated with mutant genes in mice. In Gordon, A. S. ed., *Regulation of Hematopoiesis*. New York: Appleton-Century-Crofts, 1970.

13. Snell, J. D., Dausset, J., and Nathenson, S. *Histocompatibility*. New York: Academic Press, 1976.

14. Thomas, E. D., Fefer, A., Buckner, C. D., and Storb, R. Current status of bone marrow transplantation for aplastic anemia and acute leukemia. *Blood, 49*:671, 1977.

31

Hemopoietic Stem Cells in Experimental Hypoplastic Marrow Failure of the Mouse

Kazuo Kubota,
Hideaki Mizoguchi,
Yasusada Miura, Shogo Kano,
and Fumimaro Takaku

Aplastic anemia is a syndrome characterized by a failure of the supply of erythrocytes, leukocytes, and platelets to the peripheral blood, owing to a decreased production of these cells. In this disease, the number of hemopoietic cells in the bone marrow usually fall, and they are replaced by fatty cells.

The pathogenesis of this disease is still unknown. The recent development of *in vitro* culture systems, however, has revealed that the frequency of occurrence of both granuloid and erythroid progenitor cells is depressed markedly in this disease and tends to return to normal levels in remission (8,9,12,15,18–22). From these findings some abnormalities in the hemopoietic stem cells have been presumed to be pathognomonic to this disease.

Recently, some cases of aplastic anemia were attributed to a lymphocyte-mediated injury to stem cells (1,10,11), although this observation is still controversial. This observation may supply a promising clue to the elucidation of the pathogenesis of aplastic anemia, however, if animal models of aplastic anemia, induced by immunologic methods, were available. Barnes and Mole (3) described a fatal marrow aplasia in mice given allogeneic lymph node cells after sublethal irradiation.

The present experiments were designed to study the changes in hemopoietic stem cells in experimentally induced bone marrow hypoplasia and to determine if the mechanisms of this experimental hypoplasia could be used to elucidate those of aplastic anemia in man.

MATERIALS AND METHODS

Eight to fourteen-week-old inbred female C3H/He (H-2^k, Mls^c) and B10 × BR (H-2^k, Mls^b) mice were used throughout the experiments. The C3H/He mice were obtained from the Funabashi Animal Center, Chiba, Japan, whereas the B10 × BR mice were bred in our laboratory by strict brother-sister mating. Cages, shavings, diet, and water for the mice were sterilized before use.

The C3H/He mice were randomized into four groups of four to 12 animals. Group A was not given any treatment. Group B was given 1×10^7 lymph node cells prepared from B10 × BR mice. Group C received 450 or 600 rad of wholebody X-irradiation. Group D was given 1×10^7 lymph node cells obtained from B10 × BR mice immediately after 450 or 600 rad wholebody X-irradiation.

Body weight was determined and hematologic examinations were performed individually before treatment and 7, 14, 21, 28, and 35 days after treatment. Hematocrit (Ht) readings were obtained with microhematocrit tubes, the leukocyte (WBC) count with a standard hemocytometer, and the platelet count with a phase-contrast microscope.

Mice were sacrificed 14, 17, 19 and 21 days after treatment and the number of nucleated cells, CFU-s, and CFU-c in the femoral bone marrow was determined. The marrow from the femurs was examined histologically 21 days after treatment. In addition, the wet weight and the number of CFU-c of the spleen were also determined.

The C3H/He mice were exposed in a plastic cage to x-ray from a Toshiba LMR-13 x-ray unit with the following physical factors: 10 MeV, target midline cage distance 100 cm and dose rate 130 rad/min. Dosimetry was checked with Victoreen ionization chambers.

The B10 × BR mice were killed by cervical dislocation. The mesenteric lymph node cells were collected and soaked in saline, cut into small pieces with scissors, and pressed with the tip of forceps. After passage through a stainless steel mesh (mesh size 250 μ), the cells were washed twice with saline, and the cell suspension was adjusted to the proper concentration for injection into mice. Ten million cells suspended in 1 ml saline were injected intravenously into mice via a lateral tail vein.

The number of nucleated cells per femur was determined using a standard hemocytometer. The number of colony-forming units in spleen (CFU-s) was determined according to the method of Till and McCulloch (25). A sample of 1×10^5 nucleated marrow cells was injected into a series of five to six lethally irradiated (900 rad) C3H/He mice. The mice

were sacrificed 9 days after injection of cells, the spleens were removed and fixed in Bouin's solution, and the macroscopically visible colonies were counted.

Colony-forming units in culture (CFU-c) were assayed essentially by the soft agar method described by Bradley and Metcalf (4), with L-cell-conditioned medium added as a colony-stimulating factor. The marrow cells were added to a final concentration of 7.5×10^4 cells/dish and the spleen cells to a final concentration of 5.0×10^5 cells/dish, respectively. After incubation for 7 days in a CO_2 incubator, the aggregates of more than 50 cells were scored as colonies.

Histologic examinations were performed 21 days after treatment. The femurs were fixed with 10% formalin, decalcified, embedded in paraffin, sectioned, and stained with hematoxylin and eosin.

RESULTS

Table 31.1 shows the typical changes in body weight and hematology after 600 rad. Groups A and B showed no significant changes in Ht readings and WBC and platelet counts during the course of treatment. Although a drop in body weight was observed in group B 21 days after treatment in this experiment, an increase was observed in six of seven previously conducted experiments. Therefore, we do not consider the loss of body weight in this experiment to be significant. In group C, there was a

TABLE 31.1 Body Weights, Hematocrit Readings, and White Blood Cell and Platelet Counts in C3H/He Mice after 600 Rad Wholebody Irradiation

GROUPS OF MICE[a]	PARAMETER[b]	BEFORE TREATMENT	WEEKS AFTER TREATMENT: 1	2	3
A	BW	26.1 ± 0.6[c]	26.1 ± 0.6	25.6 ± 1.4	26.4 ± 1.5
	Ht	50.8 ± 0.3	48.0 ± 2.1	50.0 ± 1.3	46.0 ± 3.0
	WBC	9.2 ± 0.9	10.8 ± 1.6	13.3 ± 0.2	11.1 ± 0.1
	Pl	142.2 ± 8.3	156.4 ± 12.2	140.6 ± 7.3	130.0 ± 6.8
B	BW		26.2 ± 1.1	26.0 ± 1.3	22.6 ± 1.6
	Ht		47.8 ± 1.0	48.3 ± 1.0	48.5 ± 1.5
	WBC		11.5 ± 0.9	13.4 ± 1.2	10.6 ± 1.0
	Pl		166.4 ± 7.3	159.4 ± 9.4	175.6 ± 5.2
C	BW		25.3 ± 0.8	25.5 ± 0.7	27.4 ± 0.2
	Ht		39.7 ± 1.3	29.3 ± 2.1	43.0 ± 1.0
	WBC		0.9 ± 0.1	3.6 ± 0.8	6.1 ± 1.0
	Pl		26.3 ± 1.2	41.0 ± 10.6	121.9 ± 30.6
D	BW		20.9 ± 0.6	22.2 ± 0.8	22.3 ± 0.8
	Ht		40.3 ± 1.5	20.2 ± 2.8	16.5 ± 0.5
	WBC		4.0 ± 0.3	0.6 ± 0.2	1.5 ± 0.5
	Pl		3.3 ± 0.5	4.9 ± 0.8	8.1 ± 0.5

[a]Group A was untreated (control); group B was given 10^7 lymph node cells; group C received rad irradiation; group D was given 10^7 lymph node cells after 600 rad irradiation.

[b]BW, body weight in grams; Ht, hematocrit, in per cent; WBC, white blood cell count $\times 10^3/mm^3$; Pl, platelet count $\times 10^4/mm^3$.

[c]Mean ± SE.

decrease in body weight, Ht readings, and WBC and platelet counts during the first 2 weeks after irradiation. Thereafter, weight and cell counts increased steadily and returned to normal levels 21 days after irradiation. There were no deaths in groups A, B, and C during the 21 days of treatment. In group D, Ht readings and WBC and platelet counts decreased to levels of 17%, 1,500/mm³ and 81,000/mm³, respectively, by the 21st day after treatment. The body weight was also markedly reduced, and the mice developed clinical signs of illness 14 days after treatment. Thus, the mice in group D developed severe pancytopenia 14 to 21 days after the experimental procedures and about 80% died within 21 days. These data agree with those of Barnes and Mole (3).

Table 31.2 shows the changes in body weight and the hematologic values after 450 rad. There were no deaths in groups A, C, and D within 35 days of treatment. The mice in group D developed pancytopenia, however, but normal values were observed 35 days after the experimental procedure.

The marrow from the femurs was examined histologically 21 days after various experimental treatments. There were no significant differences between groups A and B [Figure 31(A,B)]. Evidence of marked granulocytic hyperplasia was observed in group C [Figure 31(C)]. Marked hypoplasia, with severe depletion of the hemopoietic cells and an increase in fat cells, was observed in the marrow of mice from group D [Figure 31(D)].

Changes in the numbers of nucleated cells, CFU-s, and CFU-c per femur after 600 rad are shown in Table 31.3 and Figures 31.2 (A,B,C). There was no significant difference between groups A and B in these parameters. In group C, the number of nucleated cells, CFU-s, and CFU-c per femur decreased, due to irradiation damage, 14 days after

treatment. Thereafter they gradually increased and returned to normal 21 days after treatment. In group D, however, the numbers of nucleated cells, CFU-s, and CFU-c remained at very low levels. These data correlate well with the results of both the peripheral blood and the histologic examinations.

The wet weight and the number of CFU-c of the spleen after 600 rad are shown in Table 31.3. There were no significant differences between groups A and B. In group C, the wet weight of the spleen was the same as that of group A, but the number of CFU-c decreased 17 days after treatment, but at 19 days, the wet weight and the number of CFU-c of the spleen increased, compared to group A. In group D, however, the wet weight and the number of CFU-c of the spleen were markedly reduced at both 17 and 19 days after treatment.

DISCUSSION

In the present study, C3H/He mice developed severe pancytopenia and marrow hypoplasia 14 to 21 days after exposure to 600 rad wholebody x-irradiation and transplantation of 10^7 B10 × BR lymph node cells. The number of marrow CFU-s and CFU-c, as well as splenic CFU-c, were markedly reduced. Similar marked decreases in the number of both granuloid (CFU-c) and erythroid progenitor cells (CFU-e) are common to patients with aplastic anemia (8,9,-12,15,18,19,21,22). Therefore, the findings of this experimental model closely resemble those of aplastic anemia in man.

Although Barnes and Mole (3) used CBA/H mice as irradiated recipients and C3H/H as cell donors, we produced "aplastic mice" by using C3H/He mice as irradiated recipients and B10 × BR mice as cell

TABLE 31.2 Body Weight, Hematocrit Readings, and White Blood Cell and Platelet Counts in C3H/He Mice after 450 Rad Wholebody Irradiation

GROUPS OF MICE[a]	PARAMETER[b]	BEFORE TREATMENT	WEEKS AFTER TREATMENT:				
			1	2	3	4	5
A	BW	24.0 ± 0.6[c]	25.7 ± 0.8	25.7 ± 0.8	27.2 ± 1.0	27.6 ± 0.8	28.2 ± 1.0
	Ht	50.0 ± 1.4	47.0 ± 1.7	48.2 ± 1.0	48.8 ± 0.9	49.5 ± 0.9	50.5 ± 0.7
	WBC	9.0 ± 0.9	8.0 ± 1.7	14.5 ± 1.2	10.6 ± 1.6	13.5 ± 0.5	10.0 ± 1.9
	Pl	147.8 ± 7.3	153.6 ± 14.2	154.0 ± 7.9	168.4 ± 12.5	147.9 ± 7.3	162.4 ± 7.8
C	BW		24.6 + 0.7	24.1 ± 0.8	25.4 ± 0.7	26.3 ± 0.7	27.0 ± 0.7
	Ht		43.0 ± 1.7	38.8 ± 1.1	43.3. ± 0.9	49.3 ± 1.6	48.3 ± 0.4
	WBC		4.9 ± 0.7	3.7 ± 0.5	8.8 ± 1.6	5.6 ± 1.1	7. 0 ± 1.1
	Pl		43.6 ± 6.0	35.3 ± 3.9	105.2 ± 7.7	144.9 ± 5.3	165.0 ± 9.2
D	BW		21.3 ± 0.3	23.1 ± 0.4	24.8 ± 0.5	25.5 ± 0.6	26.1 ± 0.5
	Ht		35.9 ± 0.7	31.5 ± 2.1	38.6 ± 1.8	42.9 ± 1.4	47.4 ± 0.8
	WBC		5.3 ± 0.6	3.0 ± 0.9	5.7 ± 1.0	7.1 ± 0.9	7.0 ± 0.9
	Pl		8.5 ± 1.4	21.0 ± 2.9	40.3 ± 5.5	92.6 ± 7.9	144.6 ± 6.1

[a]Group A was untreated (control); group C received 450 rad irradiation; group D was given 10^7 lymph node cells after 450 rad irradiation.
[b]BW, body weight, in grams; Ht, hematocrit, in percent; WBC, white blood cell count × 10³/mm³; Pl, platelet count × 10⁴/mm³.
[c]Mean ± SE.

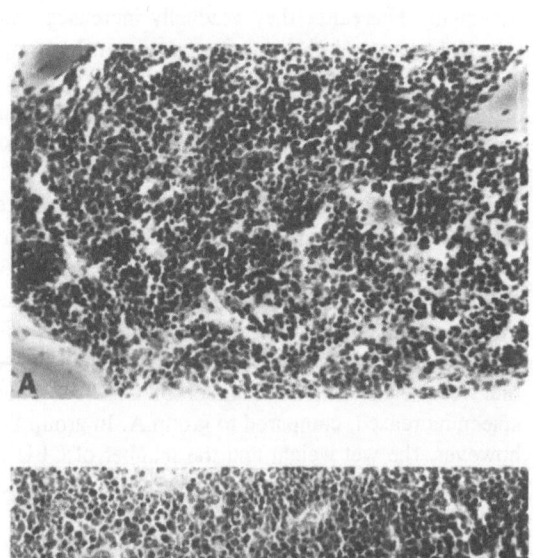
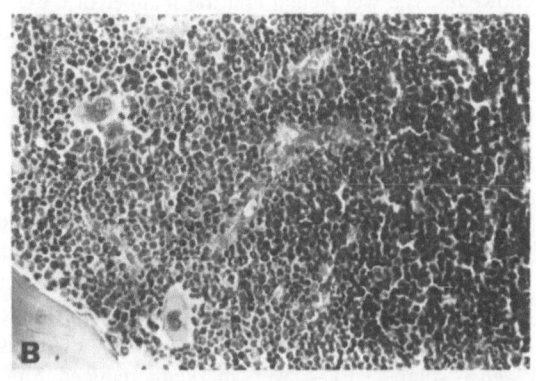
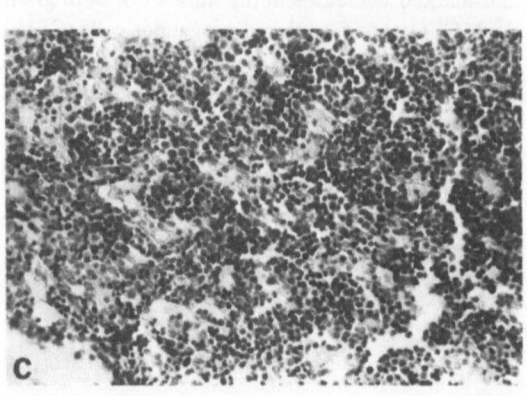
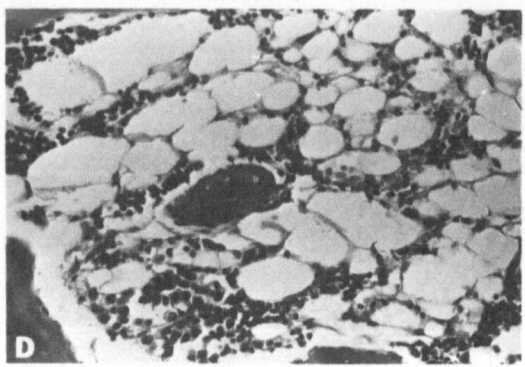

donors. Thus, the occurrence of marrow hypoplasia is not limited to a particular strain of mouse or to a limited number of strain combinations, but is a more generalized phenomenon with similar mechanisms. It is of interest to determine whether this phenomenon is due to graft-versus-host (GvH) reaction or not. It is well known that the GvH reaction has a harmful effect on hemopoietic stem cells (13,14,26). The GvH reaction is usually strongest between two strains that differ in the major histocompatibility complex (*H-2*). Both C3H/He and B10 × BR mice, however, have the same *H-2* alleles (or haplotype), and bone marrow allografts between mice of the same *H-2* haplotype are feasible (17). We confirmed these findings in our two strains of mice (data not shown). Furthermore, the signs and symptoms of the "aplastic mice" in our studies as well as in those of Barnes and Mole (3), which included marrow hypoplasia and weight loss, differed from those described for GvH disease, which include diarrhea, splenomegaly, and hair loss (24).

The C3H/He and B10 × BR mice are different in minor histocompatibility loci other than *H-2* loci. The *Mls* locus, another lymphocyte-activating locus, was originally detected through mixed lymphocyte culture reactions (5). Weak GvH reactions were also demonstrated in *H-2* identical, *Mls* locus differing combinations using a sensitive popliteal lymph node assay (16, 23). The specific *Mls* loci of C3H/He and B10 × BR mice used in our experiments are *Mls*c and *Mls*b, respectively. It has been confirmed that the *Mls* loci of CBA/H and C3H/H mice used in the experiments of Barnes and Mole (3) are also *Mls*b and *Mls*c, respectively. They are very weak or nonstimulatory toward each other (6,7). Therefore, the marrow failure observed in the present experiments might be due to a weak GvH reaction caused by the difference in the *Mls* locus as well as other minor histocompatibility loci. Barnes and Mole (3) also suggested that the mechanism of the marrow hypoplasia in their experiments was due to a GvH reaction of the engrafted, immunologically competent cells against the marrow. Experiments to character-

FIGURE 31.1. Histology of bone marrow from the femur. (A) A normal C3H/He mouse in group A; (B) a C3H/He mouse in group B 21 days after transplantation of 10₇ lymph node cells prepared from B10 BR mice; (C) a C3H/He mouse in group C 21 days after 600 rad wholebody x-irradiation; (D) a C3H/He mouse in group D 21 days after 600 rad wholebody x-irradiation followed by transplantation of 10₇ lymph node cells prepared from B10 BR mice. Marked hypoplasia with loss of hemopoietic cells and an increase in fat cells was observed (D). Original magnification X 200 (hematoxylin and eosin).

TABLE 31.3 Nucleated Cells, CFU-s, and CFU-c of Bone Marrow and the Wet Weight and CFU-c of Spleen after 450 Rad Wholebody Irradiation

EXP. NO.	DAY OF ASSAY	GROUPS OF MICE[a]	NUCLEATED CELLS/ FEMUR ($\times 10^7$)	CFU-s/10^5 MARROW CELLS	CFU-s/ FEMUR	CFU-c/7.5 $\times 10^4$ MARROW CELLS	CFU-c/FEMUR	WET WEIGHT OF SPLEEN (MG)	CFU-c/ 5.0 $\times 10^5$ SPLEEN CELLS
1	14	A	1.46 ± 0.28^b			127 ± 18	$24,000 \pm 1,300$		
		B	1.56 ± 0.36			95 ± 1	$19,800 \pm 4,900$		
		C	0.61 ± 0.16^c			3 ± 1^c	300 ± 100^c		
		D	0.09 ± 0.09			0 ± 0^c	0 ± 0^c		
2	17	A	1.15 ± 0.05	14.0 ± 2.0	$1,620 \pm 300$	111 ± 20	$17,200 \pm 3,800$	275 ± 37	84 ± 1
		B	1.14 ± 0.14	11.0 ± 2.0	$1,230 \pm 80$	91 ± 26	$13,400 \pm 2,300$	284 ± 9	85 ± 1
		C	0.59 ± 0.05^d	9.5 ± 1.5	570 ± 130	18 ± 15	$1,500 \pm 1,300$	251 ± 7	5 ± 4^c
		D	0.11 ± 0.04^e	1.0 ± 0^c	10 ± 4^e	0 ± 0^e	0 ± 0^e	34 ± 1^e	0 ± 0^c
3	19	A	1.47 ± 0.05	20.5 ± 2.5	$3,030 \pm 470$	196 ± 13	$38,400 \pm 1,300$	168 ± 23	68 ± 13
		B	1.65 ± 0.01^e	13.0 ± 2.0	$2,150 \pm 300$	152 ± 18^e	$33,400 \pm 3,700^d$	264 ± 2	61 ± 3
		C	1.15 ± 0.03^e	10.5 ± 6.5	$1,190 \pm 700$	82 ± 20^e	$12,500 \pm 2,600^d$	375 ± 47	47 ± 5
		D	0.07 ± 0.01^c	1.0 ± 0^d	7 ± 1^e	0 ± 0^c	0 ± 0^c	66 ± 1^e	0 ± 0^e
4	21	A	1.45 ± 0.15			121 ± 8	$23,600 \pm 4,000$		
		B	1.02 ± 0.15			101 ± 29	$14,200 \pm 5,600$		
		C	1.09 ± 0.13			171 ± 21	$24,500 \pm 3,200$		
		D	0.34 ± 0.02^d			13 ± 1^c	590 ± 9^c		

[a]Group A was untreated (control); group B was given 10^7 lymph node cells; group C received 600 rad irradiation; group D was given 10^7 lymph node cells after 600 rad irradiation.

[b]Mean ± SE.

[c]$p < .01$.

[d]$p < .02$.

[e]$p < .05$.

ize the engrafted lymph node cells that might be harmful to hemopoietic stem cells are under way with the object of elucidating the mechanism of this phenomenon. It will be of interest to perform the experiment using both *H*-2- and *Mls*-identical mice.

In the present experiments, the induction of marrow hypoplasia required two factors, irradiation and transplantation of lymph node cells, neither of which was sufficient to cause fatal damage alone. This may be important when one considers the pathogenesis of aplastic anemia in man. Although its cause is not clear, we can often observe the clinical onset or deterioration of aplastic anemia after irradiation, infection, drugs, or other events that, in themselves, are not usually serious in normal healthy people. If we can utilize the present experiments as a model of aplastic anemia, it may be possible that there are cases where, although one factor alone does not cause severe aplastic anemia, the addition of another factor or factors has a more than additive, seriously damaging effect on the hemopoietic cells. The fact that severe aplastic anemia was not achieved when the dose of irradiation was reduced to 450 rad suggests that there is a threshold for the induction of bone marrow aplasia. Experiments in which the dose of irradiation and the number of lymphocytes being injected were changed may be useful in analyzing the mechanisms in aplastic anemia.

It is of interest that in some cases of aplastic anemia the pathogenesis may be immunologic. Ascensao et al. (1) reported that the addition of anti-lymphocyte globulin to the bone marrow culture of a patient with aplastic anemia increased the number of CFU-c. Kagan et al. (11) demonstrated that cells present in the bone marrow of a patient with aplastic anemia actively suppress granuloid colony formation *in vitro*. Further, Hoffman et al. (10) indicated that some patients with aplastic anemia possess a population of lymphocytes capable of suppressing *in vitro* erythropoiesis. We have performed experiments in which peripheral lymphocytes derived from some patients with aplastic anemia were added to the *in vitro* cultures of both CfU-c and CFU-e. Suppression of erythroid colony formation was demonstrated in some cases, but no suppression of granuloid colony formation was observed (T. Suda et al., unpublished data). A case of successful treatment of aplastic anemia with cyclophosphamide gave further support to a immunologic mechanism (2). From these findings, it could be postulated that a cell-mediated mechanism plays an important role in the pathogenesis of aplastic anemia in some patients.

It might also be postulated that in some cases of aplastic anemia, a mutant population of lymphocytes appears *in vivo* after exposure to irradiation, infectious agents or drugs. Such lymphocytes may harm

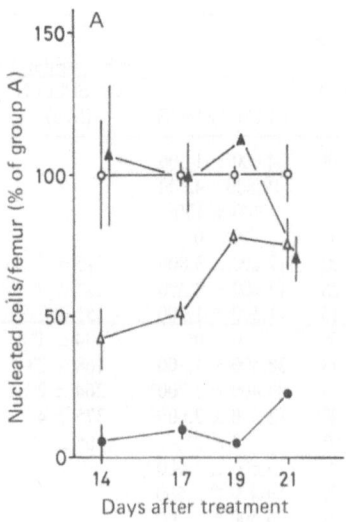

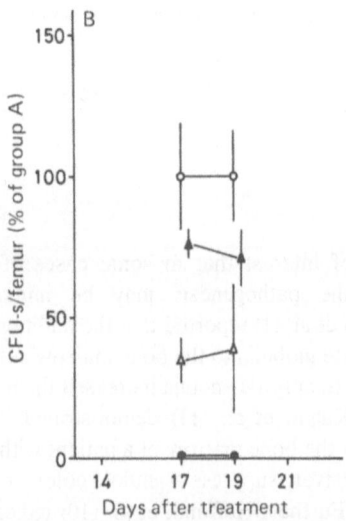

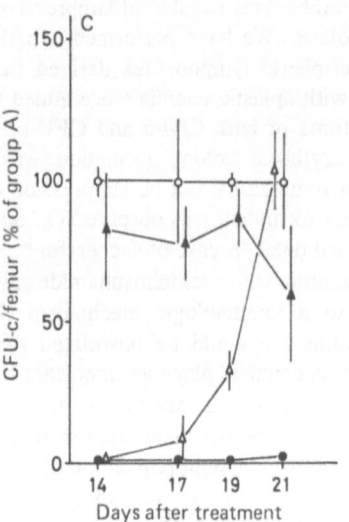

the hemopoietic stem cells, which have already been damaged; this could lead to marrow hypoplasia and pancytopenia. The present experiments lend support to this hypothesis.

Thus, the present model may be representative of at least some cases of aplastic anemia in man and should provide a useful means of investigating its pathogenesis and treatment. Means to prevent and treat this experimental aplastic anemia are being developed.

ACKNOWLEDGMENTS

We wish to thank Dr. Toshihiko Sado for his helpful participation in discussions regarding this paper and for his generous gifts of breeding pairs of B10 × BR mice, Dr. Reiko Masuda for histologic examinations, and Miss Sachiko Kurokawa for her excellent technical assistance.

This work was supported by grants from the Japanese Ministry of Education and the Ministry of Health and Welfare.

SUMMARY

Immunologically induced aplastic anemia in mice was used as a model for investigating the pathogenesis of aplastic anemia in man. Here, C3H/He (H-2^k,Mls^c) and B10 × BR (H-2^k,Mls^b) mice were used throughout the experiments because these two strains of mice share the same major histocompatibility antigen. The C3H/He mice received 600 rad wholebody X-irradiation followed by the transplantation of 10^7 lymph node cells derived from B10 × BR mice. The C3H/He mice developed severe pancytopenia and marrow hypoplasia 14 to 21 days after treatment. The total numbers of nucleated cells, CFU-s, and CFU-c per femur and the wet weight and CFU-c of the spleen decreased remarkably. Mice receiving irradiation alone suffered from a transient, nonlethal hematologic suppression; their blood cell levels returned to normal 21 days after exposure. Transplantation of lymph node cells alone did not

FIGURE 31.2. (A) Changes in the numbers of nucleated cells per femur after 600 rad; (B) changes in the numbers of CFU-s per femur after 600 rad; (C) changes in the numbers of CFU-c per femur after 600 rad. Group A, no treatment (o——o); Group B, 10₇ lymph node cells (▲——▲); Group C, 600 rad irradiation (△——△); Group D, 600 rad irradiation plus 10₇ lymph node cells (●——●). The results are compared (per cent) to the data from the corresponding experiments in Group A. Each point shows the mean with one standard error when the limits are larger than the size of the circle or the triangle.

produce marrow hypoplasia. The results of the present experiments indicating granuloid and erythroid progenitor cell decrease are consistent with those of aplastic anemia in man. Therefore, this experimental model may provide a useful means of investigating the pathophysiology of and the development of treatment for aplastic anemia.

REFERENCES

1. Ascensão, J., Pahwa, R., Kagan, W., Hansen, J., Moore, M., and Good, R. Aplastic anaemia: Evidence for an immunological mechanism. *Lancet, I*:669, 1976.

2. Baran, D. T., Griner, P. F., and Klemperer, M. R. Recovery from aplastic anemia after treatment with cyclophosphamide. *N. Eng. J. Med., 295*:1522, 1976.

3. Barnes, D. W. H., and Mole, R. H. Aplastic anaemia in sublethally irradiated mice given allogeneic lymph node cells. *Brit. J. Haematol., 13*:482, 1967.

4. Bradley, T. R., and Metcalf, D. The growth of mouse bone marrow cells *in vitro. Aust. J. Exp. Biol. Med. Sci., 44*::287, 1966.

5. Festenstein, H. Antigenic strength investigated by mixed cultures of allogeneic mouse spleen cells. *Ann. N.Y. Acad. Sci., 129*:567, 1966.

6. Festenstein, H. Pertinent features of M locus determinants including revised nomenclature and strain distribution. *Transplantation, 18*:555, 1974.

7. Festenstein, H., and Démant, P. Workshop summary on genetic determinants of cell-mediated immune reactions in the mouse. *Transplant. Proc., 5*:1321, 1973.

8. Greenberg, P. L., Nichols, W., and Schrier, S. L. Granulopoiesis in acute myeloid leukemia and preleukemia. *N. Eng. J. Med., 284*:1225, 1971.

9. Hansi, W., Rich, I., Heimpel, H., Heit, W., and Kubanek, B. Erythroid colony forming cells in aplastic anaemia. *Brit. J. Haematol., 37*:483, 1977.

10. Hoffman, R., Zanjani, E. D., Lutton, J. D., Zalusky, R., and Wasserman, L. R. Suppression of erythroid-colony formation by lymphocytes from patients with aplastic anemia. *N. Eng. J. Med., 296*:10, 1977.

11. Kagan, W. A., Ascensão, J. A., Pahwa, R. N., Hansen, J. A., Goldstein, G., Valcera, E. B., Incefy, G. S., Moore, M. A. S., and Good, R. A. Aplastic anemia: Presence in human bone marrow of cells that suppress myelopoiesis. *Proc. Nat. Acad. Sci. U.S., 73*:2890, 1976.

12. Kern, P., Heimpel, H., Heit, W., and Kubanek, B. Granulocytic progenitor cells in aplastic anaemia. *Brit. J. Haematol., 35*:613, 1977.

13. Kitamura, Y., Kawata, T., and Kanamaru, A. Parental lymph node cell dose necessary to change differentiation pattern of parental colony-forming cells in F_1 hybrid mice. *Transplantation, 14*:568, 1972.

14. Kitamura, Y., Kawata, T., Suda, O., and Ezumi, K. Changed differentiation pattern of parental colony-forming cells in F_1 hybrid mice suffering from graft-versus-host disease. *Transplantation, 10*:455, 1970.

15. Kurnick, J. E., Robinson, W. A., and Dickey, C. A. *In vitro* granulocytic colony-forming potential of bone marrow from patients with granulocytopenia and aplastic anemia. *Proc. Soc. Exp. Biol. Med., 137*:917, 1971.

16. Lilliehöök, B., and Blomgren, H. Weak graft-versus-host response of CBA lymphocytes against the H-2 identical strain C3H. *Scand. J. Immunol., 3*:637, 1974.

17. Lotzová, E., Dicke, K. A., Trentin, J. J., and Gallagher, M. T. Genetic control of bone marrow transplantation in irradiated mice: Classification of mouse strains according to their responsiveness to bone marrow allografts and xenografts. *Transplant. Proc., 9*:289, 1977.

18. Mizoguchi, H., Miura, Y., Chiyoda, S., and Takaku, F. Myeloid stem cells in various hemopoietic disorders. In Nakao, K., Fisher, J. W., and Takaku, F., eds., *Erythropoiesis: Proceedings of the Fourth International Conference on Erythropoiesis.* Tokyo: University of Tokyo Press, 1975, p. 199.

19. Mizoguchi, H., Miura, Y., Kubota, K., and Takaku, F. Hemopoietic stem cells in aplastic anemia and pure red cell aplasia. In Hibino, S., Takaku, F., and Shahidi, N. T., eds., *Aplastic Anemia: Proceedings of the First International Symposium on Aplastic Anemia.* Tokyo: University of Tokyo Press, 1978, p. 37.

20. Mizoguchi, H., Miura, Y., Takaku, F., Sassa, S., Chiba, S., and Nakao, K. The effect of erythropoietin on human bone marrow cells *in vitro*. I. Studies on nine cases of bone marrow failure. *Blood, 37*:624, 1971.

21. Moore, M. A. S., and Spitzer, G. *In vitro* studies in the myeloproliferative disorders. In Lindahl-Kiessling, K., and Osoba, D., eds., *Lymphocyte Recognition and Effector Mechanisms: Proceedings of Eighth Leucocyte Culture Conference.* New York: Academic Press, 1974, p. 431.

22. Moore, M. A. S., Williams, N., and Metcalf, D. *In vitro* colony formation by normal and leukemic human hemopoietic cells: Characterization of the colony-forming cells. *J. Nat. Cancer Inst., 50*:603, 1973.

23. Salaman, M. H., Wedderburn, H., Festenstein, H., and Huber, B. Detection of a graft-versus-host reaction between mice compatible at the H-2 locus. *Transplantation, 16*:29, 1973.

24. Simonsen, M. Graft versus host reactions. Their natural history and applicability as tools of research, *Progr. Allergy, 6*:349, 1962.

25. Till, J. E., and McCulloch, E. A. A direct measurement of the radiation sensitivity of normal mouse bone marrow cells. *Radiat. Res., 14*:213, 1961.

26. van Bekkum, D. W., Vos, O., and Weyzen, W. W. H. The pathogenesis of the secondary disease after foreign bone marrow transplantation in X-irradiated mice. *J. Nat. Cancer Inst., 23*:75, 1959.

32

Irradiation-Induced Canine Leukemia: A Proposed New Model. Incidence and Hematopathology

D. V. Tolle, T. M. Seed,
T. E. Fritz, and W. P. Norris

A leukemia model in a large animal, analogous to human myelogenous leukemia, would be invaluable for not only leukemogenic studies, but also in the design of chemotherapeutic trials and supportive therapy. The incidence of spontaneous myeloid leukemia in most domestic animals, however, is quite low (15,17,26), and this disease rarely occurs in the dog (10).

In ongoing studies at this laboratory designed to determine the late effects of protracted low-dose wholebody γ-irradiation, 26 cases of leukemia have occurred in beagles exposed to ^{60}Co at either 5, 10, or 17 R/day. Twenty-one cases [15 myelogenous leukemia (ML) five erythroleukemia (EL), and one lymphocytic leukemia (LL)] occurred in a group of 53 dogs irradiated beginning at 400 days of age for duration of life. The other five cases [four ML, one monocytic leukemia (MOL)] occurred in dogs who were exposed, as above, over long periods of time, with a specific total exposure of either 2,000 or 4,000 R and were then maintained under normal kennel conditions to be observed for late effects.

These induced canine leukemias are similar to those in humans in that they are preceded in many cases by a "preleukemic" state with dyserythropoiesis, bizarre circulating nucleated erythrocytes, oscillating platelet values, giant platelets, and atypical granule formation in the neutrophilic series. Terminally, as in the human disease, there is marked anemia, thrombocytopenia, and circulating blast cells. The bone marrow is hyperplastic, and in most cases, there is hepatosplenomegaly. Leukemic infiltrates are seen in hemopoietic tissue, as well as in a variety of non-hemopoietic tissues.

The high incidence of myeloid leukemia that can be obtained in the beagle by continuous, low-level γ-irradiation, and the ease with which clinical evaluations can be made in this relatively large animal make this system an attractive alternative to the rodent myeloid leukemia models. The objective of this chapter is to discuss both the incidence and the hematopathology of irradiation-induced canine leukemia and to propose this system as a new animal model of human disease.

MATERIALS AND METHODS

Animals and Irradiation

Groups of young to adult beagles of both sexes (~400 days old) were exposed 22 hr/day, 7 days/week at one of three different exposure rates. Animals were exposed to ^{60}Co γ-rays in a specially constructed facility (18) either for duration of life (53 dogs) or to predetermined total exposures (156 dogs) of either 2,000 or 4,000 R (Table 32.1) (12). Exposure rates in both cases were 5, 10, or 17 R/day. Total exposures of 2,000 R were given at 5, 10, and 17 R/

TABLE 32.1 Irradiation Groups

A. Continuous Irradiation until Death

EXPOSURE RATE (R/22-HR DAY)	NUMBER OF DOGS	NUMBER DEAD
5	24	24
10	16	16
17	13	13
	53	53

B. Terminated Exposures

EXPOSURE RATE (R/22-HR DAY)	NUMBER OF DOGS	TOTAL EXPOSURE (R)	NUMBER DEAD (AUGUST, 1978)
5	20	2,000	1
10	28	2,000	13
10	31	4,000	27
17	53	2,000	35
17	24	4,000	17
	156		93

day and 4,000 R at 10 and 17 R/day. During the 2-week period before radiation exposure, the animals were placed in the exposure cages and given physical examinations, which included obtaining base-line hematologic values. Control dogs were handled in an identical fashion, except they were caged for the duration of the experiment in an anteroom adjoining the radiation facility. The animals received water *ad libitum* and were given standard dog food once a day (Rockland Dog Diet, Teklad, Monmouth, Ill.).

Hematology

Hemograms were obtained at 14- and 28-day intervals for the irradiated and control animals, respectively. Blood samples were collected from the jugular vein into Vacutainer tubes containing EDTA. Erythrocyte and leukocyte counts were performed electronically. Packed cell volumes were determined by microhematocrit centrifugation methods. Peripheral blood platelets were enumerated by direct observation with phase-contrast optics (4). Hemoglobin was measured spectro-photometrically as cyanmethemogloblin at 540 nm. Differential white blood cell determinations were made by direct microscopic examination of Wright-stained thin films.

Pathology

When the leukemic animals became acutely ill, they were sacrificed by exsanguination while under Nembutal (sodium pentobarbtial, Parke Davis) anesthesia. Gross pathologic changes were recorded and tissue samples were taken for light-microscopic examination. Touch imprints were made from femoral bone marrow, spleen, liver, and various excised lymph nodes. These imprints were stained with Wright-Giemsa, and interpreted by light microscopy. On selected imprints, peroxidase (22) Sudan black (25), periodic acid-Schiff (20), and naphthol-as-*d*-acetate (1) stains were used. Bone marrow differential cell counts (1,000 cells) were performed and myeloid:erythroid (M:E) ratios were calculated.

RESULTS

Causes of Death and Incidence of Leukemia

Early nonleukemic deaths (less than 2,000 R total accumulation or within 100 days after removal from the gamma field) at the highest dose rate (17 R) were due to either aplastic anemia or septicemia in all cases. At 5 and 10 R/day, septicemia was not observed, and aplasia was the major cause of early death.

Most late occurring deaths in dogs irradiated for life at all three dose rates were due to leukemia. But, at 5 R/day, a variety of non-leukemic malignancies (e.g., osteosarcoma and mammary and ovarian carcinoma), and hepatic degeneration were the cause of death.

Under continuous exposure to ^{60}Co γ-irradiation until death at either 5, 10, or 17 R/22-hr day, 21 of 53 dogs died of either myelogenous leukemia, erythroleukemia, or lymphocytic leukemia. Erythroleukemia occurred only at the lowest exposure rate, 5 R/day. Myelogenous leukemia was the predominant type seen at all three exposure rates (Tables 32.2 and 32.3). The overall incidence of these three types of leukemia at the various exposure rates was 11 of 24 at 5 R/day (45.8%), eight of 16 at 10 R/day (50.0%), and two of 13 at 17 R/day (15.4%). Mean survival times were 1,454, 987, and 1,038 days for total accumulated exposures of 7,280, 9,870, and 17,646 R, respectively.

Of 156 dogs that received terminated exposures of either 2,000 or 4,000 R at the same exposure rates, i.e., 5, 10, or 17 R/day, 63 are still alive at this time (August 1978). Five leukemic deaths have occurred among the 93 decedents. The incidence was 1 ML at 5 R/day, 2,000 R total exposure; 3 ML at 10 R/day, 4,000 R total exposure; and 1 MOL at 17 R/day, 2,000 R total exposure. The times to death in all groups are shown in Table 32.3.

Hematology

During irradiation, all dogs developed a progressive leukopenia and thrombocytopenia, the rate of decline being directly related to dose rate. At all three exposure rates, the nadir of leukocyte and platelet values was reached when the total accumulated exposure was approximately 2,000 R (117 days at 17 R, 200 days at 10 R, and 400 days at 5 R). At this critical period, a number of dogs at 10 and 17 R/

TABLE 32.2 Summary of Hematologic Data on Dogs Dying with Hemoproliferative Disorders after Protracted Whole Body Exposure to γ-Radiation from Cobalt-60

EXPOSURE (R/DAY)	MARROW AND HISTOPATHOLOGIC INTERPRETATION	NO. DOGS	MEAN TERMINAL (RBC/MM³ × 10⁶)	MEAN NRBC[a] (CORRECTED WBC/MM³)	MEAN PLATELET COUNT (PER MM³)	MEAN 1,000 CELL MARROW (M:E)	INCIDENCE OF HPD IN DECEDENTS (%)[b]
			IRRADIATION GIVEN UNTIL DEATH				
5	Erythroleukemia (mixed type)	5	1.37	11,224	3,100	0.46	20.8 (5/24)
5	Myelogenous leukemia	6	1.70	13,035	13,500	5.93	25.0 (6/24)
10	Myelogenous leukemia	7	1.87	7,019	3,700	20.7 (N = 6, dog 1215 not included) Total	45.8 (11/24) at 5R/day 43.8 (7/16)
10	Lymphocytic leukemia	1	4.41 (1 dog)	44,589	137,000	1.4[c]	6.2 (1/16)
17	Myelogenous leukemia	2	2.86	40,606	5,500	10.1 Total	50.0 (8/16) at 10 R/day 15.4 (2/13)
	Incidence of HPD in decedents irradiated until death at 5, 10, and 17 R/day = 39.6% (21/53)						
				TERMINATED EXPOSURES			Total Accumulated Exposure (R)
5	Myelogenous leukemia	1	0.66	11,790	10,000	13.7	2,000
10	Myelogenous leukemia	3	1.64	34,169	7,500	4.8	4,000
17	Monocytic leukemia	1	1.30 (1 dog)	95,509	225,000	20.3[d]	2,000

Incidence of HPD in decedents given terminated exposures at 5, 10 and 17 R/day = 5.3% (5/93).

[a]Nucleated red blood cells.
[b]Incidence of HPD in decedents given terminated exposures at 5, 10 and 17 R/day = 5.3% (5/93).
[c]Lymphocyte/erythroid ratio.
[d]Monocyte/erythroid ratio.

TABLE 32.3 Terminal Hematology in Dogs Dying with Hemoproliferative Disorders after Protracted Wholebody Exposure to γ-Radiation from Cobalt-60

DOG NO.	EXPOSURE (R/DAY)	TOTAL DAYS IRRADIATED	DAYS BETWEEN FINAL HEMOGRAM AND DEATH	RBC/MM³ ($\times 10^6$)	WBC/MM³ (NRBC [a] CORRECTED)	PLATELETS/ MM³	1,000 CELL BONE MARROW (M:E)	MARROW AND HISTO-PATHOLOGIC INTERPRE-TATION[b]
1392	5	989	1	0.98	3,517	3,500	0.3	EL
1439	5	1,430	1	1.73	18,260	1,000	0.6	EL
1469	5	1,440	2	2.07	9,668	2,000	0.2	EL
1454	5	1,803	0	1.17	8,964	8,000	1.0	EL
1366	5	1,949	0	1.21	15,709	1,000	0.2	EL
1464	5	1,030	1	1.70	9.961	2,500	4.4	ML
1212	5	1,163	0	1.71	1,710	5,500	5.6	ML
1379	5	1,186	0	4.20	20,665	55,000	11.8	ML
1389	5	1,551	0	0.65	3,004	14,000	6.5	ML
1473	5	1,611	0	0.74	3,986	3,000	3.2	ML
1472	5	1,865	0	0.89	38,883	1,000	4.1	ML
1397	10	383	9	3.39	2,093	500	17.8	ML
1462	10	504	4	2.25	7,494	1,000	5.1	ML
1210	10	662	0	0.70	9,349	4,500	61.5	ML
1382	10	669	1	1.81	7,859	1,500	20.7	ML
1386	10	929	9	2.10	4,487	10,500	10.8	ML
1252	10	1,166	0	1.86	7,622	5,500	8.3	ML
1215	10	1,622	2	0.99	10,231	2,500	Autolysis	ML
1448	10	1,966	0	4.41	44,589	137,000	1.4[c]	LL
1394	17	1,015	1	4.49	41,316	9,000	16.2	ML
1381	17	1,061	0	1.23	37,896	2,000	4.1	ML
2775	5	(400) + 329[d]	0	0.66	11,790	10,000	12.9	ML
1678	10	(405) + 250	1	1.37	13,033	1,000	2.4	ML
1933	10	(405) + 407	2	2.46	24,916	9,500	7.7	ML
1913	10	(405) + 576	2	1.15	64,587	12,000	4.3	ML
2331	17	(117) + 1,241	0	1.30	95,509	225,000	20.3[e]	MOL

[a]NRBC, nucleated red blood cells.
[b]EL, erythroleukemia; ML, myelogenous leukemia; LL, lymphocytic leukemia; MOL, monocytic leukemia.
[c]Lymphocyte:erythroid ratio.
[d]Terminated exposure at (N) days; death occurred at + days after removal from the γ-ray field.
[e]Monocyte:erythroid ratio.

day died of marrow aplasia, which was at times complicated by septicemia (17R/day). Preselected groups of animals who received a total terminated exposure to 2,000 R were removed from the γ field and returned to the normal kennel for lifetime observation. Other preselected animals were left in the γ field for an additional 2,000 R accumulated exposure, then removed, or remained in the γ field for duration of life. Dogs that remained in the γ field after 2,000 R showed a partial recovery of hemopoietic function in which leukocyte and platelet values slowly rose to approximately 50% of their pre-irradiation values in spite of the continuing irradiation.

Dogs that developed ML, EL, or MOL showed a similar peripheral blood response (Figure 32.1). Between 200 and 600 days before death with leukemia, a rather predictable oscillation in platelet values was seen in the majority of the dogs (19 of 25 cases). At this time, various morphologic abnormalities in the peripheral blood were anisopoikilocytosis, macrocytosis, atypical granule formation, granule coalescence in the neutrophilic series, and the pres-

ence of target cells and nucleated erythrocytes. A few dogs had an increase in the number of immature neutrophils in the peripheral blood, although circulating blast cells were not observed. The one dog that died of LL did not exhibit the above morphologic abnormalities, nor did it develop severe terminal anemia or thrombocytopenia.

Fifty to 100 days before death, all dogs with ML or EL entered an acute phase of the disease. During this time, immature granulocytes including blast forms appeared in the peripheral blood. Circulating nucleated erythrocytes, which included rubriblasts and megaloblastic forms especially in the EL cases, were observed (Figure 32.2). Giant platelets and cytoplasmic fragments were common. The changes seen in cell type and morphology progressed with time and were most severe just before death.

Only nine of 25 dogs with ML, EL, or MOL had a terminal leukocytosis, defined as a total leukocyte count over 15,000/mm³. All had terminal anemia and, with the exception of the only LL and MOL case, they were thrombocytopenic (Table 32.3).

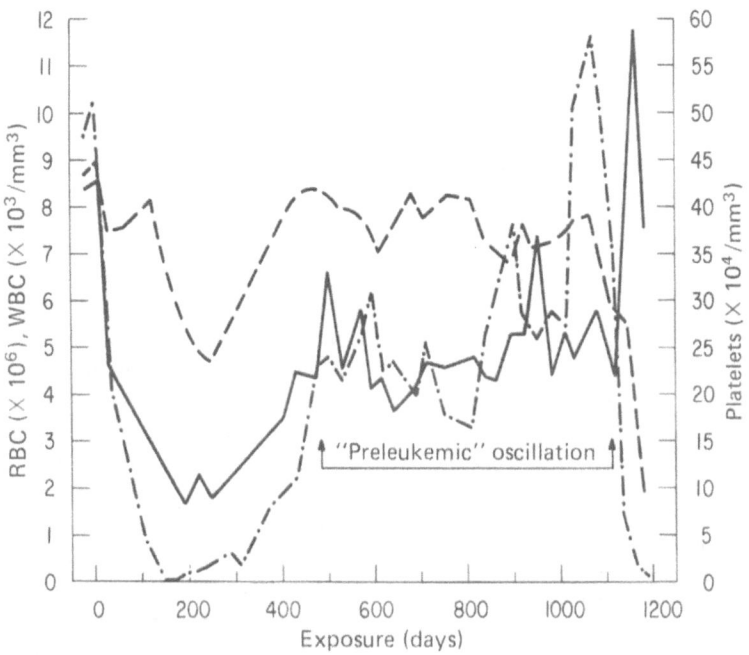

FIGURE 32.1. Peripheral blood values for a representative dog (1252) that died of myelogenous leukemia after continuous irradiation at 10 R/day until death. Erythrocytes (---); total leukocytes (——); platelets (—·—·—).

Pathology

At necropsy, hepatosplenomegaly and lymphadenopathy was present in the majority of dogs. The bone marrow was hyperplastic, dark brownish-red, and pulpy. There was little remaining fat. Touch impressions of the bone marrow showed an increased percentage of immature cells, and asyn-

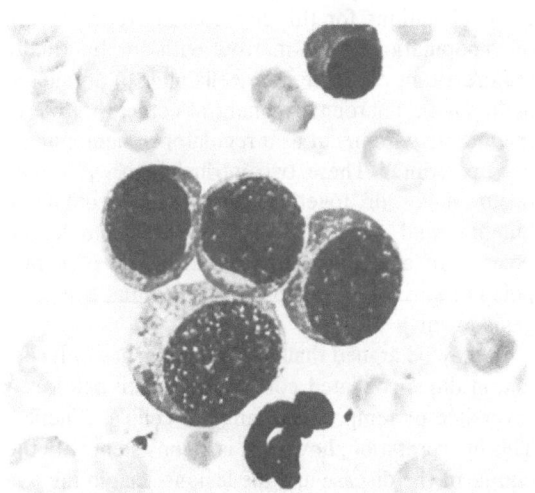

FIGURE 32.2. Peripheral blood film of dog 1469, 14 days before death with erythroleukemia. Four rubriblasts with prominent nucleoli are shown (Wright-Giemsa, 1600 X).

chrony of nuclear/cytoplasmic maturation (Figures 32.3 and 32.4). The bone marrow myeloid/erythroid ratios and differential cell counts are shown in Table 32.4.

Other Tissues

Leukemic infiltration was common in the spleen (Figure 32.5), liver (Figure 32.6), and lymph nodes (Figure 32.7). The degree of infiltration ranged from focal infiltration to massive replacement of normal tissue. Other tissues frequently involved included the gastrointestinal wall, heart, kidney, lung, and in a few cases, the retina. The hematopathology, histopathology, and electron microscopy of these canine leukemias have been previously described (11,23,27,28).

DISCUSSION

It is well known that exposure to ionizing radiation is leukemogenic to man (3,5,14,29) as well as animals (9–13,23,24,27,28). In the dog, continuous exposure to γ-irradiation until death at the appropriate exposure rate (5 and 10 R/day) results in a high incidence (~50.0%) of leukemia, with a mean period to death of ~3.5 years (383 to 1,966 days). Continuous irradiation to predetermined total exposures also induces leukemia, although not as effectively, and

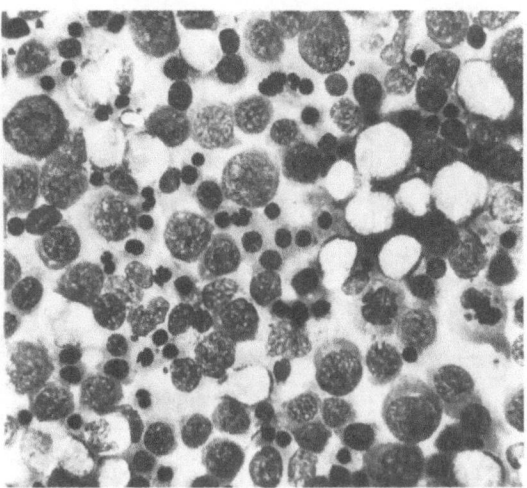

FIGURE 32.3. Terminal bone marrow impression of dog 1366 that died of erythroleukemia. There is proliferation of rubriblasts, with few intermediate stage cells (prorubricyte, rubricyte). Note the prominent blast nucleoli (Wright-Giemsa, 640 X).

the latent period may be as long as 3.4 years (1,241 days) after removal from the radiation field. In man, the estimated latent period in radiation-induced leukemia is 7 years (16).

In these studies, both total exposure and exposure rate of irradiation are critical factors in determining the incidence of leukemia. At 17 R/day, death during (under 2,000 R total exposure) or shortly following the irradiation period was due to either aplasia or septicemia. Those dogs that survive 2,000

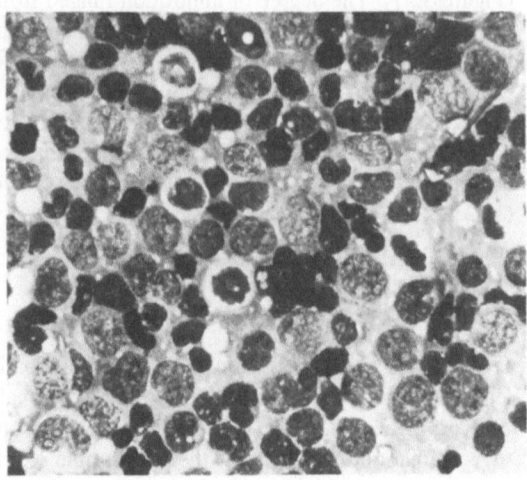

FIGURE 32.4. Terminal bone marrow impression of dog 1394 that died of myelogenous leukemia. There is myeloblast proliferation, occasional mitotic figures, and depletion of erythroid precursors (Wright-Giemsa, 640 X).

R total exposure accommodate to the irradiation stress and show hemopoietic recovery and, somewhat later, develop overt leukemia. At 5 and 10 R/day, early deaths during irradiation were due solely to aplasia. As with the 17 R/day dogs, those that survive 2,000 R show hemopoietic recovery and die later of either leukemia (~50%) or other non-hemopoietic causes. The results of our studies are consistent with the hypothesis presented by Cronkite (6), Dameshek (7), and others (2,9), that is, continuous, low-level irradiation of the bone marrow does not completely prevent hemopoietic function, but places the marrow under continuous stress, forcing the marrow to accommodate by producing clones of abnormally radioresistant hemopoietic cells.

The importance of exposure rate is striking in that although approximately 50% of the dogs irradiated for life at 5 and 10 R/day developed leukemia, erythroleukemia was observed only in the 5 R/day group (five of 11 cases). This suggests that 5 R/day represents an optima erythroleukomogenic stimulus for the beagle. Although the significance is less clear, it is important to note that the leukemias occurring in dogs irradiated at 5 R/day all occurred by 1,949 days of irradiation. All dogs dying at later times died of other causes.

The period of hemopoietic recovery seen in animals that survive an initial exposure of 2,000 R, irrespective of exposure rate and irrespective of whether they receive terminated exposures or are irradiated for life, is potentially the most informative for studying the early events of leukemogenesis. During this period, the hemopoietic system accommodates to the irradiation stress as shown by increasing leukocyte and platelet values. Two possible explanations for this hemopoietic recovery are (a) repopulation of the marrow with an abnormally radioresistant clone of stem cells and (b) an altered hemopoietic microenvironment affecting the role of the microenvironment as a regulator of hemopoiesis via "poietins." These two intrinsic processes may occur singly or together, with as yet unknown humoral and cellular events that regulate hemopoiesis. In any event, this initial period of hemopoietic recovery may represent the initial leukemogenic event.

It may be argued that the high incidence of leukemia in dogs irradiated continuously until death is a reversible or temporary change in cellular kinetics. This interpretation, however, is inconsistent with the course of the disease and the lesions seen in the five dogs that developed overt leukemia several hundred days following removal from the radiation field.

It appears that approximately 2,000 R at 5, 10, or 17 R/day is the minimal total exposure to induce leukemia. We have yet to see a single case of ML,

TABLE 32.4 Marrow Differential Cell Counts of Dogs Dying with Hemoproliferative Disorders (%) after Protracted Wholebody Exposure to γ-Radiation from Cobalt-60[a]

	DOG NO.	Myeloblast	Promyelocyte	Myelocyte	Metamyelocyte	Band	Segmented	Rubriblast	Prorubricyte	Rubricyte	Metarubricyte	1000 Cell (M:E ratio)
1.	1392	5.0	7.0	11.0	2.0	2.0	0	13.0	13.0	16.0	23.0	0.3:1
2.	1439	1.5	1.0	4.0	4.5	12.5	9.0	22.5	10.0	16.0	11.5	0.6:1
3.	1469	0.5	3.0	6.5	5.5	1.5	0	32.5	23.0	13.0	8.0	0.2:1
4.	1453	2.0	4.5	26.5	8.5	5.5	1.5	10.0	11.	13.5	11.5	1.0:1
5.	1366	0.5	3.0	5.0	5.5	4.5	1.0	18.5	12.0	18.5	27.5	0.2:1
6.	1464	0	5.0	13.0	12.0	40.0	10.0	0	1.0	5.0	12.0	4.4:1
7.	1212	40.0	11.5	13.5	5.5	10.5	2.0	0	1.5	4.0	8.0	5.6:1
8.	1379	0.5	2.0	10.5	18.5	31.0	22.5	0	1.0	3.0	3.0	11.8:1
9.	1389	7.5	3.5	22.5	8.5	28.5	8.0	1.0	3.0	3.5	5.0	6.5:1
10.	1473	11.0	7.5	26.0	9.5	8.0	2.5	0	2.0	6.0	13.5	3.2:1
11.	1472	2.0	6.0	17.5	28.0	10.0	2.0	0.5	1.5	9.0	5.5	4.1:1
12.	1397	59.0	14.0	10.0	3.0	2.5	2.0	0	0	1.0	4.0	17.8:1
13.	1462	5.0	4.0	37.0	13.0	10.0	12.0	0	0	8.0	8.0	5.1:1
14.	1210	21.5	9.0	26.0	5.5	12.5	14.5	0	0	1.0	0.5	61.5:1
15.	1382	21.0	48.0	15.0	5.0	1.0	0	0	2.0	4.0	0	20.7:1
16.	1386	2.0	12.0	23.0	19.0	20.0	13.0	0	0	3.0	6.0	10.8:1
17.	1252	14.5	5.5	9.0	9.5	40.5	6.5	0	0	3.0	7.0	8.3:1
18.	1215	Autolysis does not permit differential cell count										
19.	1448	Lymphocytic leukemia—75% of leukocytes are lymphoid—lymphocyte/erythroid ratio = 1.4:1										
20.	1394	8.0	24.0	28.0	10.0	15.0	9.0	1.0	4.0	3.0	2.0	16.2:1
21.	1381	9.5	14.0	35.5	7.0	6.0	8.0	1.0	4.0	8.0	5.5	4.1:1
22.	2775	6.0	7.0	18.5	17.5	31.0	10.0	0	0.5	2.0	4.5	12.9:1
23.	1678	8.0	4.0	19.0	10.0	18.0	13.0	0	2.0	10.0	13.0	2.4:1
24.	1933	4.0	6.5	30.5	19.0	15.0	8.5	0	0	3.0	8.0	7.7:1
25.	1913	29.5	5.0	9.5	13.5	10.0	8.5	1.5	4.0	7.0	5.5	4.3:1
26.	2331	Monocytic leukemia—95% of luekocytes are monoblasts and promocytes—monocyte/erythroid ratio = 20.3:1										
	Control[b] Mean ± SE	0.4 ± 0.1	1.1 ± 0.2	10.6 ± 0.6	9.1 ± 0.5	15.6 ± 0.8	20.3 ± 1.2	0.2 ± 0.1	2.9 ± 0.3	16.2 ± 0.3	20.6 ± 0.8	1.5 ± 0.1

[a]Values do not total 100% because all cell types were counted. Only granulocytic and erythrocytic cells are included here, unless otherwise indicated.

[b]35 untreated clinically normal dogs.

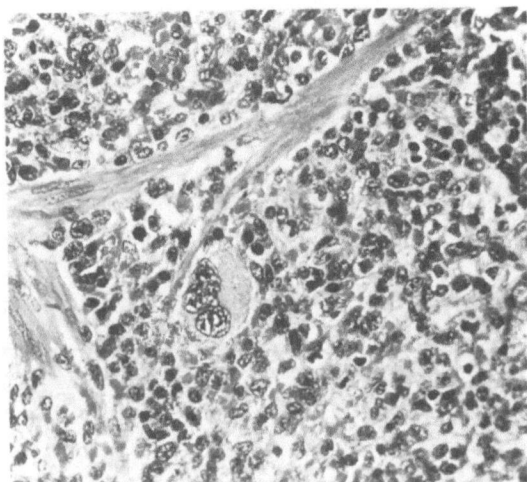

FIGURE 32.5. Spleen section of dog 1913 that died with myelogenous leukemia. The normal tissue is completely replaced with immature myeloid cells. Note the megakaryocyte (hematoxylin-eosin, 400 X).

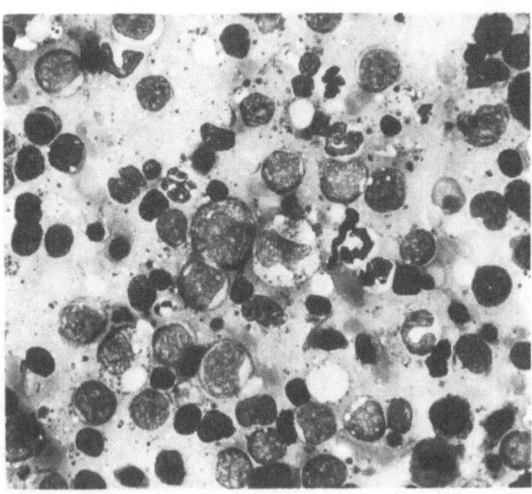

FIGURE 32.6. Liver impression of dog 1381 that died with myelogenous leukemia. There is a predominance of blast cells and other immature myeloid cells (hematoxylin-eosin, 640 X).

EL or MOL in any dog irradiated at under 2,000 R. Further, no cases of these leukemias have been seen at any exposure rate greater than 17 R/day (total decedent dogs in ⁶⁰Co irradiation study = 378). One case of LL has been observed in an untreated colony control dog. But ML, EL, or MOL have never been observed in any other decedent control animals (*N* = 97). One case of ML occurred in an earlier study in which a single IV injection of ¹⁴⁴Ce (10) was used to produce protracted irradiation that is related to the deposition of the radionuclide and to its physical and biological half-life.

In the majority of these canine leukemias, a "pre-leukemic" state is seen (24), which is similar to that in man (8,19,21) and is characterized by refractory anemia, oscillation of the platelet values, and various morphologic abnormalities, that is, macrocytosis, megaloblastoid maturation, and giant platelets. The canine myeloid leukemias are similar in many ways to the terminal state, in man, acute myelogenous leukemia or the "blast crisis" in chronic myelogenous leukemia. Clinically, there is anorexia, progressive weakness, and loss of weight. The terminally ill and severely anemic, thrombocytopenic, and more often than not leukopenic. Petechial and ecchymotic hemorrhages are seen in a variety of tissues, and hepatosplenomegaly and lymphadenopathy are common clinical findings.

Histologically, leukemic infiltrates are seen in a variety of tissues. The bone marrow, spleen, and liver are most severely and frequently affected. The bone marrow is hypercellular with little or no fat remaining. The degree of shift to more immature

cells varies from moderate to extreme (a marrow packed with blasts). Severe maturation defects, such as asynchronous nuclear/cytoplasmic maturation, atypical granule formation, and in some cases agranularity, are common morphologic observations. The degree of leukemic infiltration in other tissues varies from focal infiltration (gastrointestinal tract, heart, kidney, and lung) to massive infiltration and replacement of normal tissues (spleen, liver, and lymph nodes).

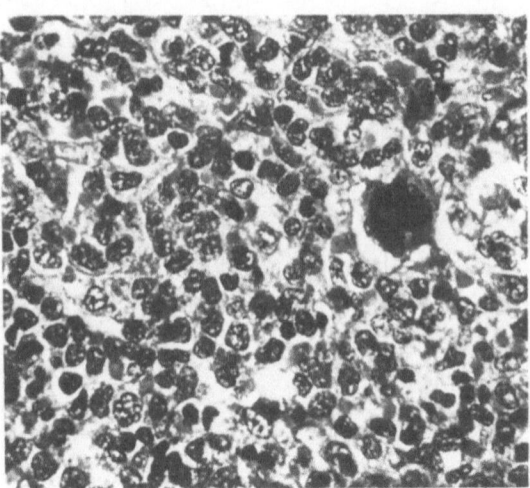

FIGURE 32.7. Mesenteric lymph node section of dog 1913 that died with myelogenous leukemia. There is complete replacement of the parenchyma with immature myeloid cells. (hematoxylin-eosin, 400 X).

SUMMARY

With precise control of the leukemogenic stimulus (^{60}Co gamma irradiation) we have produced a high incidence of both myelogenous and erythroleukemia in the dog that is clinically, hematologically, and histologically similar to the human disease. Because these radiation-induced canine leukemias are reproducible, and occur in high incidence, and because the dog is a large animal, which is easily manipulated clinically, it is an attractive alternative animal model for the hematologist-oncologist interested in leukemogenesis and especially in developing new clinical therapies.

ACKNOWLEDGMENTS

The authors wish to thank C. Poole, D.V.M., L. Lombard, D.V.M., J. Angerman, L. Bell, S. Cullen, D. Doyle, L. Kaspar, W. Keenan, and P. Polk, as well as the animal care personnel for valuable assistance. We also thank Ms. L. Purdy for typing the manuscript.

Supported by the U.S. Department of Energy.

REFERENCES

1. Bennett, J. M., and Reed, C. E. Acute leukemia cytochemical profile: Diagnostic and clinical implications. in *Blood cells. 1, Inserm symposium, Paris*. Berlin: Springer-Verlag, 1975, p. 101.

2. Bierman, H. R. The leukemias-proliferative or accumulative?. *Blood., 30*:238, 1967.

3. Bizzozero, O. J., Johnson, K. G., Ciocco, A., Kawasaki, S., and Toyoda, S. Radiation-related leukemia in Hiroshima and Nagasaki 1946–1964. II. Observations on type-specific leukemia, surviorship, and disease behavior. *Ann. Intern. Med., 66*:522, 1967.

4. Brecher, G., and Cronkite, E. P. Morphology and enumeration of human blood platelets. *J. Appl. Physiol., 3*:365, 1950.

5. Court-Brown, W. M., and Doll, R. Leukemia and aplastic anemia in patients irradiated for ankylosing spondylitis. *Med. Res. Council. Special Report Ser.* London. 1957, p. 295.

6. Cronkite, E. P. Kinetics of leukemic cell proliferation. In Dameshek, W., and Dutcher R. M., eds., *Perspectives in Leukemia*. New York: Grune & Stratton, 1968, p. 158.

7. Dameshek, W. Riddle: What do aplastic anemia, paroxysmal nocturnal hemoglobinuria (PNA) and hypoplastic leukemia have in common? *Blood., 30*:251, 1967.

8. Dörmer, P., Hegemann, R., and Brinkmann, W. Proliferation and production of hemopoietic cells in two stages of disease: Preleukemia and overt leukemia. *Klin. Wschr., 54*:461, 1976.

9. Dungworth, D. L., Goldman, M., Switzer, J. W., and McKelvie, D. H. Development of a myeloproliferative disorder in beagles exposed continuously to Sr-90. *Blood, 34*:610, 1969.

10. Fritz, T. E., Norris, W. P., Rehfeld, C. E., and Poole, C. M. Myeloproliferative disease in beagle dogs given protracted whole-body irradiation or single doses of ^{144}Ce. In Clarke, W. J., Howard, E. G., and Hacket, P. L., eds., *Myeloproliferative Disorders of Animals and Man*. Oak Ridge: USAEC CONF-680529, 1970, p. 219.

11. Fritz, T. E., Norris, W. P., and Tolle, D. V. Myelogenous leukemia and related myeloproliferative disorders in beagles continuously exposed to ^{60}Co γ-radiation. In Dutcher, R. M., and Chieco-Bianchi, L., eds. *Unifying Concepts of Leukemia, Bibl. Haematol*. Basel: Karger, 1973, p. 170.

12. Fritz, T. E., Norris, W. P., Tolle, D. V., Seed, T. M., Poole, C. M., Lombard, L. S., and Doyle, D. E. Relationship of dose rate and total dose to responses of continuously irradiated beagles. *International Symposium on the Late Biological Effects of Ionizing Radiation*. Vienna: IAEA-SM-224/206, 1978.

13. Howard, E. B., and Clarke, W. J. Strontium-90-induced hematopoietic neoplasms in minature swine. In Clarke, W. J., Howard, E. G., and Hackett, P. L., eds., *Myeloproliferative Disorders of Animals and Man*. Oak Ridge: USAEC CONF-680529, 1970, p. 379.

14. Lawrence, J. W. Irradiation leukemogenesis, *JAMA, 190*:93, 1964.

15. Meier, H. Neoplastic diseases of the hematopoietic system (so-called leukosis-complex) in dogs. *Zentrabl. Veterinaermed., 4*:633, 1957.

16. Modan, B., and Lubin, E. Radiation induced leukemia in man. In Gunz, F. W., ed., *The Etiology of Leukemia. Ser. Haematol*. Vol. 7: 1974, p. 192.

17. Moulton, J. E. *Tumors in Domestic Animals*. Berkeley: University of California Press, 1961, p. 85.

18. Norris, W. P., and Fritz, T. E. Interactions of total dose and dose rate in determining tissue responses to ionizing radiations. In Stover, J., and Jee, W. S. S., eds., *Radiobiology of Plutonium*. Salt Lake City: J. W. Press, 1972, p. 243.

19. Pierre, R. V. Preleukemic states. *Semin. Hematol., 11*:73, 1974.

20. Quaglino, D., and Hayhoe, R. G. J. Periodic acid-Schiff positivity in erythroblasts with special references to DiGuglielmo's disease. *Br. J. Haematol., 6*:26, 1960.

21. Saarni, M. I. and Linman, J. W. Preleukemia: The hematologic syndrome preceding acute leukemia. *Am. J. Med., 55*:38, 1973.

22. Sato, A., and Sekiya, S. A simple method for differentiation of myeloid and lymphatic leukocytes of the human blood. *Tohoku J. Exp. Med., 7*:111, 1956.

23. Seed, T. M., Tolle, D. V., Fritz, T. E., Devine, R. L., Poole, C. M., and Norris, W. P. Irradiation-induced erythroleukemia and myelogenous leukemia in the beagle dog: Hematology and ultrastructure. *Blood., 50*:1061. 1977.

24. Seed, T. M., Tolle, D. V., Fritz, T. E., Cullen, S. M.,

Kaspar, L. V., and Poole, C. M. Hemopathological consequences of protracted gamma irradiation in the beagle: Preclinical phases of leukemia induction. *International Symposium on the Late Biological Effects of Ionizing Radiation.* Vienna: IAEA-SM-224/308, 1978.

25. Sheehan, H. L., and Storey, G. W. An improved method of staining leukocyte granules with Sudan black. *Br. J. Pathol. Bacteriol., 59*: 336, 1947.

26. Squire, R. Hematopoietic tumors of domestic animals. *Cornell Vet.,54*:97, 1964.

27. Tolle, D. V., Fritz, T. E., and Norris, W. P. Radiation-induced erythroleukemia in the beagle dog. *Am. J. Pathol., 87*:499, 1977.

28. Tolle, D. V., Seed, T. M., Fritz, T. E., Lombard, L. S. Poole, C. M., and Norris, W. P. Acute monocytic leukemia in an irradiated beagle. *Vet. Pathol.* (in press).

29. Vaughan, J. M. *The effects of irradiation in the skeleton.* Oxford: Clarendon, 1973, p. 119.

33

Lymphocyte/ Marrow Co-Cultures as a Tool to Detect Transfusion-Induced Sensitization and Predict Marrow Graft Rejection in Dogs

Beverly J. Torok-Storb,
Rainer Storb, H. Joachim Deeg,
Theodore C. Graham,
Cathy M. Wise, Paul L. Weiden,
and John W. Adamson

Recent reports from several investigators make it reasonable to assume that lymphocytes, specifically T-cells, play a role in stimulating hemopoiesis (7,10,-11,13,26). In keeping with these observations, we have reported that peripheral blood lymphocytes (PBL) from normal dogs significantly increase the number of in vitro erythroid colonies (EC) grown from DLA-identical littermate marrow (22). The magnitude of stimulation observed depended upon the ratio of lymphocytes to EC precursors cultured, suggesting that lymphocytes might interact directly with erythroid colony-forming units in some capacity.

In contrast to the stimulating ability of normal dog lymphocytes, we found that lymphocytes from transfusion-sensitized dogs had a significantly reduced ability to stimulate and that they actually inhibited EC growth from marrow of the DLA-identical transfusion donor. This was in keeping with earlier observations made in vivo that showed that dogs given blood transfusions frequently rejected subsequent marrow grafts from the transfusion donor The purpose of the present study was to investigate whether the inhibition of in vitro growth of donor marrow by "sensitized" recipient lymphocytes might predict graft rejection in vivo.

MATERIALS AND METHODS
Dogs
Twenty pairs of DLA-identical canine littermates of various breeds were identified by serotyping and mutual nonreactivity in mixed leukocyte cultures as described previously (1,19). Litters were obtained from kennels in the states of Washington, Oregon, and Virginia.

Transfusions
Prospective recipients were sensitized by transfusion of donor blood products. Six dogs were injected intravenously with 50 ml of heparinized whole blood on each of three occasions, 24, 17, and 10 days before transplantation. Five dogs were given 50 ml of donor blood on day 10 before transplantation. Nine dogs were given infusions of platelets ($5.5 \pm 1.5 \times 10^9$) on each of three occasions, 24, 17, and 10 days before transplantation. Platelets were separated as described previously.

Bone Marrow Transplantation
Recipients were conditioned for transplantation by exposure to 1,200 R total-body irradiation, delivered by two opposing ^{60}Co sources. Marrow transplantation and post-transplant supportive care were carried out as previously described (19). Successful engraftment was indicated by a prompt and sustained rise in peripheral blood counts, and further

documented by marrow histology, the development of graft-versus-host disease (GvHD) and, when possible by the demonstration of donor sex chromosomes in cells from marrow and peripheral blood. Graft rejection was documented by the demonstration of a severely hypocellular marrow in those dogs either without recovery of peripheral blood counts after the post-irradiation nadir or, after an initial recovery, a subsequent decline of white blood cell counts to less than 300/mm^3 and platelets to less than 10,000/mm^3.

Bone Marrow and Lymphocyte Preparation

Donor bone marrow cells (BMC) were obtained for culture prior to sensitization of the recipient and again on the day of transplantation. Cells were aspirated from the humoral head into a 10-ml syringe containing 5 ml TC-199 tissue culture medium and 20 units of preservative-free heparin. Buffy coat cells were separated, washed three times, and suspended in supplemented α-medium. Mononuclear cells were obtained by layering 10 ml of heparinized blood over Ficoll-Hypaque ("Lymphoprep," Nyegaard Co., Oslo, Norway) (2). The cells were washed three times and then suspended in supplemented α-medium. All cells were counted on a hemocytometer using trypan blue exclusion to determine viability.

Erythroid Colony Assay

Cells were cultured in 0.1 ml plasma clots using the technique described by Stephenson et al. as modified in this laboratory for dog cells (3,15). In all experiments, 2×10^4 donor BMC were cultured alone and in the presence of 5×10^4 recipient PBL, with a final concentration of 1.0 unit erythropoietin/ml culture (Step III, sheep plasma, Connaught Laboratories, Ltd., Willowdale, Ontario). The clots were harvested after 72 hr, fixed with 5% glutaraldehyde on glass slides, stained with benzidine, and aggregates of eight hemoglobinized cells or more were counted as EC (8). There were at least six replicates of each culture.

Data Analysis

The results were expressed as ratios derived by dividing the mean number of EC grown in the presence of lymphocytes by the mean number obtained in control cultures without lymphocytes. Ratios greater than one indicated that lymphocytes increased the number of EC, whereas ratios less than one indicated a reduced number of EC. The distribution of ratios >1 versus ≤1 between dogs with successful engraftment or dogs with graft rejection was analyzed by Fisher's Exact Probability test (14). The mean ratios of EC obtained with lymphocytes were compared to control ratios by the Student's t test.

RESULTS

In our previous study, normal dog PBL gave stimulation ratios of greater than one in 50 out of 52 experiments. In the current study, co-culturing with normal dog PBL increased EC numbers in all 20 experiments. The degree of stimulation ranged from 1.02 to 9.00 times control values and was found to be inversely correlated with the number of EC in control cultures. This is in keeping with our previous findings, which indicate that as the number of targets or EC increases, the ability of a set number of PBL (5×10^5/ml) to stimulate growth decreases (22).

For this reason, we hypothesized that the expected effect of normal PBL was stimulation of allogeneic EC. To further test this theory, we studied five pairs of dogs, each tested on three separate occasions. As shown in Table 33.1, stimulation of allogeneic EC by normal PBL is a consistent phenomenon. For this reason, ratios over one are considered normal and those equal to or under one are grouped together as abnormal. Data can then be analyzed as the distribution of ratios over one versus ratios equal to or under one in relation to engraftment versus graft rejection.

The PBL obtained from recipients after transfusion failed to stimulate EC in 14 of 20 cases and in 11 instances, it decreased the number of EC. The degree of inhibition varied from 0.89 to 0.55 (Table 33.2). Fourteen of the transfused dogs rejected the subsequent marrow graft from the DLA-identical littermate, whereas six did not. Of the 14 dogs with graft rejection, 13 were predicted by ratios equal to or less than one. Of the six dogs with sustained engraftment, five were predicted by ratios greater than one. The correlation of *in vitro* and *in vivo* results was significant at $p = 0.002$ (Table 33.3)

DISCUSSION

We have previously shown that normal dog lymphocytes cultured with littermate marrow increase EC numbers (22). Stimulation of an *in vitro* erythropoiesis by allogeneic PBL was not a surprising observa-

TABLE 33.1 Consistent Stimulation (Ratios > 1) of Erythroid Colonies by Normal DLA-Identical Littermate Peripheral Blood Lymphocytes Tested on Three Separate Occasions.

DOG	EXP. I	EXP. II.	EXP. III
1	2.84	1.84	1.10
2	1.38	1.20	1.20
3	1.52	1.81	1.69
4	1.51	1.59	1.57
5	2.09	1.94	1.12

TABLE 33.2 *In Vitro* Data from Donor Marrow Cultured Alone (Control) and With Transfusion-Sensitized Recipient Peripheral Blood Lymphocytes. Correlation with the *In Vivo* Results of Subsequent Marrow Transplants.

DOGS	MEAN ± SE NUMBER OF ERYTHROID COLONIES/2 × 10⁴ MARROW CELLS		RATIO	RESULT OF MARROW TRANSPLANT
	Without PBL	With PBL		
1	43 ± 1.6	43 ± 2.4	1.00	Rejection
2	53 ± 4.8	29 ± 2.3	0.55	Rejection
3	42 ± 3.1	33 ± 2.6	0.79	Rejection
4	93 ± 6.2	82 ± 4.1	0.88	Rejection
5	23 ± 1.4	20 ± 1.1	0.89	Rejection
6	76 ± 4.2	76 ± 5.6	1.00	Rejection
7	89 ± 4.3	55 ± 4.6	0.62	Rejection
8	103 ± 4.9	78 ± 1.5	0.76	Rejection
9	199 ± 5.5	161 ± 8.2	0.81	Rejection
10	28 ± 1.7	28 ± 2.3	1.00	Rejection
11	51 ± 1.5	37 ± 1.4	0.73	Rejection
12	43 ± 2.1	34 ± 1.7	0.79	Rejection
13	149 ± 2.5	123 ± 3.5	0.83	Rejection
14	16 ± 1.3	56 ± 3.6	3.29	Rejection
15	47 ± 1.4	74 ± 0.6	1.57	Engraftment
16	17 ± 0.5	15 ± 0.8	0.88	Engraftment
17	29 ± 0.7	36 ± 1.3	1.24	Engraftment
18	26 ± 0.7	38 ± 0.7	1.46	Engraftment
19	49 ± 1.5	98 ± 1.7	1.99	Engraftment
20	26 ± 1.9	95 ± 1.6	3.67	Engraftment

tion, therefore, since previous reports from several investigators suggest that lymphocytes play a role in stimulating hemopoiesis. Nathan et al. (10) have shown that T-cells from human peripheral blood enhance *in vitro* growth of erythroid burst-forming units (BFU-e) present in human blood. Other workers have found that the addition of thymocytes promotes hemopoietic regeneration in transplanted mice (7,11). We have reported enhancement of allogeneic marrow engraftment in irradiated dogs after the addition of peripheral blood buffy coat cells to the marrow inoculum (16). The addition of thoracic duct lymphocytes was similarly effective (4), whereas irradiated buffy coat cells were ineffective (25). Additional evidence is provided by the observation in mice that treatment of donor marrow with anti-theta serum results in a failure of the marrow to reconstitute the genetically anemic *W/W*ᵛ mouse

TABLE 33.3 2 × 2 Contingency Table Showing Distribution of *In Vitro* Ratios in Relation to Outcome of Transplantation

RESULTS OF MARROW TRANSPLANT	NUMBER OF DOGS WITH RATIOS	
	> 1	≤ 1
Fourteen dogs with graft rejection	1	13
Six dogs with engraftment	5	1

(26). Subsequent addition of thymocytes to the treated marrow reduced the effect of anti-serum.

The exact nature of the lymphocyte-marrow interaction responsible for the *in vitro* stimulation has not been defined. Lymphocyte-conditioned medium has been shown to increase *in vitro* erythropoiesis (6,10) and to increase the number of stem cells in mice (5). These findings suggest that lymphocytes produce a factor that mediates this response. The demonstration of such a factor does not preclude the possibility that lymphocytes can interact directly with bone marrow cells. Our studies in dogs have shown that the degree to which a given number of lymphocytes enhance colony growth is related to the number of bone marrow cells in culture (22), suggesting that there is an optimal effector-target cell ratio that can maximize this response. This is in keeping with, but does not prove cell to cell interaction.

In contrast to the stimulating ability of normal dog lymphocytes, we found that PBL from transfusion-sensitized dogs had a significantly reduced ability to stimulate and that they actually inhibited EC growth from the marrow of the DLA-identical littermate transfusion donor. These findings suggested that *in vivo* sensitization against minor histocompatibility antigens could be detected *in vitro* by marrow-lymphocyte co-cultures. Since marrow graft rejection in the dog has been shown to occur if the recipient has been previously exposed to minor antigens (17,18), we sought to determine whether the *in vitro* inhibition of donor marrow growth would predict marrow graft rejection *in vivo*. The results clearly indicate that transfusion-induced sensitization and subsequent rejection of a DLA-identical marrow graft are predicted by reduced EC growth of donor marrow when it is cultured with recipient lymphocytes.

These results are of potential importance for the multiply transfused human patient with severe aplastic anemia who is a marrow transplant candidate. Marrow graft rejection has been observed in 20 to 30% of these patients and is presumably related to transfusion-induced sensitization (12,20). The availability of tests documenting sensitization and predicting rejection is important, since sensitized patients can then be treated with different immunosuppressive conditioning regimens aimed at preventing rejection (20). Results of two tests currently in use, one a positive relative response index in mixed leukocyte culture (9), the other a positive lymphocyte-mediated ^{51}Cr release assay (23), have been found to correlate significantly with graft rejection. Unfortunately, they both have a 15% incidence of "false negatives." Hence, additional and perhaps more accurate *in vitro* tests are needed. The current studies in dogs indicate that results obtained from culturing recipient PBL with donor marrow are highly predictive of graft rejection or sustained engraftment. Whether or not the technique will be applicable to human patients with aplastic anemia remains to be determined.

SUMMARY

We have previously shown that peripheral blood lymphocytes from normal dogs co-cultured with DLA-identical littermate marrow result in an increase of the number of erythroid colonies over that obtained with marrow alone. In addition, blood transfusion from a matched littermate reduces the ability of lymphocytes to enhance erythroid colony growth. We are currently reporting the correlation of these *in vitro* findings with the observations made *in vivo* following marrow transplantation. Twenty-five pairs of DLA-identical littermates were studied. One member of each pair served as the blood and marrow donor and the other as the recipient. Recipients were conditioned for transplantation by 1,200 R total-body irradiation. Marrow was transplanted within 4 hr of irradiation. All recipients received blood products from their littermate marrow donor before transplantation: 11 received three transfusions of 50 ml whole blood and nine were given three transfusions of platelets obtained from 50 ml of blood on days -24, -17, -10; five received one transfusion of whole blood on day -10. On the day of transplantation, 2×10^5 donor marrow cells were cultured alone and together with 5×10^5 recipient lymphocytes in 1-ml plasma clots. The number of erythroid colonies grown with 1.0 unit of erythropoietin was determined on day 3 of culture. In 18 of the 24 experiments, "sensitized" lymphocytes either failed to stimulate or inhibited erythroid colony growth. In the remaining seven, significant stimulation was observed. Marrow graft rejection occurred in 19 of the 25 dogs. Seventeen of these 19 rejections were predicted by the *in vitro* observation of either absent stimulation or inhibition of donor marrow growth by recipient lymphocytes. Successful engraftment occurred in six dogs. Five of the six cases were associated with significant stimulation of donor marrow in co-culture. The results are significantly different ($p = .002$), as analyzed by the Fisher exact probability test. We suggest that transfusion-induced sensitization and subsequent marrow graft rejection can be detected *in vitro* as reduced erythroid colony growth of donor marrow cultured with recipient lymphocytes.

ACKNOWLEDGMENTS

Thanks are expressed for the excellent technical assistance of S. DeRose, R. Raff, R. Colby, L. Cook, and G. Maul. We also thank J. Schroeder for help in preparing the manuscript.

Supported in part by Grant Numbers CA 18047, CA 18221, Am 19410, awarded by the National Institutes of Health, DHEW, and by Grant Number AM 05356, a Research Service Award to Dr. Torok-Storb.

REFERENCES

1. Albert, E. D., Erickson, V. M., Graham, T. C., Parr, M., Templeton, J. W., Mickey, M. R., Thomas, E. D., and Storb, R. Serology and genetics of the DLA system. I. Establishment of specificities. *Tissue Antigens,* 3:417, 1973.
2. Boyum, A. Isolation of mononuclear cells and granulocytes from human blood. *Scand. J. Clin. Lab. Invest.,* 21(Supp 97)77, 1968.
3. Brown, J. E., and Adamson, J. W. Modulation of *in vitro* erythropoiesis. Enhancement of erythroid colony growth by cyclic nucleotides. *Cell Tissue Kinet.,* 10:289, 1977.
4. Deeg, H. J., Storb, R., Weiden, P., Torok-Storb, B., Graham, T., Thomas, E. D. Resistance to marrow grafts in dogs mediated by antigens close to but not identical with DLA-A, B, and D and overcome by infusion of thoracic duct lymphocytes. (Abst.) *Exp. Hematol.* 6(3):23, 1978.
5. Hamano, T., and Nogai, K. Effects of allogeneic stimulations on the proliferation of differentiation of the hemopoietic stem cell. *Transplantation,* 25(1):23, 1978.

6. Johnson, G. R., and Metcalf, D. Pure and mixed erythroid colony formation *in vitro* stimulated by spleen conditioned medium with no detectable erythropoietin. *Proc. Nat. Acad. Sci. U.S., 74*:3879.

7. Lord, B. I., and Schofield, R. The influence of thymus cells in hematopoiesis: Stimulation of hematopoietic stem cells in a syngeneic, *in vivo,* stituation. *Blood, 42*:395, 1973.

8. McLeod, D. L., Shreeve, M. M., and Axelrad, A. A. Improved plasma culture system for production of erythrocytic colonies *in vitro:* Quantitative assay method for CFU-E. *Blood, 44*:517, 1974.

9. Mickelson, E. M., Fefer, A., Storb, R., Thomas, E. D. Correlation of the relative response index with marrow graft rejection in patients with aplastic anemia. *Transplantation, 22*:294, 1976.

10. Nathan, D. G., Chess, L., Hillman, D. G., Clarke, B., Breard, J., Merler, F., and Housman, D. E. Human erythroid burst forming unit (BFU-E) T-cell requirement for proliferation *in vitro*. *J. Exp. Med., 147*:324, 1978.

11. Pritchard, L., Shinpock, S. G., and Goodman, J. W. Augmentation of marrow growth by thymocytes separated by discontinuous albumin density gradient centrifugation. *Exp. Hematol., 3*:94, 1975.

12. Report from ACS/NIH Bone Marrow Transplant Registry: Bone marrow transplantation from histocompatible, allogeneic donors for aplastic anemia. *JAMA, 236*:113, 1976.

13. Resnitzky, P., Zipou, D., and Trainin, W. Effect of neonatal thymectomy on hematopoietic tissue in mice. *Blood, 37*:364, 1971.

14. Siegel, S. *Nonparametric Statistics*. New York: McGraw-Hill, 1956, p. 96.

15. Stephenson, J. R., Axelrad, A. A., McLeod, D. L., and Shreeve, M. M. Induction of colonies of hemoglobin-synthesizing cells by erythropoietin *in vitro*. *Proc. Nat. Acad. Sci. U.S., 68*:1542, 1971.

16. Storb, R., Epstein, B., Bryant, J., Ragde, H., and Thomas, E. D. Marrow grafts by combined marrow and leukocyte infusions in unrelated dogs selected by histocompatibility typing. *Transplantation, 6*(4):587, 1968.

17. Storb, R., Epstein, R. B., Rudolph, R. H., and Thomas, E. D. The effect of prior transfusion on marrow grafts between histocompatible canine siblings. *J. Immunol., 105*:627, 1970.

18. Storb, R., Rudolph, R. H., Graham, T. C. and Thomas, E. D. The influence of transfusions from unrelated donors upon marrow grafts between histocompatible canine siblings. *J. Immunol., 107*:409, 1971.

19. Storb, R., Rudolph, R. H., Kolb, H. J., Graham, T. C., Michelson, E., Erickson, V., Lerner, K. G., Kolb, H., and Thomas, E. D. Marrow grafts between DLA matched canine littermates. *Transplantation, 15*:92, 1973.

20. Storb, R., Prentice, R. L., and Thomas, E. D.: Marrow transplantation for treatment of aplastic anemia. An analysis of factors associated with graft rejection. *New Eng. J. Med., 296*:61, 1977.

21. Torok-Storb, B., Storb, R., Weiden, P., and Adamson, J. The effects of lymphocytes from non-transfused and transfused dogs on the growth of erythroid colonies (EC) from DLA-identical littermates. (Abst.) *Exp. Hematol., 5*(2):97, 1977.

22. Torok-Storb, B., Storb, R., Graham, T. C., Prentice, R. L., Weiden, P. L., and Adamson, J. W. *In vitro* erythropoiesis: The effect of normal versus "transfusion-sensitized" mononuclear cells. (In press) *Blood*.

23. Warren, R. P., Storb, R., Weiden, P. L., Michelson, E. M., and Thomas, E. D. Direct and antibody-dependent cell-mediated cytotoxicity against HLA-identical sibling lymphocytes. Correlation with marrow graft rejection. Brief communication. *Transplantation, 22*:631, 1976.

24. Weiden, P. L., Storb, R., Slichter, S., Warren, R. P., and Sale, G. E. Effect of six weekly transfusions on canine marrow grafts: Tests for sensitization and abrogation of sensitization by procarbazine and antithymocyte serum. *J. Immunol., 117*:143, 1976.

25. Weiden, P., Storb, R., Graham, T., Sale, G., and Thomas, E. Resistance to DLA-nonidentical marrow grafts in lethally irradiated dogs. *Transplant. Proc., 9*(1):285, 1977.

26. Wiktor-Jedrzejczak, W., Sharkis, S., Ahmed, A., and Sell, K. W. Theta-sensitive cells and erythropoiesis: Identification of a defect in *W/W*v anemic mice. *Science, 196*:313, 1977.

Index